Springer Proceedings in Physics 70

Springer Proceedings in Physics

Managing Editor: H. K. V. Lotsch

44 *Optical Fiber Sensors*
Editors: H. J. Arditty, J. P. Dakin, and R. Th. Kersten

45 *Computer Simulation Studies in Condensed-Matter Physics II: New Directions*
Editors: D. P. Landau, K. K. Mon, and H.-B. Schüttler

46 *Cellular Automata and Modeling of Complex Physical Systems*
Editors: P. Manneville, N. Boccara, G. Y. Vichniac, and R. Bidaux

47 *Number Theory and Physics*
Editors: J.-M. Luck, P. Moussa, and M. Waldschmidt

48 *Many-Atom Interactions in Solids*
Editors: R .M. Nieminen, M. J. Puska, and M. J. Manninen

49 *Ultrafast Phenomena in Spectroscopy*
Editors: E. Klose and B. Wilhelmi

50 *Magnetic Properties of Low-Dimensional Systems II: New Developments*
Editors: L. M. Falicov, F. Mejía-Lira, and J. L. Morán-López

51 *The Physics and Chemistry of Organic Superconductors*
Editors: G. Saito and S. Kagoshima

52 *Dynamics and Patterns in Complex Fluids: New Aspects of the Physics–Chemistry Interface*
Editors: A. Onuki and K. Kawasaki

53 *Computer Simulation Studies in Condensed-Matter Physics III*
Editors: D. P. Landau, K. K. Mon, and H.-B. Schüttler

54 *Polycrystalline Semiconductors II*
Editors: J. H. Werner and H. P. Strunk

55 *Nonlinear Dynamics and Quantum Phenomena in Optical Systems*
Editors: R. Vilaseca and R. Corbalán

56 *Amorphous and Crystalline Silicon Carbide III, and Other Group IV-IV Materials*
Editors: G. L. Harris, M. G. Spencer, and C. Y. Yang

57 *Evolutionary Trends in the Physical Sciences*
Editors: M. Suzuki and R. Kubo

58 *New Trends in Nuclear Collective Dynamics*
Editors: Y. Abe, H. Horiuchi, and K. Matsuyanagi

59 *Exotic Atoms in Condensed Matter*
Editors: G. Benedek and H. Schneuwly

60 *The Physics and Chemistry of Oxide Superconductors*
Editors: Y. Iye and H. Yasuoka

61 *Surface X-Ray and Neutron Scattering*
Editors: H. Zabel and I. K. Robinson

62 *Surface Science: Lectures on Basic Concepts and Applications*
Editors: F. A. Ponce and M. Cardona

63 *Coherent Raman Spectroscopy: Recent Advances*
Editors: G. Marowsky and V. V. Smirnov

64 *Superconducting Devices and Their Applications*
Editors: H. Koch and H. Lübbig

65 *Present and Future of High-Energy Physics*
Editors. K.-I. Aoki and M. Kobayashi

66 *The Structure and Conformation of Amphiphilic Membranes*
Editors: R. Lipowsky, D. Richter, and K. Kremer

67 *Nonlinearity with Disorder*
Editors: F. Abdullaev, A. R. Bishop, and S. Pnevmatikos

68 *Time-Resolved Vibrational Spectroscopy V*
Editor: H. Takahashi

69 *Evolution of Dynamical Structures in Complex Systems*
Editors: R. Friedrich and A. Wunderlin

70 *Computational Approaches in Condensed-Matter Physics*
Editors: S. Miyashita, M. Imada, and H. Takayama

71 *Amorphous and Crystalline Silicon Carbide IV*
Editors: C. Y. Yang, M. M. Rahman, and G. L. Harris

Volumes 1–43 are listed on the back inside cover

S. Miyashita M. Imada H. Takayama (Eds.)

Computational Approaches in Condensed-Matter Physics

Proceedings of the
6th Nishinomiya-Yukawa Memorial Symposium,
Nishinomiya, Japan, October 24 and 25, 1991

With 125 Figures

Springer-Verlag
Berlin Heidelberg New York
London Paris Tokyo
Hong Kong Barcelona
Budapest

Professor Dr. Seiji Miyashita

Department of Physics, CLAS, Kyoto University, Kyoto 606, Japan

Professor Dr. Masatoshi Imada

Institute for Solid State Physics, University of Tokyo,
Tokyo 106, Japan

Professor Dr. Hajime Takayama

Institute of Physics, University of Tsukuba,
Tsukuba 305, Japan

ISBN-13:978-3-642-84823-0 e-ISBN-13:978-3-642-84821-6
DOI: 10.1007/978-3-642-84821-6

Library of Congress Cataloging-in-Publication Data. Nishinomiya-Yukawa Memorial Symposium (6th : 1991 : Nishinomiya-shi, Japan) Computational approaches in condensed matter physics : proceedings of the 6th Nishinomiya-Yukawa Memorial Symposium, Nishinomiya, Japan, October 24 and 25, 1991 / S. Miyahita, M. Imada, H. Takayama, eds. p. cm. – (Springer proceedings in physics ; v. 70) Includes bibliographical references and index. ISBN-13:978-3-642-84823-0 1. Condensed matter–Congresses. 2. Spin glasses–Congresses. 3. Numerical calculations–Congresses. I. Miyashita, S. (Seiji), 1954- . II. Imada, M. (Masatoshi), 1953- . III. Takayama, H. (Hajime), 1945- . IV. Title. V. Series.
QC173.4.C65N57 1991 530.4'1–dc20 92-
28269

Softcover reprint of the hardcover 1st edition 1992

Typesetting: Camera ready by authors/editors
54/3140 - 5 4 3 2 1 0 - Printed on acid-free paper

Preface

This volume combines the proceedings of two related meetings held in Japan: the 6th Nishinomiya-Yukawa Memorial Symposium on "Computational Approaches in Condensed Matter Physics" held on October 24 and 25, 1991, in Nishinomiya, and the accompanying international meeting on "Computational Physics for Condensed Matter Phenomena–Methodology and Applications (CPCMP)" held from October 21 to 23, 1991, in Osaka.

In the recent several years, much progress and numerous advances have been made in computational physics in condensed matter research, particularly in the fields of strongly correlated electronic systems, quantum spin systems, and random and/or nonlinear dynamical systems (complex systems). The rapid growth of our computing facilities has certainly been one reason for this progress, but more important has been the huge development of algorithms, which has been closely related to our understanding of such complicated systems. Algorithm developments provide the possibility of combining different methods and concepts from different fields to create a new efficient algorithm whose applicability can then be extended further to different problems in various fields. Among the subjects that were extensively discussed in these two meetings were: quantum simulations in strongly correlated systems and quantum spin systems (in particular, the negative sign problem in quantum Monte Carlo methods); first principles calculations of electronic structure with applications to correlated systems; molecular dynamics calculations, including those with quantum systems; and some important simulations on complex systems. Recent developments of computer facilities are also reviewed.

The Nishinomiya-Yukawa Memorial Symposium was sponsored by Nishinomiya City, where Prof. Hideki Yukawa was living when he wrote his celebrated paper on mesons that was later honored by a Nobel prize. The symposium was intended to promote the interest of young scientists in various fields of physics and consisted of lectures with an introductory character.

The international meeting CPCMP, on the other hand, was devoted to discussions on specialized problems in the same field as that from which the topic of the 6th Nishinomiya-Yukawa Memorial Symposium was chosen. The CPCMP meeting was organized as one of the activities of a research project on "Computational Physics as a New Frontier in Condensed-Matter Physics", which was supported by a Grant-in-Aid for Scientific Research on Priority Areas from the Ministry of Education, Science and Culture of Japan.

The present proceedings contain most of the invited and contributed papers of these two meetings. We are very grateful to all authors for their efforts in preparing such excellent manuscripts. Although we were forced to limit the length of the manuscripts, we believe that this volume will aid readers in understanding current developments in the field.

The symposium and the international meeting were made possible by the generous support of Nishinomiya City. We express sincere thanks to Mr. Yoneji Yagi, mayor of Nishinomiya City, Prof. Masato Morita, chairman of the Steering Committee of the Nishinomiya-Yukawa Memorial Activities and all the members of the Nishinomiya Comprehensive Education Center.

Nishinomiya
October 1991

Seiji Miyashita
Masatoshi Imada
Hajime Takayama

Contents

Part IV **Critical Phenomena**

Part V Spin Glasses

Part VI Other Complex Systems

Introduction

S. *Miyashita*[1], *M. Imada*[2], *and H. Takayama*[3]

[1]Department of Physics, CLAS, Kyoto University, Kyoto 606, Japan
[2]Institute for Solid State Physics, University of Tokyo, Roppongi 7-22-1, Minato-ku, Tokyo 106, Japan
[3]Institute of Physics, University of Tsukuba, Tsukuba 305, Japan

In condensed-matter physics, the subjects of research are always interacting many-body systems. Many-body effects are the source of both the interest and the difficulty in this field. In order to understand macroscopic properties of matter in terms of microscopic knowledge, many analytic and approximate methods have been developed. For example, the Hartree-Fock approximation for interacting electrons and the mean-field approximation for phase transitions provide a way of understanding many-body systems from the viewpoint of "one-body systems" surrounded by an average field, and certainly have succeeded in clarifying the essential physics of various phenomena. Perturbation expansion is another fruitful method if the interaction is weak enough. However, it has become increasingly clear that various difficulties debar many fundamental problems in condensed-matter physics from reliable analytic treatments. These so far inescapable difficulties may be attributed to the intrinsic nature of systems often represented by nonlinearity, randomness, quantum mechanical effects, strong correlation, and so on. A naive use of uncontrolled approximations often yields results in which it is unclear whether the properties obtained are intrinsic and relevant to the "systems" or "models" we started with or simply artifacts due to the approximations.

Numerical calculations have been long and widely used to partly avoid these difficulties and to allow an analytic approach in such cases as numerical integration and numerical solution of coupled self-consistent equations including differentiation and/or integration. In fact, the rapid growth of computer facilities has made it possible to substantially increase the scale of calculations of such cases as perturbation expansion, series expansion and coupled nonlinear equations, which allows not only quantitative but also qualitative changes in their applicabilities.

In addition, the recent developments of numerical approaches have opened a conceptually new way of understanding materials and phenomena in them. These are methods of obtaining a correct answer by excluding uncontrolled approximations as much as one can with the help of computational facilities when the system or the model is set up. For example, the development of supercomputers and parallel processors enables direct methods, such as molecular dynamics, Monte Carlo methods and exact diagonalization of quantum systems, to be acknowledged more and more as practical tools. These methods provide a way of grasping many-body effects as they are without resorting to ad hoc and often prejudiced assumptions, which, of course, may be understood more thoroughly with the help of various analytic concepts and methods in the next step. These computational approaches have revealed many common features in various different disciplines and have begun to assert their identity as a common field. This is called "computational physics".

Among the various computational approaches and fields, we have paid special attention to three different problems in this proceedings volume: (1) first principles calculations of electronic structure, (2) strongly correlated fermion and spin systems, and (3) random and complex systems. Although various interesting results of physics obtained from numerical approaches are the main concern in these proceedings, recent developments of methodology and algorithms are also extensively discussed, especially for ab initio molecular dynamics simulations, fermion simulations, quantum spin simulations, and algorithms for random and complex systems.

The papers are grouped into seven parts according to their physical subjects. In Part I, **First Principles Calculations**, recent developments and applications of local density approximations (LDA) and their extension to the so-called Car-Parrinello method are included. Based on a more or less established method – the local density approximation – a new experimental technique and new materials have been analyzed: Tsukada discusses a systematic way of analyzing images from scanning tunneling microscopes using the knowledge of the electronic structure of the sample-probe system, while Saito et al. report results of the calculated electronic structure of new materials, carbon clusters C_{60} and K_3C_{60}.

Recently, Car and Parrinello have introduced a new method of molecular dynamics simulation to discuss structure stability and optimization of materials. In this method, equations of motion for atoms are solved using the potential determined from the charge distribution of electrons while the electronic state evolves in fictitious time introduced so as to reproduce the LDA result in the steady state. Morikawa et al. and Uda and Ihara show recent applications of this method to the Si surface. Ad-atom structure and the reconstruction of the surface are discussed. Yonezawa and Sakamoto also apply this method to SiH. From the viewpoint of taking electron correlation more seriously than LDA, Hamann reviews three different Monte Carlo methods, namely, variational, Green's function and auxiliary field Monte Carlo methods, together with a discussion of possible future progress.

Strongly Correlated Fermion Systems, one of the most challenging and fundamental problems in recent condensed-matter theory, is the subject of Part II. Among others, the quantum Monte Carlo method is a powerful numerical method for treating larger systems. However, in fermion simulations, we are faced with the sign problem, which is a consequence of the fact that the measure in the path integral is not positive definite in interacting fermion systems. It leads to exponentially increasing statistical errors of measured results with decreasing temperatures or increasing system size if the computational time is fixed. This difficulty is a central issue in developing the methodology of fermion simulations. The sign problem is discussed and reviewed in the articles by Hamann, Sorella, and Imada and Furukawa. Various prescriptions and improvements of algorithms such as the optimization of trial wave functions and the positive projection method are discussed in these three articles. The nature of the Hubbard model near half-filling is known to be a challenging and fundamental but difficult problem of strongly correlated systems. Several recently clarified aspects of correlation and single-particle properties of spin and charge are discussed by Imada and Furukawa and Sorella. Exact diagonalization studies of the t-J model with two planes are reported by Dagotto with the emphasis placed on a possible mechanism of superconductivity. The numerical renormalization group (RG) method is another challenging subject of this part. Shimizu and Sakai report its application to the excitation spectra of the impurity Anderson model. White discusses the possibility of extending the RG method to lattice fermion models such as the Hubbard model.

The dynamical properties of quantum systems are known to be more difficult to obtain than thermodynamic quantities. Gubernatis discusses basic principles and recent applications of the maximum entropy method developed to obtain dynamical properties.

Quantum spin systems in insulators provide a field in which to examine various concepts obtained from perspectives and insights connected to those of field theories and statistical mechanics. The interplay of quantum mechanical effects and the symmetric nature of the ordering is a key for the understanding of the systems. This issue is closly related to the subjects discussed in Part II in the sense that the quantum spin systems may be realized as fermion systems under local constraints. Part III, **Quantum Spin Systems**, discusses various topics. Makivic reviews the quantum Monte Carlo method based on Suzuki-Trotter decomposition on the hypercubic parallel supercomputer. This article also includes an application of the maximum entropy method for the determination of dynamical properties of the system, which is explained in detail by Gubernatis in Part II. Miyashita and Behre study impurity effects in ordering of the 2D Heisenberg antiferromagnet and discuss "quantum interference phenomena" in spin ordering. The one-dimensional spin-1 antiferromagnetic system is believed to have an energy gap, known as the Haldane gap. Numerical calculations have been very powerful both for confirming its existence and clarifying its nature, as reviewed by Takahashi. Nishimori summarizes the finite size scaling procedures done to obtain exactly diagonalized results on small clusters of several quantum spin systems and points out a difficulty in the extrapolation to the thermodynamic limit. He emphasizes the importance of combining with analytic studies. Gubernatis discusses numerical and symmetry breaking and its restoration in quantum Monte Carlo results by studying coupled double-well oscillators.

Critical phenomena and the nature of ordered phases have been long-lived problems in condensed-matter physics and statistical physics. The recent trend of this field is toward understanding new types of phase transitions in complex systems such as random systems with frustrated interactions, whose prototype system is a spin glass. Even in homogeneous systems, phase transitions associated with order parameters having a certain internal structure have been extensively studied. In Part IV, **Critical Phenomena**, Suzuki et al. discuss basic concepts and applications of the coherent anomaly method, which has been introduced to determine critical points and the exponents of various phase transitions in a systematic way. Okabe and Kikuchi examine the distribution of order parameters to discuss the order of phase transitions in various Potts models and propose new analyses for critical properties.

Recent topics on spin glasses and related systems are presented in Part V, **Spin Glasses**. Recently, research is particularly focused on the investigation of phase transitions and the nature of the low temperature phase in spin glasses with short-ranged interactions. Young reviews, with emphasis on these problems, the most up-to-date results obtained by Monte Carlo simulations on Ising spin glasses in two to four dimensions. Matsubara et al. introduce a new hybrid algorithm which very efficiently simulates Heisenberg spin glasses and emphasize the importance of anisotropic interactions for these spin glasses in three dimensions to realize the spin glass phase. Kawamura proposes the possibility of a novel ordered phase associated with discrete degrees of freedom in the local order parameter (chirality) in vector spin glasses. The extension of concepts obtained in spin glasses to other related systems is also of current interest. Binder discusses the nature of glass transitions in orientational glasses, which exhibit similar behaviors, such as slow relaxational dynamics, to those observed in spin glasses.

For systems having more complex interactions and/or structures, more powerful computational approaches are necessary to investigate their cooperative nature. In fact, there are many interesting phenomena for which the computational method is, at present, a unique tool to open the possibility of grasping the essential mechanism of the phenomena. Various numerical methods applicable to complex systems are surveyed in Part VI, **Other Complex Systems**. Binder reviews results of extensive Monte Carlo simulations on polymer materials and emphasizes the role of computational methods in polymer physics, in particular those based on the coarse-grained model for polymers. Kaneko introduces the "Coupled Map Lattice" as a prototype model of nonlinear dynamical systems with many degrees of freedom and shows that it exhibits several fundamental universal behaviors, such as the occurrence of spatio-temporal chaos, expected in complex dynamical systems.

Finally, recent developments of computer facilities are reviewed in the last part, **Computer Facilities**. Loh reviews studies done with the help of the "Connection Machine" and Ito et al. give an explanation of their special purpose computer for personal use, the "Ising Machine". The Hypercubic machine is also briefly explained by Makivic in Part III.

Most of the chapters include several contributed papers. Although we could not introduce their contents here because of lack of space, we believe they will aid the reader, as much as the invited papers will do, in understanding current developments in the field.

Part I

First Principles Calculations

First Principles Methods for Correlated Ground States of Solids

D.R. Hamann

AT&T Bell Laboratories, Murray Hill, NJ 07974, USA

Abstract. This paper introduces three quantum Monte Carlo methods which are being explored for application to improved treatment of electron correlation in solids. These include the variational, diffusion, and auxiliary-field methods. The fundamental theory of each approach is reviewed, and recent progress in each is discussed.

1. Introduction

The correlated motion of electrons plays an important role in the bonding of atoms into molecules and solids. Most calculations of the ground state properties of real materials have only incorporated these correlations in an averaged way, using the so-called local density approximation. While this approach has been very successful for many properties and many classes of solids, it cannot be systematically improved. It typically gives good results for structural properties, but makes 10 to 20% errors in cohesive energies, and predicts qualitatively incorrect ground states for some materials such as Fe and La_2CuO_4. With the rapid growth of available computational power, it is of interest to explore methods which deal directly with electron correlations, and can in principle give exact results.

This talk will introduce three methods which are being explored today to accomplish this goal. They are the variational Monte Carlo method, the diffusion or greens function Monte Carlo method, and the auxiliary field Monte Carlo method. All three methods have seen only limited application to solids to date. They are all useful for extended periodic systems, and should all have reasonable scaling properties of computational effort with system size. (It is these two considerations that cause us to reject consideration of the configuration interaction method and related methods developed to treat correlations in small molecules.) The Monte Carlo methods are all good candidates for the "next generation" first-principles approach, and the relative merits of each should become better known as research proceeds over the next few years.

2. Variational Monte Carlo Method

The variational quantum Monte Carlo method employs an explicit correlated wave function. It was first introduced by McMillan in an application to liquid He.[1] The most widely used form for a system of M electrons can be written

$$\Phi(\vec{r}_1 \cdots \vec{r}_M) = \prod_{i=1}^{M} g(\vec{r}_i) \prod_{i,j=1}^{M} f(|\vec{r}_i - \vec{r}_j|)\Psi(\vec{r}_1 \cdots \vec{r}_M) , \quad (1)$$

where Ψ is a Slater determinant, f is a pair correlation function, and g is a one-particle function to be discussed below. We wish to evaluate the expectation values of various quantities, most typically the energy, using this wave function. We then vary parameters characterizing the wave function to minimize the energy. The integrals to be evaluated,

$$<H> = \frac{\int dR \Phi(R) H \Phi(R)}{\int dR |\Phi(R)|^2} \approx \frac{1}{N} \sum_{k=1}^{N} \Phi^{-1}(R_k) H \Phi(R) , \quad (2)$$

where we have abbreviated the M 3-dimensional electron coordinates (and, implicitly, spin) as R, are not separable as in the Hartree-Fock approximation, and cannot be done analytically. The goal of the method is to use statistical sampling to estimate the normalized integral as efficiently as possible. This is done by generating a random set of N coordinates $\{R_k\}$ with a probability distribution $|\Phi(R)|^2$. The sum in Eq. (2) over members of this set then approximates the integral. The better the wave function, the lower the energy and the smaller the variance, but the variance always decreases as $N^{-1/2}$.

It is not obvious that it is possible to generate such a distribution. This may, however, be accomplished in several ways. Consider an arbitrary probability distribution of a single variable $p(x)$, which is non-zero for $a<x<b$, and bounded by $p_M > p(x)$. Suppose we have a computer program available which generates pseudorandom numbers u uniformly distributed over the interval $[0,1)$. Let us generate a u, and a trial value $x_t = a+u(b-a)$. Now we generate a new u. If $p(x_t)/p_M > u$, we keep x_t as part of our sample. Otherwise we reject it and generate a new x_t. While it is easy to see that this approach accomplishes the desired objective, it is extremely inefficient when generalized to a multi-dimensional situation. $|\Phi(R)|^2$ is usually highly peaked in R-space, and almost all trial values would be rejected.

A better approach is to take our last accepted x_k, move it a random distance to get our trial $x_t = x_k + c(u-½)$, and compute $\eta = p(x_t)/p(x_k)$. We then obtain a new random u, and accept x_t as x_{k+1} if $\eta > u$ (including $\eta > 1$). If we reject x_t, we set $x_{k+1} = x_k$. This approach is the Metropolis algorithm, and is related to the principle of detailed balance in a system in thermal equilibrium.[2] We can adjust the parameter c at will to achieve a compromise between high acceptance (desirable) and high correlations between sample points (undesirable). This method is efficient for highly peaked multi-dimensional distributions, and is what is used in practice. We have discussed sampling in this case in great detail because similar considerations of strategy, acceptance, and correlation enter into all the Monte Carlo methods, but are most simply introduced here.

The first (and only) application of this variational method to solids was recently carried out by Fahy et al.[3] There were many details to be developed and tested. They chose Ψ to be a determinant of local density functional eigenfunctions, expecting these to be better than the Hartree-Fock functions usually used for atoms and molecules. Because of the large range of energies and distance scales involved in treating core electrons, they introduced pseudopotentials so as to deal with the valence electrons only. Otherwise, statistical noise from the cores would overwhelm the calculation of energies of interest for cohesive properties. To be accurate, first-principles pseudopotentials must be non-local operators, in the sense that different angular momentum components of the electron wave function experience different potentials at the atom cores.[4] Generalizing the statistical sampling method in a way to incorporate such operators was a major accomplishment.[3]

The purpose of the 2-body term f in Φ is to keep electrons apart, to reduce their repulsive coulomb energy. Fahy et al. followed the usual practice of using a form that exactly satisfies the 2-electron Schrödinger equation at short distances, and has a range which can be varied. Since this smooths the electron charge to an unphysical extent, they used the 1-electron term g in Φ to return this quantity towards the local density shape by a variable amount. Quantum many-body simulations must involve supercells rather than single unit cells of the solid to allow for the long range of the coulomb interaction. They used up to a 56 atom supercell, and developed scaling methods to extrapolate to the infinite-supercell limit.[3]

The error bars in their energy calculations are comparable to the change in energy accompanying a change in lattice parameter of a few percent, and the equilibrium lattice parameter and energy are extracted from a smooth fit. Their lattice parameter is within 0.03 Å of experiment for both diamond and silicon, and their predicted cohesive energy is within ~0.1 eV. This is an order of magnitude better than the local density approximation. Curiously, they find that the local density energy is very close to their variational energy for the solids, and that most of the improvement in the cohesive energy is from their variational free atom calculations.[3]

The major limitation of the variational Monte Carlo method is that there is no way to systematically improve the form of the guessed wave function. Even for a fixed form, it is difficult to vary many parameters, and not even practical to optimize the one electron states in the determinant.

3. Diffusion Monte Carlo Method

Strategies for systematically improving a trial wave function $|\psi_t\rangle$ can be based on the Schrödinger equation in imaginary time τ,

$$-\frac{d}{d\tau}|\Psi(\tau)\rangle = H|\Psi(\tau)\rangle \ ; \ |\Psi(0)\rangle = |\psi_t\rangle \ . \tag{3}$$

The formal solution to this equation expanded in exact many-body eigenstates Φ_i,

$$|\Psi(\tau)\rangle = \sum_i e^{-\varepsilon_i \tau} |\Phi_i\rangle\langle\Phi_i|\psi_t\rangle \ , \tag{4}$$

shows that in the large τ limit, $|\Psi(\tau)\rangle$ must approach the ground state $|\Phi_0\rangle$, since all excited states decay at faster exponential rates. The solution of Eq.(3) can be considered to project the exact ground state component out of the trial state.

One realization of such a projection approach is to make use of Eq. (3) directly in real space,

$$[-\frac{d}{d\tau} + \frac{1}{2}\nabla_R^2 - V(R)]\ \Psi(R) = 0\ , \qquad (4)$$

where R is the many-body coordinate introduced previously, and the potential V contains both the 1-body and 2-body potentials. This is a diffusion equation with branching. While it cannot be solved in closed form, an iterative approach in which the solution is advanced over a small time slice $\Delta\tau$ can be given,

$$\Psi(R,\tau+\Delta\tau) = \int dR'\, e^{-|R-R'|^2/\Delta\tau}\ e^{-\Delta\tau V(R')}\ \Psi(R',\tau)\ , \qquad (5)$$

which becomes exact as $\Delta\tau \to 0$. In effect, the local growth or decay part of Eq. (4) and the free-space diffusion part are treated as if they were independent.[5]

The multi-dimensional gaussian convolution in Eq. (5) faces the same difficulties we encountered for the variational method. A Monte Carlo approach can be devised based upon representing Ψ as a sum of δ functions,

$$\Psi(R',\tau) = \sum_i \delta[R'-R_i(\tau)]\ , \qquad (6)$$

so that the density of the set $\{R_i'\}$ approximates the amplitude of the wave function. A random process which produces suitable sets is to kill or replicate members of the set with relative probability $e^{-\Delta\tau V(R')}$, and then generate a new set $\{R_i(\tau+\Delta\tau)\}$ with a gaussian distribution of moves. (There are well known algorithms for this step.) In practice, this process is made more efficient if the trial function is used to "bias" or "guide" this random walk towards high-probability regions.[5]

While the process outlined above works fine for bosons, electrons are fermions, and their wave functions must be antisymmetric. The needed generalization of Eq. (6) is

$$\Psi(R) = \sum_i \delta[R-R_i^+] - \sum_i \delta[R-R_i^-]\ , \qquad (7)$$

where separate sets of "random walkers" R have been introduced to represent positive and negative regions of the wave function. Unfortunately, the separated $\{R^+\}$ and $\{R^-\}$ distributions started according to an approximate antisymmetric state quickly collapse to a common distribution, that representing the symmetric ground state, when evolved in τ. This effect, and the resulting exponential decay of the statistical "signal to noise ratio" is dramatically illustrated by Figs. 3 and 4 of Arnow et al.[6] It is known as the "fermion sign problem," and constitutes a major obstacle to developing exact simulation methods for electrons.

Antisymmetry alone cannot determine the location of the nodes of $\Psi_0(R)$. If the nodes of the trial function $\psi_t(R)$ are considered to be a reasonable approximation to the exact nodes, it makes sense to modify the random walk process to confine the $\{R^+\}$ and $\{R^-\}$ sets to the positive and negative regions of ψ_t. This so-called "fixed node" approximation was introduced by Anderson,[6] and later shown to give a variational upper bound on the energy.[5] In a representative recent application to what is considered a rather large molecular system in these circles, methane (CH_4), Garmer and Anderson showed that this method could obtain 89% of the experimental correlation energy with a moderate quality ψ_t, and 97% with their best ψ_t.[7]

Only very recent developments have permitted the application of the fixed-node diffusion Monte Carlo method to solids. The core electron problem must be dealt with, and standard non-local pseudopotentials [4] are incompatible with the method. Bachelet et al. have devised a way of incorporating many of the the angular-momentum dependent scattering properties of the pseudopotential into the kinetic energy operator as position-dependent "effective mass" functions.[8] Using such operators, Li et al. have studied everyone's favorite solid, silicon, and preliminary reports have indicated excellent results for the cohesive energy.[9]

Our limited ability to improve ψ_t (and hence its nodal structure) impose limitations on the fixed-node diffusion method that echo those of the variational method. In addition, the kinetic energy approach to the pseudopotential has constraints which apparently restrict its applicability to certain atoms. Considerably more experience will have to be gained to evaluate the accuracy and general utility of this approach for solids.

4. Auxiliary Field Simulations

The auxiliary field method is considered to be good for strongly correlated electrons, and had been widely applied to simple models. It offers several potential advantages for real materials calculations. Well-developed tools from applications of the local density approximation, such as basis sets and the best available pseudopotentials, may be incorporated. Fermion antisymmetry is built in analytically, so detailed nodal constraints are not needed. The formalism is suitable for the development of systematic approximations, which might supplement all-out simulations. However, it is also known to display "the" (more correctly, "a") fermion sign problem.

To introduce this method, let us drop back from a full materials Hamiltonian to the simplified Hubbard model, with a hopping form of kinetic energy on a lattice of sites, and strictly local interactions. (This is a tutorial rather than a real restriction, since we have already explored the path back to general interactions.[10]) The Hubbard model hamiltonian is

$$H = \sum_{ij\sigma} t_{ij} a^{\dagger}_{i\sigma} a_{j\sigma} + U \sum_{i} n_{i\uparrow} n_{i\downarrow} \, , \tag{8}$$

where $a^{\dagger}_{i\sigma}$, $a_{i\sigma}$, and $n_{i\sigma}$ are the second-quantized fermion creation, annihilation, and number operators, t_{ij} is the hopping term, and U the local interaction. Following the ground state projection strategy, we write the formal solution to the imaginary-time Schrödinger equation in the form

$$|\Psi(\tau)> = e^{-\Delta\tau H} \cdots e^{-\Delta\tau H} e^{-\Delta\tau H} |\psi_t> \, , \tag{9}$$

where we have already broken τ into small slices.

The interaction term can be rewritten as a square and a one-electron term using operator identities,

$$U n_{\uparrow} n_{\downarrow} = -\tfrac{1}{2} U (n_{\uparrow} - n_{\downarrow})^2 + \tfrac{1}{2} U (n_{\uparrow} + n_{\downarrow}) \, , \tag{10}$$

where site indices are dropped for simplicity. Eq. (9) is then transformed using the following gaussian integral identity,

$$e^{\Delta\tau U(n_\uparrow - n_\downarrow)^2} = \frac{1}{\sqrt{2\pi}} \int_{-\infty}^{\infty} dx e^{-\frac{1}{2}x^2 + \sqrt{U\Delta\tau}\,(n_\uparrow - n_\downarrow)} . \tag{11}$$

The result of this transformation is

$$|\Psi(\tau)> = \int \prod_{i=1}^{N} \prod_{j=1}^{\ell} dx_{ij} e^{-\frac{1}{2}\sum_{ij} x_{ij}^2} e^{-\Delta\tau H_\ell} \cdots e^{-\Delta\tau H_2} e^{-\Delta\tau H_1} |\psi_t> , \tag{12}$$

where the individual time-slice hamiltonians,

$$H_j = \sum_{ik\sigma} t_{ik} a_{i\sigma}^\dagger a_{k\sigma} + \sqrt{U/\Delta\tau} \sum_i x_{ij}(n_{i\uparrow} - n_{i\downarrow}) , \tag{13}$$

are one-electron operators.[11] The application of each of the "time slice propagators" $\exp(-\Delta\tau H_j)$ to a Slater determinant produces another single determinant, and these calculations can be carried out exactly.[12]

For any particular set of values of the auxiliary field x_{ij}, Eq. (12) represents non-interacting electrons propagating under the influence of a space and time varying external field. An expectation value like $<\psi_t|\Psi(\tau)>$ could be expanded in powers of this external field, represented by Feynman diagrams, with only electron loops and x's on them indicating interactions with the field. For any given diagram, performing the gaussian integration joins up the x's with interaction lines in all possible ways.

An attempt to evaluate the gaussian integrals of some expectation value by a Monte Carlo sampling approach quickly reveals noise diverging with τ, a fermion sign problem in another context. Aware that this was an obstacle which would block any application of these methods to solids, Fahy and I have recently reformulated the problem in a way which has led to some new insights. The gaussian distribution of x_{ij}'s in Eq. (12) can be considered to generate a distribution of Slater determinants, so that

$$|\Psi(\tau)> = \int d\psi f(\psi,\tau)|\psi> , \tag{14}$$

where the integral is over the manifold of Slater determinants, and f is the distribution.[13] A numerical exploration of this distribution for an extremely simple case is illustrated in Fig. 1. Here, we have a Hubbard model for an H_2 molecule. The determinants can be characterized by 2 continuous variables, represented here by the angles θ_1 for the spin up electron and and θ_2 for the spin down. $\theta=0$ corresponds to site 1, and $\theta=\pi/2$ to site 2. The system is started in a ψ_t corresponding to both electrons on site 1, and a set of points distributed according to $f(\psi,\tau)$ is generated at the indicated τ values. This distribution is seen to undergo diffusion, drift, and branching (ie. growth and decay), and evolves towards a Heitler-London like set of states at the largest τ.

The behavior in Fig. 1 is suggestive of a result that can be proved to be mathematically exact: $f(\psi,\tau)$ obeys a diffusion equation with a second-derivative ψ operator, a ψ-dependent force, and a ψ-dependent growth potential, all of which can be calculated explicitly.[13] The diffusion is caused strictly by the electron-electron interactions, not by the kinetic energy as in the method studied in Sec. 3. In general, this diffusion equation is much too complicated to solve directly. However, because

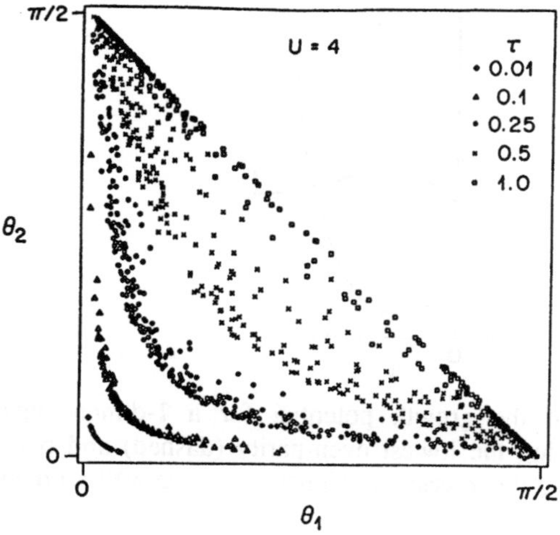

Fig. 1. Evolution in τ of points distributed according to the function $f(\psi,\tau)$ in Eq. (14) for a 2-site Hubbard model. ψ is the determinant $(cos\theta_1 a_{1\uparrow}^{\dagger} + \sin\theta_1 a_{2\uparrow}^{\dagger})$ $(cos\theta_2 a_{1\downarrow}^{\dagger} + \sin\theta_2 a_{2\downarrow}^{\dagger})|0>$, and the distribution is started as a delta function at $\theta_1 = \theta_2 = 0$. (From reference 13.)

it is symmetric under the parity operation $\psi \rightarrow -\psi$, we know from the general properties of such equations that the dominant solution at large τ must have even parity. Such an even-parity solution, f^+, gives $|\Psi>$ identically zero when substituted in Eq. (14). The leading odd-parity solution, f^-, gives the dominant non-vanishing contribution to the integral. In attempting to evaluate the integral by statistical sampling, we fail to achieve the exact cancellation of the f^+ contributions which we know must occur in an analytical solution, and the noise from this incomplete cancellation overwhelms the non-canceling f^- contributions.[13]

The above analysis suggested a new approximation procedure, which we have dubbed "positive projection."[14] Consider a simple diffusion equation with a one-dimensional ψ manifold and a double-well growth potential, as sketched in Fig. 2. The lowest eigenfunctions f^+ and f^- are sketched in (a). The modified problem in (b), in which an infinite wall (absorbing barrier) has been placed at the position of the f^- node, has as its leading solution f^*, which is identical to f^- in the allowed region.

The new approximation, then, consists of introducing a single nodal surface cutting the manifold of determinants, and modifying our procedures for sampling the auxiliary fields x_{ij} so that any field configurations which allow $|\Psi(\tau)>$ to cross this surface are rejected. This sampling does not prevent all cancellation in the integrations for physical expectation values and their normalization, but effectively places a finite lower bound on the signal to noise ratio at large τ.[14] While this is reminiscent of the fixed-node approximation discussed earlier, and was in fact inspired by it, it is a vastly less restrictive constraint.

Tests of the positive projection approximation have yielded expectation values that are difficult to distinguish from exact diagonalization results within the

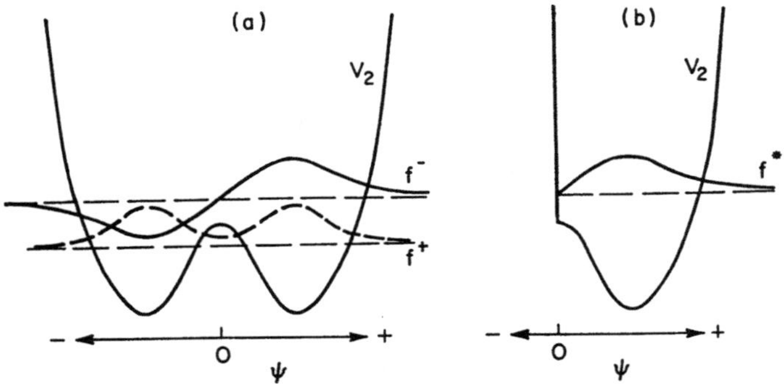

Fig. 2 Schematic representation of the growth potential for a 1-dimensional diffusion equation for $f(\psi,\tau)$, showing the lowest even-parity (dashed) and odd-parity (solid) eigenfunctions in (a), and the lowest eigenfunction for the well with an added infinite barrier in (b). (From reference 14.)

Table 1. Monte Carlo (3F-PP) and exact (DIAG) results and their difference Δ for an 8-site model described in the text. Energy, kinetic energy, and parallel and antiparallel pair correlation functions for neighbor shells (n) are given.

	3F-PP	DIAG	Δ
E	1.0217(40)	1.0218	0.0001
KE	-1.4668	-1.4668	0.0
ppc(1)	0.1771	0.1774	-0.0003
ppc(2)	0.2597	0.2607	-0.0010
ppc(3)	0.1896	0.1859	0.0037
apc(0)	0.2175	0.2161	0.0014
apc(1)	0.2531	0.2534	-0.0003
apc(2)	0.2490	0.2488	0.0002
apc(3)	0.2764	0.2774	-0.0010

uncertainties of statistical noise and $\Delta\tau \to 0$ extrapolation. A representative set of results for a model consisting of 8 sites arranged in a cube is given in Table 1. In this case, we have generalized from the simple local interaction and are treating an interaction that has values 2, 1, and 0.5 between electrons on the local, first neighbor, and second neighbor sites, respectively, in units where the hopping t is 1.[10] There are 8 electrons, and since the non-interacting problem has an s-like ground state and 3-fold degenerate p-like first excited states, this could be considered a "toy model" for electron correlations in the valence shell of a neon atom. The energy and kinetic energy per electron are given, along with the parallel and anti-parallel spin pair correlation functions for the several neighbor shells. While the analysis of the statistical errors is difficult and still incomplete, it is not clear that any of the departures from exact results are statistically significant.

5. Summary and Prospects

The three methods discussed are each at different stages of development for applications to solids. In the near term, the variational Monte Carlo method should be tried on a variety of different classes of materials. We need to know how large a fraction of the correlation energy it can be expected to recover. Are the excellent silicon and carbon results just a lucky accident? Only such testing will establish for which types of materials it is best, and for which it is worst.

The diffusion or green's function Monte Carlo method should develop in the medium term. The most important questions here are the limitations of the "kinetic pseudopotential," and the possibility of incorporating other types of pseudopotentials through controlled approximations. In conjunction with this, the question of the fixed node approximation and its relative effects in different classes of materials needs to be answered. Issues of statistical error estimation, time slice convergence, supercell scaling, and computational effort scaling must also be explored.

Auxiliary field calculations have never been applied to solids, and are a long-term (and perhaps long-shot) approach. All the issues discussed above plus questions of extensions from lattice models to the continuum limit, introduction of basis sets, and the robustness of the positive projection approximation need to be investigated. These questions will have to be answered before it is established that a solid-state calculation of useful precision can even be performed. The relative advantages among the various approaches I have discussed will probably not emerge for several years.

References

1. W. L. McMillan, Phys. Rev. A **138**, 442 (1965).
2. N. Metropolis and S. Ulam, J. Amer. Stat. Assoc. **44**, 335 (1949).
3. S. Fahy, X. W. Wang, and S. G. Louie, Phys. Rev. B **42**, 3503 (1990).
4. D. R. Hamann, M. Schlüter, and C. Chiang, Phys. Rev. Lett. **43**, 1494 (1979).
5. P. J. Reynolds, D. M. Ceperley, B. J. Alder, and W. A. Lester, Jr., J.Chem. Phys. **77**, 5593 (1982);
   D. M. Arnow, M. H. Kalos, M. A. Lee, and K. E. Schmidt, J. Chem. Phys. **77**, 5562 (1982).
6. J. B. Anderson, J. Chem. Phys. **65**, 4121 (1976).
7. D. R. Garmer and J. B. Anderson, J. Chem. Phys. **86**, 4025 (1987).
8. G. Bachelet, D. Ceperley, and M. Chiocchetti, Phys. Rev. Lett. **62**, 2088 (1988).
9. X.-P. Li, D. Ceperley, and R. M. Martin, Bull. Am. Phys. Soc. **36**, 875 (1991).
10. D. R. Hamann and S. B. Fahy, unpublished.
11. J. Hubbard, Phys. Rev. Lett. **3**, 77 (1959); R. L. Stratonovich, Dokl. Akad. Nauk. SSSR **115**, 1097 (1957) [Soviet Phys. Doklady **2**, 416 (1957)].
12. G. Sugiyama and S. E. Koonin, Annals of Phys. **168**, 1 (1986).
13. S. B. Fahy and D. R. Hamann, Phys. Rev. B **43**, 765 (1991).
14. S. B. Fahy and D. R. Hamann, Phys. Rev. Lett. **65**, 3437 (1990).

Computational Physics Approach to Scanning Tunneling Microscopy and Spectroscopy

M. Tsukada[1], *K. Kobayashi*[1], *H. Kageshima*[2], *N. Isshiki*[3], and *S. Watanabe*[4]

[1]Department of Physics, University of Tokyo, Hongo 7-3-1, Bunkyo-ku, Tokyo 113, Japan
[2]NTT LSI Laboratories, Morinosato-Wakamiya 3-1, Atsugi-shi, Kanagawa-ken 243-01, Japan
[3]Institute for Knowledge and Intelligence Science, Kao Corporation, Bunka 2-1-3, Sumida-ku, Tokyo 113, Japan
[4]Aono Atomcraft Project, ERATO, JRDC, Kaga 1-7-13, Itabashi-ku, Tokyo 173, Japan

Abstract. Bardeen's formula of the tunneling current is extended to include atomic features of the tip-surface system of scanning tunneling microscopy (STM). The electronic states of the sample surface and the tip is calculated by the local density functional approach (LDA) and they are used as the input data for the numerical simulation of the STM image. Normal images are resulted by the tip with a single apex atom, while abnormal images tend to be caused by the tip with several atoms on the tip apex. For the former type of the tips, tunnel current concentrates dominantly on a single apex atom. Some numerical results are presented, which demonstrate the importance of the first-principles simulation resolving puzzles appeared in STM process.

1. Introduction

Scanning tunneling microscopy (STM) and spectroscopy (STS) opened a new epoch of surface science, by providing very clear atomistic images of surface in a direct way[1]. Mechanism of STM is rather simple, i.e., the image is obtained by scanning the sample surface with a sharp metal tip attached to a piezoelectronic device with monitoring the tunnel current. It is surprising that very clear image of atoms on the surface can be obtained with this seemingly simple mechanism. However, in spite of the rapid progress in experimental technique, microscopic feature of the electron tunneling between the tip and the surface is far from trivial. How quantitatively can experimental data of STM/STS be analysed to deduce the information of the surface in a microscopic level? And how does the microscopic state of the tip influence the experimental data? To resolve these fundamental problems and to establish a firm basis of STM/STS, numerical simulations based on the computational physics approach play very essential role. In this article, we present the LDA theory of STM/STS simulation developed in our group recently[2], with some interesting examples showing crucial role of the first-principles simulation.

Springer Proceedings in Physics, Vol. 70 Computational Approaches in Condensed-Matter Physics
Editors: S. Miyashita · M. Imada · H. Takayama

2. The Formalism of the Simulation

Based on the Bardeen's perturbation theory[3], the tunnel current can be expressed as [4]

$$I = \frac{2\pi e}{\hbar} \int dE[f(E) - f(E + eV)]$$
$$\times \int_{\Omega_T} d\mathbf{r} \int_{\Omega_T} d\mathbf{r}' V_T(\mathbf{r}) V_T(\mathbf{r}') \mathcal{G}^T(\mathbf{r}', \mathbf{r}; E + eV) \mathcal{G}^S(\mathbf{r} + \mathbf{R}, \mathbf{r}' + \mathbf{R}; E). \quad (1)$$

In the above, $\mathcal{G}^S$ and $\mathcal{G}^T$ are the imaginary part of the Green's function of the surface and the tip, respectively. The quantity V is the bias voltage, $\mathbf{R}$ represents the tip position and V_T is the potential of the tip. To obtain the Green's functions, the electronic states of the sample surface and the tip are calculated by the local density functional methods (LDA) in the separated limit. To facilitate the numerical calculations, slab model consisted of several atomic layers are used for the sample surface, while cluster models are used for the tip. If the distance between the tip and the surface is not very large, the LCAO construction of the surface wave function can be utilized. For the case of large distance between the tip and the surface, the so called connected vacuum tail method[5] is utilized to amend shortage of the LCAO in representing very tail part of the wave function. In order to reduce the effect due to the discreteness of the cluster levels, each level is broadened by the Lorenzian function.

3. STM Image of Graphite and the Tip Effect

The numerical current images of graphite with the cluster model of the tip, $W_{10}[111]$, $W_9[111]$, $W_{14}[110]$ and $W_{13}[110]$ are shown in Fig.1(a)-(d), respectively[6]. The tip-surface distance is 2.6 Å and the tip bias voltage is −0.5V. By detailed analysis of tunnel current distribution, it is found that more than 90% of the tunnel current is concentrated on a single apex atom for the case of $W_{10}[111]$ and

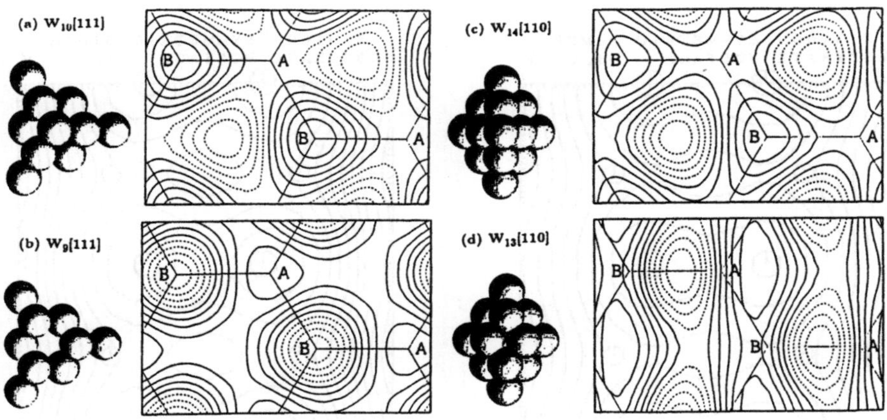

Fig.1 Calculated current images of graphite with cluster models of (a) $W_{10}[111]$, (b) $W_9[111]$, (c) $W_{14}[110]$, and (d) $W_{13}[110]$ tip. The area enclosed by the solid lines shows higher current and the dashed-lines shows lower current. A and B are the A- and B-site of graphite lattice, respectively.

W_{14}[110] tip. These STM images are considered normal, since the image hardly changes with the rotation of the tip around its axis. In the normal STM image, the strongest current region, the saddle point region, and the weakest current region appear at the B site, A site and the hexagon center of the graphite lattice, respectively. The image obtained by the W_9[111] tip shows a uniform shift relative to the graphite lattice structure. This is because, for example, when the tip axis is located on B site, the three W atoms on the top layer come very close to the hexagon center, where the tunnel current is the weakest. In the image by W_{13}[110] tip, the contour line flows linearly to the y direction and the STM image is quite dissimilar to the graphite lattice. This abnormal image is caused by a strong interference effect between the current elements contributed by the two W atoms on the short diagonal of the rhombus of the top tip plane[7].

4. STM/STS of Si(100) Reconstructed Surfaces

Fig.2(a) and 2(b) show the calculated image of the Si(100) 2×1 symmetric dimer surface for the tip height 5.3Å for the occupied states and unoccupied states, respectively. On the other hand, Fig.3 are the corresponding images for the tip height, 10.6Å[5]. The images of the occupied states are calculated for the surface bias −1.0V, and those of the unoccupied states are for the surface, 1.0V. As for the tip, W_{10}[111] cluster model is used. For the shorter tip-surface distance(5.3Å) we found no significant difference between the simulated image of the occupied and the unoccupied surface states. On the other hand for the longer distance(10.6Å), a remarkable difference appears in the STM image by the bias polarity. Namely for the occupied state, the maximum position of the tunnel current runs through the center of each dimer row, while for the unoccupied state a large valley develops along the center of each dimer row. Moreover with the increase of the tip-surface distance, the ridges of maximum current zones for the unoccupied states are pushed outwards of the dimer row, until finally the strong current region locates on the trough between the dimer rows. This trend corresponds well with the experimental observation[8].

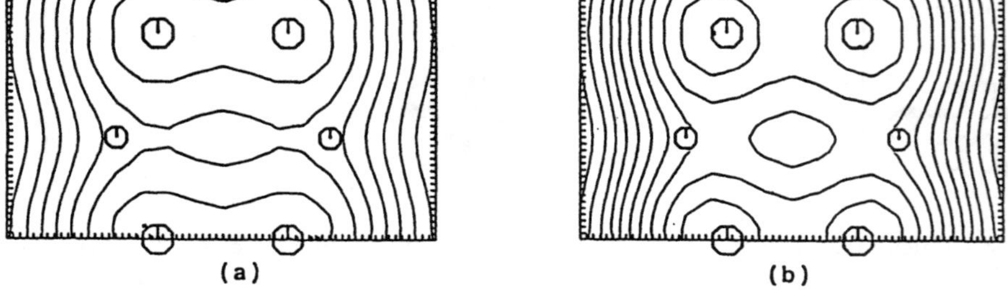

Fig.2 Calculated STM image of Si(100) 2×1 symmetric dimer surface. The tip height is 5.3Å. Surface bias is (a) −1.0V and (b) +1.0V.

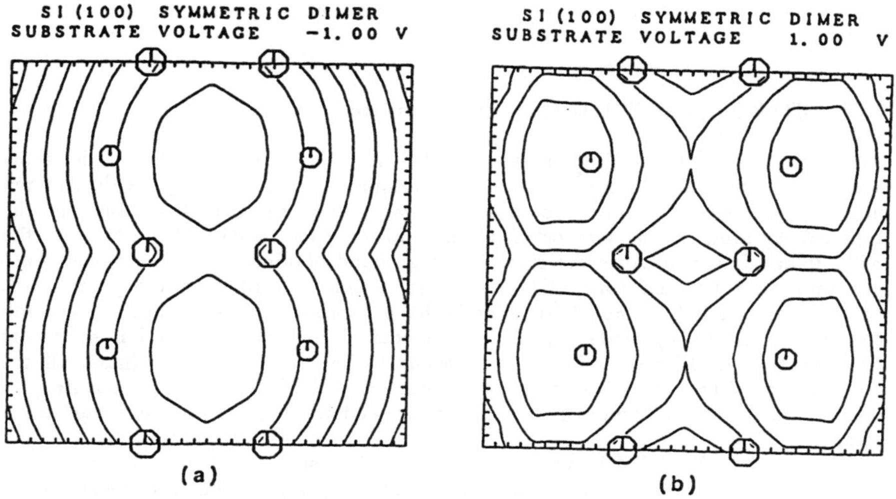

Fig.3 Calculated STM image of Si(100) 2×1 symmetric dimer surface. The tip height is 10.6Å. Surface bias is (a) −1.0V and (b) +1.0V.

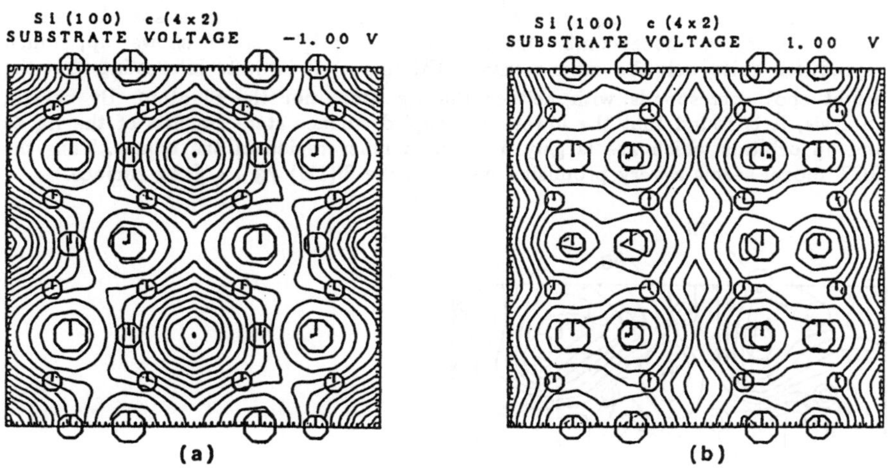

Fig.4 Calculated STM image of Si(100) c(4×2) symmetric dimer surface. The tip height is 5.3Å. Surface bias is (a) −1.0V and (b) +1.0V.

For the smaller tip distance(Fig.2), the node of the unoccupied surface states cannot be seen. There are two reasons for this. Firstly, the overlap between the two surface bands with the bonding and the antibonding character results in the coexistence of the tunnel current elements from both the bands irrespective of the bias polarity. Secondly, the tunnel current components via the tip orbitals as p_x, p_y, d_{xz}, d_{yz}, d_{xy} become maximum at the nodal lines of the surface wavefunctions. The STM images of the asymmetric Si(100) c(4×2) dimer surface are shown in Fig.4 for the surface bias −1.0V, 1.0V and the tip height 5.3Å. In this case, the images do not change so much with the tip-surface distance. The simulated image reproduces the characteristic features of the observed zigzag pattern near the step or defect regions.

5. The STM Image of Si(111)$\sqrt{3} \times \sqrt{3}$-Ag Surface

Although the observed STM image revealed the distribution of the bright spots arranged in the honeycomb structure, the atomic structure can not be determined uniquely. It seemed quite reasonable to assign the bright region either to the topmost Ag atom or the topmost Si atom. However such naive interpretation of the STM image failed, because the surface energy band obtained by either model could not reproduce the semiconducting character, which is experimentally established. On the other hand, Aono et al recently have proposed the so called honeycomb chained trimer(HCT) model based on the coaxial impact collision ion scattering spectroscopy (CAICISS)[9]. This model seems to be consistent with almost all the structure sensitive surface experiments except the STM. Thus it is interesting to clarify the electronic structure of this surface and the theoretically expected STM image.

We calculate the STM image of the HCT model using a slab model consisted of six atomic layers[10]. For the model of the tip W_{10}[111] is used. As the results of the calculation, a distinct energy gap of 0.54 eV at the Fermi level is revealed. The overall electron density of states corresponds fairly well with the UPS observation and the inverse photoemission. A typical STM current image calculated is shown in Fig.5. The surface bias voltage is 2.1V and the surface-tip distance is 3.7Å. As seen in the figure, a honeycomb structure of the bright spots shows up, which reproduces quite well the experimental STM image. It is quite interesting that the bright spot corresponds with neither the Ag atom nor the Si atom. By careful analysis, it is found that the lowest unoccupied surface state consists of the Ag 5s and 5p orbitals which are largely distributed inner region of the trimer. Thus the HCT model seems a very promising model of the Si(111) $\sqrt{3} \times \sqrt{3}$-Ag surface.

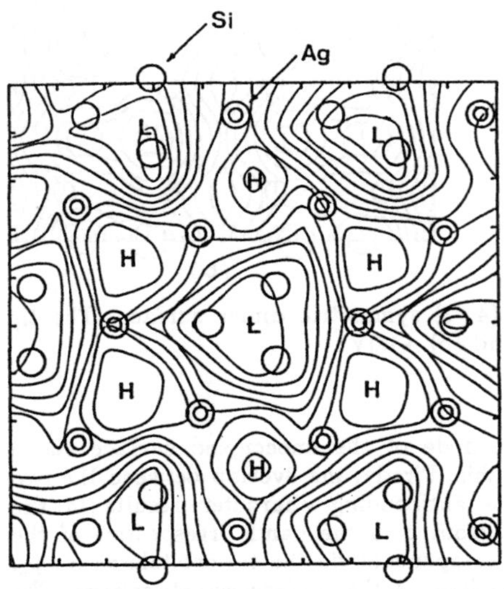

Fig.5 Calculated STM image of Si(111) $\sqrt{3} \times \sqrt{3}$-Ag surface. Double circle and single circle denote atom positions of Ag and Si, respectively. H and L indicate the maxima and minima of the tunneling current, respectively.

6. Summary

The method of a theoretical simulation of the STM image is presented based on the first-principles LDA theory of the tip/sample electronic states. Remarkable fact found here is that the tunnel current strongly concentrated on a single apex atom of the tip. This fact explains the atomic resolution of the STM, even though the macroscopic curvature of the tip usually exceeds several hundreds of Angstroms. On the other hand, if the apex of the protrusion of the tip is formed by several atoms, abnormal images tend to be formed.

The examples of the simulation of the STM image of the graphite surface and Si(111) $\sqrt{3} \times \sqrt{3}$-Ag surface tell us that interpretation of the image is, even qualitatively, often not trivial. In such cases, the theoretical simulation provide essential clue for decoding the information of the STM image. There are a number of difficult problems in STM and related technology, which awaits theoretical studies, in particular by means of the approach of computational physics.

References

[1] for example, Proceedings of the STM conferences, J. Vac. Sci. and Technol. **A8**, (1990), ibid **B9**, (1991).
[2] M.Tsukada, K.Kobayashi, N.Isshiki, and H.Kageshima, Surface Sci. Reports **13**, 265 (1991).
[3] J.Bardeen, Phys. Rev. Lett. **6**, 57 (1961).
[4] M.Tsukada and N.Shima, J. Phys. Soc. Jpn. **56**, 2875 (1987).
[5] H.Kageshima, Doctor Thesis, University of Tokyo, 1991.
[6] N.Isshiki, K.Kobayashi, and M.Tsukada, J. Vac. Sci. and Technol. **B9**, 475 (1991).
[7] N.Isshiki, K.Kobayashi, and M.Tsukada, Surface Sci. **238**, L439 (1990).
[8] T.Sakurai and T.Hashizume, private commun.
[9] M.Katayama, R.S.Williams, M.Kato, E.Nomura, and M.Aono, Phys. Rev. Lett. **67**, 267 (1991).
[10] S.Watanabe, M.Aono, and M.Tsukada, Phys. Rev. **B**, in press.

Electronic Structures of Fullerenes and Fullerides

S. Saito, A. Oshiyama, and Y. Miyamoto

Fundamental Research Laboratories and
Microelectronics Research Laboratories, NEC Corporation,
Miyukigaoka, Tsukuba, Ibaraki 305, Japan

Abstract. We report the electronic structures of C_{60} cluster, solid C_{60}, K_3C_{60}, and $C_{60}Br$ obtained by the density-functional theory. The C_{60} cluster has a closed shell electronic structure with relatively large energy gap, and solid C_{60} becomes a semiconductor. K_3C_{60} is found to be an exotic *ionic metal* and have high density of states at the Fermi energy, which seems responsible for the observed superconductivity. On the other hand, the high electronegativity of C_{60} prevents hole injection in the valence band and instead the midgap state consisting of Br $4p$ state appears in $C_{60}Br$.

1. Introduction

Since Kroto *et al.* [1] proposed truncated-icosahedron structure for C_{60}, C_{60} and related clusters ("fullerenes"), C_N and C_NX (X is a trapped atom or molecule in the cage of C_N), have been expected to be a possible constructing unit of new materials. Actually, the macroscopic quantities of C_{60} have been produced recently and C_{60} is found to form crystalline solid [2]. Moreover, K-doped solid C_{60}, K_3C_{60}, is found to be a superconductor with the transition temperature T_c=18K [3]. T_c has been found to vary with dopant, and the highest T_c is now 33K in $Cs_2Rb_1C_{60}$ [4]. In the present paper, we report the electronic structures of C_{60} cluster, solid C_{60}, and doped solid C_{60} ("fullerides"), K_3C_{60} and $C_{60}Br$, obtained by using the state-of-the-art density-functional theory with the local-density approximation [5].

2. Method

In the present calculation, norm-conserving pseudopotentials [6] with the Gaussian basis functions [7] are used. The electronic structures of fullerenes are expected to be described well since it has successfully given the electronic structures of both graphite (sp^2 C atoms) and diamond (sp^3 C atoms) and since fullerene C atoms are considered to have mainly

Springer Proceedings in Physics, Vol. 70 Computational Approaches in Condensed-Matter Physics
Editors: S. Miyashita · M. Imada · H. Takayama

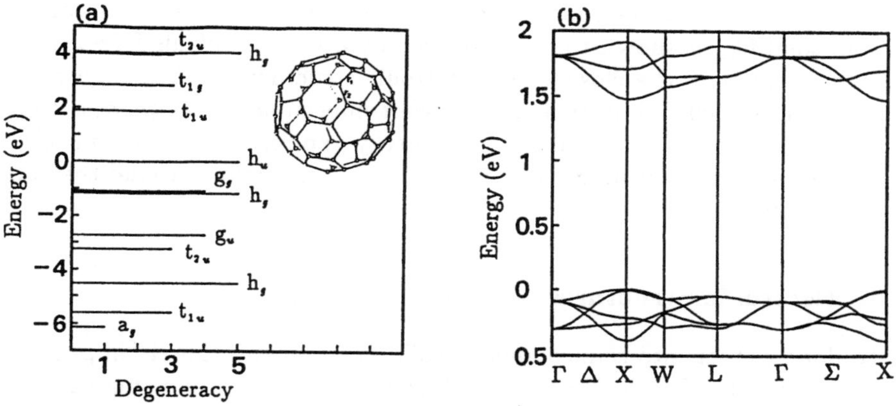

Fig.1(a) π states of the C_{60} cluster. Energies are measured from the highest-occupied state. (b) Energy bands of the fcc solid C_{60} around the gap. The valence-band top is defined as the zero energy.

sp^2 character with small amount of sp^3 component [8]. Throughout the calculation, C–C bond lengths are fixed at r_1=1.46 Å and r_2=1.40 Å (see Fig.1), which have been given by the NMR experiment [9]. The orientation of the C_{60} cluster in solid C_{60} and fullerides is chosen so that the point group has the highest symmetry.

3. Electronic Structures of C_{60}

Since C atoms in C_{60} mainly has sp^2 character, C_{60} electronic states can be classified into σ and π states. σ states are found to have large energy gap between bonding and antibonding states and most of the π states are located at the energy gap of σ states [8]. Because π states play a main role in the physical properties of fullerenes and fullerides, π-state distribution of the C_{60} cluster is shown in Fig.1(a). A rather large energy gap between the bonding and the antibonding states is also found in π states, indicating the high stability of the truncated-icosahedron structure. In addition, shell structure to be expressed by spherical harmonics Y_{lm} expected for electrons on a sphere can be clearly seen. The deepest state, a_g, corresponds to l=0 (s state). The next two states, t_{1u} and h_g, correspond to l=1 and 2 (p and d) states, respectively. The seven-fold degenerate l=3 (f) state splits into t_{2u} and g_u states because of the icosahedral potential field. The nine-fold degenerated l=4 state shows very small splittings into g_g and h_g states. Both the highest-occupied state, h_u, and the lowest-unoccupied state, t_{1u}, are the substates of l = 5 state.

From the cohesive energy and the valence-electron density distribution, solid C_{60} is shown to be condensed by a van der Waals force, since the C_{60} cluster has a closed-shell electronic structure [8]. Although both the highest-occupied state and the lowest-unoccupied state possess considerable dispersion in solid C_{60}, the energy gap remains finite and solid C_{60} becomes a semiconductor. In Fig.1(b), energy bands around the gap are shown.

4. Electronic Structures of Doped Solid C_{60}

To dope carrier into semiconductor solid C_{60}, substitutional "impurity" $C_{60}K$ cluster (K is inside the cage) and the bare K atom impurity at interstitial sites have been proposed [8]. (In $C_{60}K$, the valence electron of K atom is found to transfer to the outer C_{60} cage [10]). Actually, K-doped solid C_{60} has been found to become conductive [11] and even superconducting [3]. There are two tetrahedral interstitial sites and one octahedral interstitial site per fcc unit cell. K_3C_{60}, where K atoms occupy all the intersitial sites, has been identified as a superconducting phase and the C_{60} orientation is found to be consistent with our assumption [12].

The calculated electronic structure [13] shows that K_3C_{60} is an ionic solid where K valence electrons transferred to C_{60}. However, unlike ordinary ionic crystal, K_3C_{60} is a *metal* since transferred electrons partially fill conduction bands. Metallic kinetic energy gain as well as the Madelung energy gain is found to contribute to the cohesion of this unusual solid [13]. In Fig.2(a), the density of states (DOS) of K_3C_{60} is shown. As we can see from the figure, the Fermi level is located very close to the peak of the DOS. This rather high DOS at the Fermi energy, $N(E_F)$, may be responsible for the superconductivity. As is well known in the BCS theory, higher $N(E_F)$ usually gives higher T_c. It is consistent with

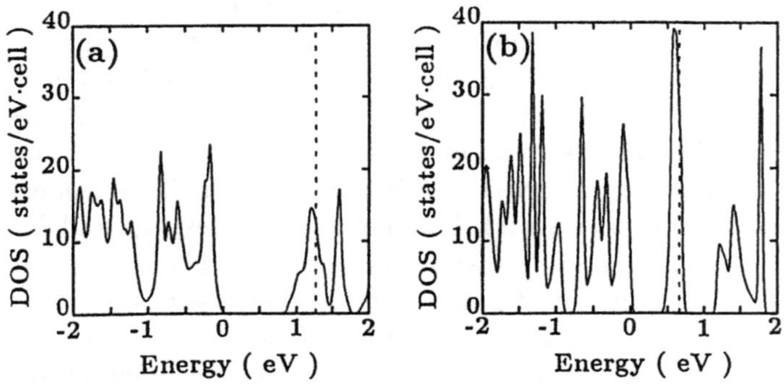

Fig.2(a) K_3C_{60} density of states. (b) $C_{60}Br$ density of states.

the observed T_c in Rb_3C_{60} (29K) [14] and $Cs_2Rb_1C_{60}$ (33K) [4]. The bigger dopants, Rb and Cs, make distance between C_{60} clusters longer, giving less intercluster overlap and less dispersion of energy bands, and, consequently, the higher $N(E_F)$. The photoemission spectra of K_xC_{60} have been reported and show that K_3C_{60} has a metallic electronic structure [15]. The agreement between our calculated DOS and photoemission spectra has been found to be excellent [16].

Since the electron-doped C_{60} has shown conductivity and superconductivity, it is worth examining hole doping to solid C_{60}, which is also an important step towards the application of C_{60} as a semiconductor material. We therefore studied the electronic structures of Br-doped solid C_{60} [17], since halogen atoms are known to give holes to the graphite-intercalation compounds. In Fig.2(b), we have shown the DOS of $C_{60}Br$, where Br is on the tetrahedral interstitial site. The Br 4*p* state, however, is found to appear in the midgap. The high electronegativity of C_{60} prevents the hole injection in the valence band.

5. Concluding Remarks

The present electronic-structure studies on C_{60} and its solid-state derivatives have revealed their novel properties. A great variety of different-size fullerenes will be available in the near future. Fullerides consisting of them will also show many interesting aspects. Larger fullerenes will have smaller energy gap and halogen atoms will be able to induce holes in their valence band. In addition, C_N X, fullerenes with trapped molecules, will also be available soon. Electronic-structure calculations willl continue to provide important and essential infromation in this new material field.

Acknowledgements

We would like to thank useful discussions with Dr. N. Hamada and Dr. S. Sawada.

References

1. H. W. Kroto, J. R. Heath, S. C. O'Brien, R. F. Curl and R. E. Smalley: Nature **318**, 162 (1985).
2. W. Krätschmer, L. D. Lamb, K. Fostiropoulos, and D. R. Huffman: Nature **347**, 354 (1990).
3. A. F. Hebard *et al.*, Nature **350**, 600 (1991).

4. K. Tanigaki, T. W. Ebbesen, S. Saito, J. Mizuki, J. S. Tsai, Y. Kubo and S. Kuroshima: Nature **352**, 222 (1991).
5. P. Hohenberg and W. Kohn: Phys. Rev. **136**, B864 (1964); W. Kohn and L. J. Sham: Phys. Rev. **140**, A1133 (1965).
6. G. B. Bachelet, D. R. Hamann and M. Schlüter: Phys. Rev. B **26**, 4199 (1982).
7. Y. Miyamoto and A. Oshiyama: Phys. Rev. B **41**, 12680 (1990).
8. S. Saito and A. Oshiyama: Phys. Rev. Lett. **66**, 2637 (1991).
9. R. D. Johnson *et al.*, in *Clusters and Cluster-Assembled Materials*, ed. R. S. Averback *et al.* (Mat. Res. Soc. Proc. **206**, Pittsburgh, 1991) p.175.
10. S. Saito: in *Clusters and Cluster-Assembled Materials* (ref.9) p.115.
11. R. C. Haddon *et al.*: Nature **350**, 320 (1991).
12. P. W. Stephens *et al.*: Nature **351**, 632 (1991).
13. S. Saito and A. Oshiyama: Phys. Rev. B **44**, 11 536 (1991).
14. K. Holczer *et al.*: Science **252**, 1154 (1991); M. J. Rosseinsky *et al.*: Phys. Rev. Lett. **66**, 2830 (1991).
15. P. J. Benning *et al.*, Science **252**, 1417 (1991); G. K. Wertheim *et al.*, Science **252**, 1419 (1991); C. T. Chen *et al.*, Nature **352**, 603 (1991).
16. N. Hamada, S. Saito, Y. Miyamoto, and A. Oshiyama, Jpn. J. Appl. Phys. **30**, L2036 (1991).
17. Y. Miyamoto, A. Oshiyama, and S. Saito: (to be published).

Structure Optimization and Dynamics of Fine Particles in LCAO Methods

C. Satoko

Department of Physics, College of Humanities and Sciences, Nihon University, Sakurajousui, Setagaya-ku, Tokyo 156, Japan

Abstract. We apply the local density functional method to perform the structure optimization of microclusters. The kinetic energy is taken to be the Thomas-Fermi and Weizsaker terms with modified coefficients. We vary the total energy with respect to the geometric structure and the charge density, which is approximated by the linear combination of atomic charge basis set. Using this procedure we are able to obtain the minimum total energy output of potassium chlorides, magnesium oxides and sodium-potassium mixed microclusters. With this information we can analyze the stability of these clusters.

1. Introduction

Due to the development of supercomputers, we are able to calculate precisely electronic structures in semiconductor systems using ab-initio molecular dynamics. Almost all of these calculations are based on a plane wave basis set technique. These electronic structure calculations are possible because the pseudo-potential is compatible with the semiconductors. But, it is not easy for the plane wave sets to be adapted for metals, especially transition metals because their orbitals are not broad, but are local. On the other hand, it is possible to describe the local orbitals and the chemical properties such as bonds via the atomic basis set. Force analyses of various materials have been developed. However, the integration method used in the atomic orbital basis is too time consuming to calculate the electronic structure in more than the ten thousand geometries.

So we have developed a total energy calculation method within the local density functional formalism. However, it is not known whether the charge density description of kinetic energy is applicable. In this paper we assume the kinetic energy terms(Thomas-Fermi and Weizsaker terms) and apply this method to alkali-halides, magnesium oxides and sodium-potassium mixed metal clusters.

2. Calculation methods and results

We assume the kinetic energy in the local density functional energy to be the sum of the Thomas- Fermi and Weizsaker term in the following form:

$$E_{kin} = \frac{3(3\pi^2)^{\frac{2}{3}}}{10} C_T \int \rho(r)^{\frac{5}{3}}\, dv + \frac{1}{72} C_w \int \frac{|\nabla\rho(r)|^2}{\rho(r)}\, dv$$

where $\rho(r)$ is the charge density of the system and the coefficients c_T and c_W are introduced for the adjustable parameters of the kinetic energy. The exchange and correlation energies are given by the Wigner interpolation formula.

We assume the general charge density to be the linear combination of some bases which are chosen by the square of the atomic orbital functions. The charge density coefficients are varied just like the usual LCAO methods. So we calculate the binding energy of the clusters to give us the minimum of the density functional total energy. The kinetic and correlation energies are calculated by using Gaussian quadrature integration. The other terms are analytically integrated.

We calculate a potential curve of KCl using the parameters C_T=1.0 and C_W=0.0. The equilibrium bond distance is near the experimental value, but the binding energy is negative. This is due to an overestimation in the kinetic energy. When we scale the Thomas-Fermi term to get the correct electron ionization energy and affinity of each atom(the scaled C_T are 1.200 for K, 1.131 for Na, and 0.797 for Cl), the calculated binding energy value corresponds with the experimental values. Choosing the scaled value of the Thomas Fermi term as 1.143 for Mg and 0.895 for O and the Weizsaker term as 0.006 for Mg and 0.01 for O, we achieve a reasonable equilibrium bond distance of 3.3 Bohr for the magnesium oxides and an orbital population $Mg^{+0.6}O^{-0.6}$.

We used this calculation method to see if we could find the "magic numbers" of sodium clusters. Here we assume the following to save computer time. First, the electron densities of sodium metal clusters are fixed at the sum of each neutral atom. This assumption is pretty good considering how strong the screening effect is. Secondly, the modified parameters of the kinetic energy are not scaled. We minimize the total energy about the geometries. The numerical derivative force is used for the search of the minimum points. The calculated total energies per atom show the minimum to be at the magic number of the clusters. The stabilities correlate with the kinetic energy of the cluster. The calculated bond distance is near the experimental values, but the binding energy is about half the experimental values. When a potassium atom is mixed with a sodium cluster, the potassium atom is apt to be inside of the sodium cluster.

In conclusion, our assumption for the kinetic energy terms works well for various materials. Our results in detail will be published in the future.

Electronic Structures of Hetero-Crystalline Semiconductor Superlattices

T. Nakayama and M. Murayama

Department of Physics, Faculty of Science, Chiba University, Yayoi, Chiba 260, Japan

Abstract. Crystal structures of hetero-crystalline superlattices are proposed and band structures of (hexagonal-diamond Si)/(cubic-diamond Si) and (wurzite ZnS)/(zinc-blend ZnS) superlattices are calculated using the first-principles pseudopotential method within LDA. It is shown that, in both superlattices, the states around the fundamental gap show charge densities localized in one constituent layers, indicating that these superlattices give good quantum well systems. Band offsets are also evaluated.

1. Introduction

In this paper, we propose the crystal structure of a superlattice (SL) of a new category; the hetero-crystalline superlattice. Hetero-crystalline SL is a superlattice which is made of materials with different crystal structures. Although hetero-crystalline SL has not been synthesized in laboratories yet, hetero-crystalline interfaces are sometimes observed as a stacking-fault in the crystals of semiconductors. From the results of experiments [1], it is found that Si has a stacking-fault which contains a hetero-crystalline interface between the hexagonal-diamond and cubic-diamond structures. On the other hand, the epitaxial growth of the materials whose structure is different from that of the substrate is one of the recent topics in the field of the crystal growth. In this view, in order to clarify the electronic structure of hetero-crystalline SL, we select Si and ZnS as candidates of constituents of SLs and calculate the band structures of (hexagonal-diamond Si)/(cubic-diamond Si) and (wurzite ZnS)/(zinc-blend ZnS) SLs using the first-principles pseudopotential method within a Local Density Approximation.

2. Results and Discussions

First, we summarize the optical properties of the hetero-crystalline SLs obtained from the calculated results of band structures. Irrespective of the layer thickness, Si-SLs are semiconductors with indirect band gap of energy around 1.1eV.[2] (Fig.1) In a Si-SL, there appears a *pseudo-direct* optical transition of energy around 1.9eV far below the direct interband transition of energy 3.3eV in bulk Si. However, the strength of that transition is small. This is because the lowest conduction band state originates from the folded state at L in the fcc Brillouin zone and there is little intervalley mixing between states at Γ and L points. On the other hand, ZnS-SLs are semiconductors with direct band gap of energy around 3.8eV, where the strength of absorption spectrum is as large as that in bulk.

Although the topology of the crystal structure is different between constituent materials of hetero-crystalline SL, a *quantum well picture* is applicable to states around the fundamental gap. For example, the heavy-hole state is strongly localized in hexagonal-diamond layers in Si-SLs (Fig.2) or zinc-blend layers in ZnS-SLs. Calculating the band

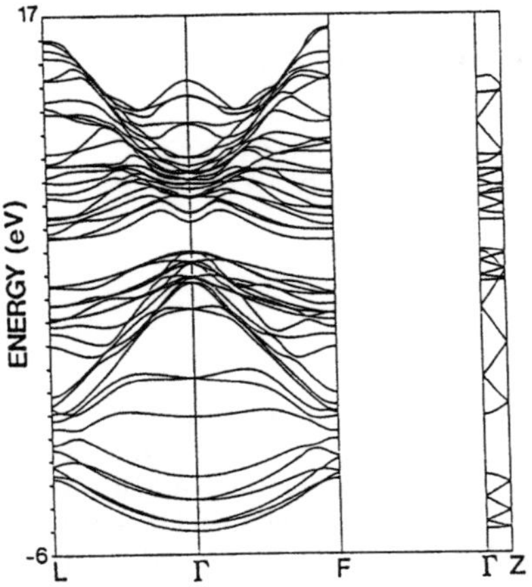

Fig.1. Calculated band structure of (hexagonal – diamond Si)$_2$/(cubic–diamond Si)$_4$. The Brillouin zone is trigonal.

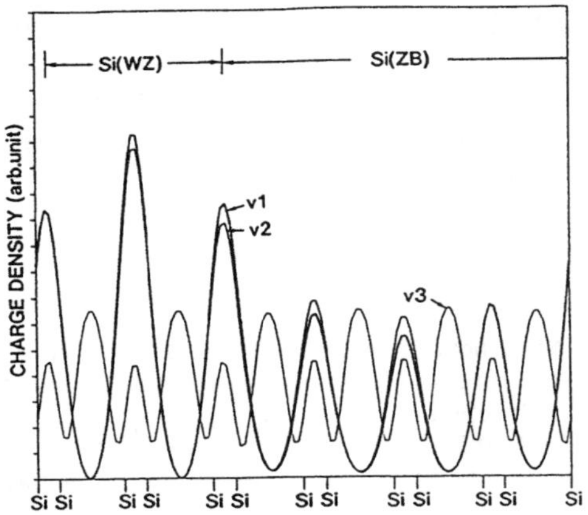

Fig.2. Charge densities of the heavy-hole (v1 and v2) and light-hole (v3) states in (hexagonal – diamond Si)$_2$/(cubic – diamond Si)$_4$. Si(WZ) indicates hexagonal layers and Si(ZB) cubic ones.

offset in a familiar way [3], Si-SLs are of type I in real space with hexagonal layers as a well and of type II in momentum space with the lowest conduction band valley at X in the fcc Brillouin zone. On the other hand, ZnS-SLs are of type II in real space with wurzite and zinc-blend layers as wells for the hole and electron states, respectively, and of type I in momentum space.

This work was partly supported by a Grant-in-Aid from the Ministry of Education, Science and Culture, Japan.

References

[1] M.Saito, K.Hiraga and K.Sumino, Jpn. J. Appl. Phy. **19**, L-195 (1980); K.Hiraga, M.Hirabayashi, M.Sato and K.Sumino, Crystal Research and Technology **17**, 189 (1982).

[2] In order to correct the band gap energies obtained in LDA, all the band gap energies are constantly shifted by +0.7eV in Si systems and +0.4eV in ZnS systems using a scissor operator. See for example, D.M.Bylander and L.Kleinman, Phys. Rev. **B34**, 5280 (1986).

[3] See for example, C.G. van de Walle and R.M.Martin, J. Vac. Sci. Tech. **B3**, 1256 (1985).

An Application of the Car-Parrinello Method to a Study of Alkali-Metal Adsorbed Si(001) Surface

Y. Morikawa[1], *K. Kobayashi*[2], *and K. Terakura*[1]

[1]Institute for Solid State Physics, University of Tokyo, Roppongi 7-22-1, Minato-ku, Tokyo 106, Japan
[2]National Institute for Research in Inorganic Materials, Namiki 1-1, Tsukuba-shi, Ibaraki 305, Japan

The Car-Parrinello method with a slight modification was applied to a study of alkali-metal (Na and K) adsorbed Si(001) surface. The nature of bond between adatoms and the substrate and the coverage dependence of adsorption energy are discussed.

1. Introduction

Several examples have demonstrated that the electronic structure calculation based on the local-density approximation (LDA) in the density functional theory may be a potentially very powerful method to provide reliable information about the structural stability of a given material [1]. However, the structural optimization based on such calculations has been formidably difficult for systems with more than three degrees of freedom. The situation has been changed by a new method proposed by Car and Parrinello [2]. The method is particularly powerful for cases where the plane wave expansion of the electronic eigenfunction is efficient. It is now possible to optimize the atomic positions of a system with more than 100 atoms. We have applied the method to the alkali-metal(AM)-adsorbed Si(001) surface [3,4].

2. Model and Calculational Method

The present calculation is based on LDA with the Wigner interpolation formula for the exchange and correlation [5], and on the norm-conserving pseudopotential [6] with a separable form proposed by Kleinman and Bylander [7]. The partial core correction (PCC) developed by Louie et al. [8] is incorporated to the pseudopotential for Na and K. PCC is very important for Na and K but can be neglected for Si.

We adopt a repeating slab geometry and plane-wave basis. The substrate is represented by a ten-layer Si(001) slab and a vacuum region of the same thickness (13.6Å) is inserted in between Si slabs. The AM atoms are introduced on each surface of the Si slab. The cutoff energy of the plane-wave basis is 6.25 Ry. The number of k points for sampling in the surface Brillouin zone is 32 and 4 for the (2×1) and (2×3) surface unit cells, respectively. The optimization of the electronic and ionic degrees of freedom is performed by a modified version of the first-principles molecular dynamics (FPMD). We adopt a steepest descent type of algorithm for the electronic degree of freedom [9] and the ordinary molecular dynamics for the ionic motion. The convergence criterion for forces acting on atoms is $3.0 \times 10^{-3}\mathrm{Ry}/a_\mathrm{H}$ with a_H denoting the Bohr radius (=0.5292Å).

As for the Lagrange multiplier, only the diagonal part $\lambda_{i,\vec{k}}$ is taken into consideration to gurantee the normalization of the wave function. The suffices i and $\vec{k}$ denote the band index and the wave vector, respectively. The wave functions are mutually orthogonalized by the Gram-Schmidt method. $\lambda_{i,\vec{k}}$ should be calculated by

$$\lambda_{i,\vec{k}}(t+\Delta t) = \int \Psi^*_{i,\vec{k}}(\vec{r}, t+\Delta t) H(t+\Delta t) \Psi_{i,\vec{k}}(\vec{r}, t+\Delta t) d^3 r, \qquad (1)$$

where $H(t)$ is the Hamiltonian in the Kohn-Sham equation at time t and $\Psi_{i,\vec{k}}(\vec{r}, t)$ is its wave function. We are only concerned with search for the optimum atomic structure and therefore the system becoms stationary at the final stage of the calculation. Therefore replacement of $H(t+\Delta t)$ with $H(t)$ in eq.(1) still gurantees the correct final solution. We can then derive the following relation for $\lambda_{i,\vec{k}}$:

$$\begin{aligned}\lambda_{i,\vec{k}}(t+\Delta t) = \lambda_{i,\vec{k}}(t) - \nu \sum_{\vec{G}} \{ & \dot{\Psi}^*_{i,\vec{k}}(\vec{G}, t) \Psi_{i,\vec{k}}(\vec{G}, t+\Delta t) \\ & - \Psi^*_{i,\vec{k}}(\vec{G}+\Delta t) \dot{\Psi}_{i,\vec{k}}(\vec{G}, t) + \Psi^*_{i,\vec{k}}(\vec{G}, t) \dot{\Psi}_{i,\vec{k}}(\vec{G}, t) \} \\ & + O(\Delta t^2), \end{aligned} \qquad (2)$$

where $\Psi_{i,\vec{k}}(\vec{G}, t)$ is defined by

$$\Psi_{i,\vec{k}}(\vec{r}, t) = \sum_{\vec{G}} \Psi_{i,\vec{k}}(\vec{G}, t) e^{i(\vec{k}+\vec{G})\cdot\vec{r}} \qquad (3)$$

with $\vec{G}$ denoting the reciprocal lattice vector. Equation (2) is convenient because only single summation over $\vec{G}$ is required.

3. Results and Discussion

Among calculated results [4], we discuss two aspects: one is the nature of bond between the adlayer and the substrate and the other is the coverage dependence of adsorption energy. Figure 1 shows the difference-electron-density map $\Delta\rho(\vec{r})$ for the half coverage case of K adsorption. K atoms are adsorbed on top of the third layer Si atoms along the trough between the surface Si-Si dimer rows. (The adsorption site is called T3 site in refs. 3 and 4.) $\Delta\rho(\vec{r})$ gives us information about the redistribution of the electron density due to the interaction between the AM adlayer and the Si substrate. The cut-planes are perpendicular to the surface and the dimer row but they cut different three positions of the dimer row, which are indicated by arrows in the insets of Fig.1. The most important aspect is that the strongest electron accumulation and depletion occurs within the cut-plane containing the Si-Si dimer rather than the one midway

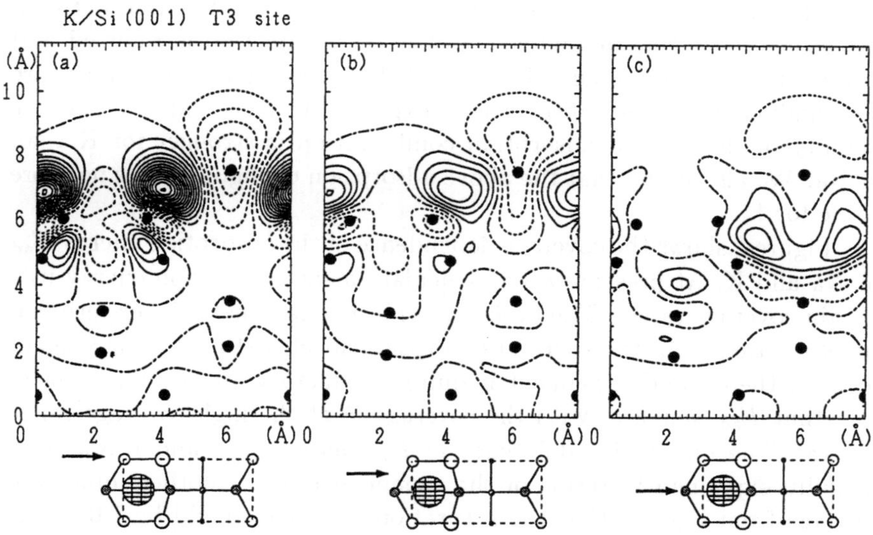

Figure 1. Difference charge density maps for K adsorption with $\Theta = 0.5$ within a cut-plane perpendicular to the surface. In each case of (a), (b) and (c), the cut-plane is defined so that it contains the arrow shown in each attached small figure. Solid (broken) curve contour denotes positive (negative) difference charge density. Along the dot-dashed lines, the difference charge density vanishes. The difference between the adjacent contour is 0.0005 a.u.. The filled circles denote atomic positions projected onto the cut-plane. In the inset, a large hatched circle denotes K and smaller circles the substrate Si atoms. (From Ref.4)

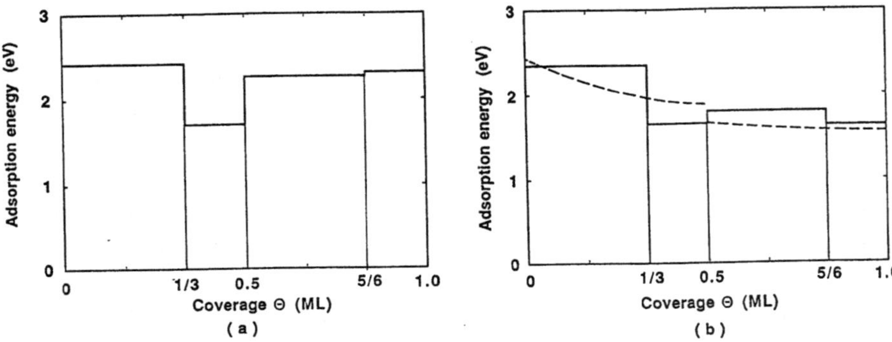

Figure 2. Coverage dependence of adsorption energy for (a) Na and (b) K. The broken curves in (b) denote the experimental estimation by the thermal desorption spectra [10]. (From Ref.4)

between the dimer and the adatom. This implies that the dangling bond of the surface Si atom interacts with the metallic bond charge between AM atoms. For Na, the extension of its wave function is reduced and thereby the electron accumulation and depletion mentioned above are less significant. These observations explain the fact that the relative stability of the present adsorption configuration is stronger for K than for Na. We also found that the above adsorption configuration is no more stable for Li.

Figure 2 shows the coverage dependence of the adsorption energy E_{ad} for Na and K. At present we can calculate mean adsorption energy in a certain coverage range. The broken lines in Fig.2a denote the experimental estimation from the thermal desorption spectra [10]. The agreement between theory and expeiment is quite satisfactory. It is interesting to note that E_{ad} increases with the coverage Θ at the high coverage range for Na. The qualitative difference in E_{ad} between Na and K was successfully explained in terms of the atomic size [4]. Another important message from Fig.2 is that the saturation coverage should be 1.0 rather than 0.5 or 0.68 [11] both for Na and K. The recent experiments support this conclusion [12,13].

Acknowledgement

The present work was partially supported by a Grant-in-Aid for Scientific Research in Priority Areas from Ministry of Education, Science and Culture. The numerical calculations were performed at the computer centers of University of Tokyo and Institute for Molecular Science. We thank Dr. S. Blügel for the collaboration at the early stage of the present work.

References

1. As a recent review of the density functional theory and its applications, R. O. Jones and O. Gunnarsson: Rev. Mod. Phys. **61**, 689(1989).
2. R. Car and M. Parrinello: Phys. Rev. Lett. **55**, 2471(1985).
3. Y. Morikawa, K. Kobayashi, K. Terakura and S. Blügel: Phys. Rev. B **44**, 3459(1991).
4. K. Kobayashi, Y. Morikawa, K. Terakura and S. Blügel: to be published in Phys. Rev. B.
5. E. Wigner: Phys. Rev. **46**, 1002(1934).
6. G. B. Bachelet, D. R. Hamann and M. Schlüter: Phys. Rev. B **26**, 4199(1982).
7. L. Kleinman and D. M. Bylander: Phys. Rev. Lett. **48**, 1425(1982).
8. S. G. Louie, S. Froyen and M. L. Cohen: Phys. Rev. B **26**, 1738(1982).
9. A. R. Williams and J. Soler: Bull. Am. Phys. Soc. **32**, 562(1987).
10. S. Tanaka, N. Takagi, N. Minami and M. Nishijima: Phys. Rev. B **42**, 1868(1990).
11. G. S. Glander and M. B. Webb: Surf. Sci. **222**, 64(1989).
12. A. J. Smith, W. R. Graham and E. W. Plummer, Surf. Sci. **243**, L37(1991).
13. S. Kono: private communications.

Ab-Initio Molecular Dynamics Study of the Si(001) Surface

T. Uda[1] *and S. Ihara*[2]

[1] Advanced Research Laboratory, Hitachi Ltd.,
Hatoyama, Saitama 350-03, Japan
[2] Central Research Laboratory, Hitachi Ltd., Kokubunji, Tokyo 185, Japan

Abstract. Using *ab initio* molecular dynamics, the structural and electronic properties of Si(001) surface are investigated. The c(4×2) and the c(2×2) reconstructions are generated in a few picoseconds starting from an ideal (1×1) surface. In addition to standard (n×m) geometries, characteristic defects appear, giving novel interpretations for single-missing and double missing dimer.

1. Introduction

Owing to the development of electronic computers and of a rapidly converging algorithm for *ab initio* molecular dynamics, computer simulation is expected to be an efficient tool for predicting the structural and electronic properties of molecules and solids. In this report, the dynamic behavior of Si(001) surface is studied with the hope of visualizing atomic relaxation into standard dimer reconstructions and of finding other complicated structures on this surface.

2. Model and Method

The surface is modeled by a slab geometry. The slab thickness should be large to ensure that the interaction between surfaces is negligible. By the same reason, the vacuum region between slabs must also be large. We used a slab containing 7 layers of Si atoms and a mono-layer of H atoms. The Si atoms are placed on the nominal lattice sites of a diamond structure. Two H atoms are attached to each Si atom on the bottom layer to eliminate dangling bonds leading to the di-hydride (1×1) structure. The unit cell area parallel to the surface is 4×2 in the largest case. Using this (1×1) surface as an initial structure, trajectories of all atoms in the specimen were traced via the *ab initio* molecular dynamics calculation proposed by Car and Parrinello [1]. During the lattice relaxation, the temperature of the system is kept constant at 300 K by rescaling atomic velocities.

3. Results and Discussion

Several simulation runs were conducted. Within a few picoseconds, reconstructions were generated in most cases. Reflecting a variety of local minimum coordinates, the obtained structures are not unique. Representative geometries are shown

Springer Proceedings in Physics, Vol. 70 Computational Approaches in Condensed-Matter Physics
Editors: S. Miyashita · M. Imada · H. Takayama

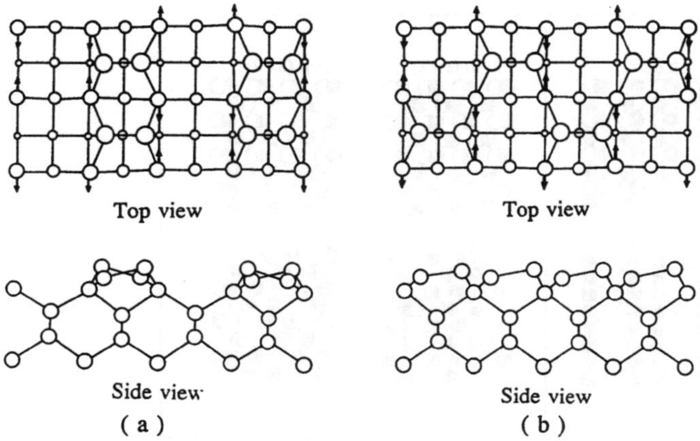

Fig. 1. Calculated reconstructions of Si(001) surface, (a) the c(4×2) and (b) the c(2×2) structures. Arrows show displacements of the second layer atoms.

in Fig. 1, where the left (a) and the right (b) figures correspond, respectively, to so-called c(4×2) and c(2×2) reconstructions. Both structures are composed of buckled (asymmetric) dimers. For the c(4×2) surface, the dimer bond length is 2.25 Å and the tilting angle is 14°, similar to a previous energy minimization calculation [2]. The dimer bond length of c(2×2) symmetry (2.29 Å) is a little longer than that of the c(4×2), but is still shorter than the bulk value (2.35Å). The tilting angle of the surface dimer is 9°, which is smaller than that of the c(4×2) surface.

Since the pioneering work by Schlier and Farnsworth [3], there have been reported various experimental and theoretical studies suggesting the formation of dimers on the Si(001) surface. Further, the energy minimization calculations show that surface energy can be lowered by allowing the dimers to buckle out of the surface plane [2,4]. Thus, it is now widely accepted that the building block of the surface is the buckled (asymmetric) dimer. Based on this observation, there are eight reconstruction patterns depending on the different arrangements and orientations of asymmetric dimers. As is shown in Fig. 2, they are classified into (I) "2×1" and (II) "c(2×2)" families [5].

In our calculation, dimer reconstruction of either type (I) or type (II) symmetry was observed within 0.5 psec after the onset of relaxation. At this stage, the tilting angles of surface dimers are random. Thus, the resultant structure cannot be assigned to a particular (n×m) geometry. With the lapse of time, there appear certain orders in the tilting angles. After a few picoseconds, from the surface with type (I) symmetry, the c(4×2) reconstruction is preferably generated, in which the direction of buckling alternates from dimer to dimer along the row. From the structure with type (II) symmetry, the c(2×2) surface is generated, in which all the dimers are tilted uniformly. These relationships between arrangements and orientations of buckled dimers can be explained by observing the displacements of atoms in the second layer, shown by arrows in Fig. 1. As is clear

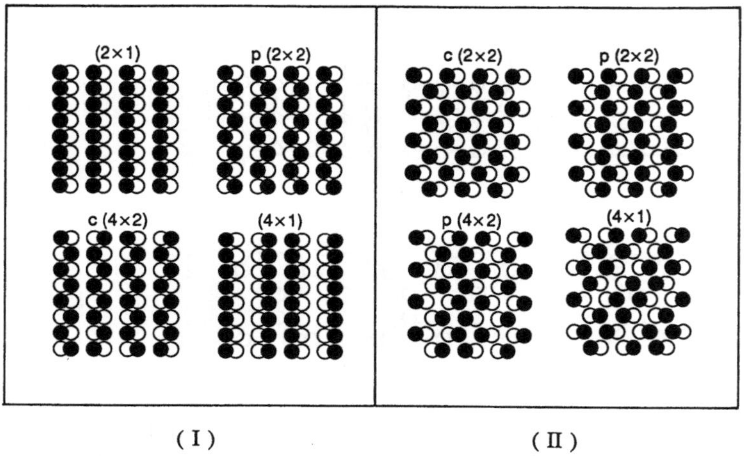

Fig. 2. Possible reconstruction geometries of the Si(001) surface for (I) "2×1" and (II) "c(2×2)" families. Black and white circles show up and down dimer atoms, respectively.

from the figure, atoms in the second layer move toward the up asymmetric dimer atoms and away from the down asymmetric dimer atoms in the direction of dimer rows. As is pointed out by Zhu et al. [2], this movement keeps the bond length between dimer atoms and the atoms in the second layer close to the bulk value, suppressing the energy increase due to bond length distortion. This relaxation is allowed only for c(2×2) geometry in the type (II) family. In contrast to this, the relaxation is allowed both for c(4×2) and p(2×2) geometries in the type (I) family, suggesting that the p(2×2) surface is another possible structure, although we have not yet obtained this structure.

The surface images by scanning tunneling microscopy (STM) show clearly the c(4×2) and the p(2×2) regions near defects. However, the c(2×2) geometry is not observed. Since the energy of the c(2×2) is 0.05 eV/dimer higher than that of the c(4×2) in this calculation, it may be transformed into the other two geometries after a very long time in the simulation scale but a very short time in the laboratory scale. A more serious problem is that most dimers in the STM images appear to be symmetric. One possible interpretation may be that the dimers are actually vibrating between the two configurations [6].

Our simulation occasionally yields structures different from any of the (n×m) geometries [7]. One of such defects is illustrated in Fig. 3. As in an ordinary reconstruction, surface atoms form dimers, but one dimer is relaxed into the surface by 1.7 Å and is located even lower than the second layer. This structure, termed 'interstitial dimer' would, therefore, give nearly the same STM image as a dimer vacancy in the dimer rows [6]. It should be noted that the dimer in question and the adjacent surface dimers are all symmetric. In Fig. 4, the density of states of the interstitial dimer is shown. A characteristic feature is that a band appears near -14 eV from the Fermi level. The charges with this energy are illus-

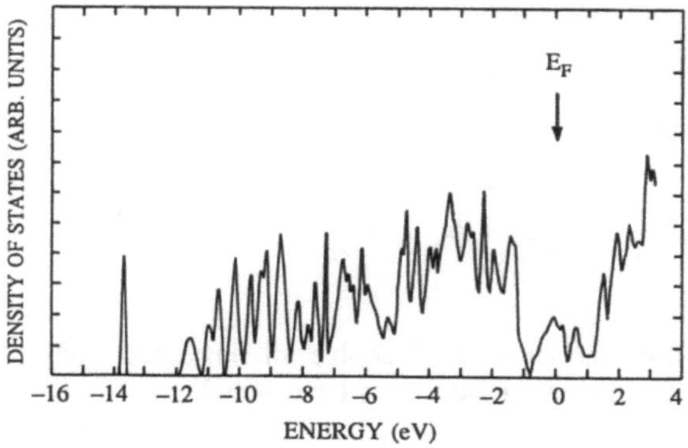

Fig. 3. Structure of the interstitial dimer. Dots show the charge distribution of the band near -14 eV from the Fermi level shown in Fig. 4.

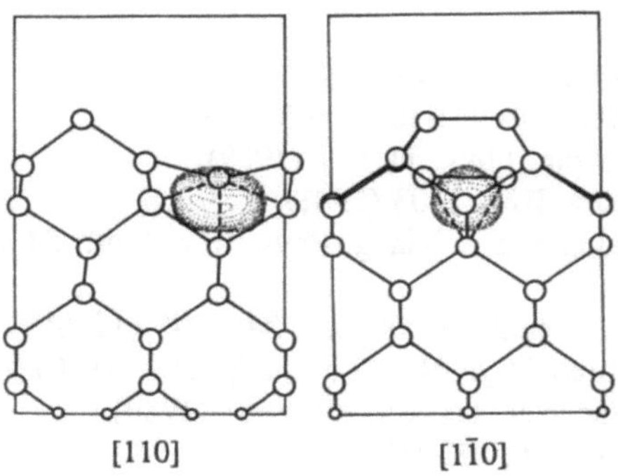

Fig. 4. Density of states of the interstitial dimer.

trated in Fig. 3. They are distributed just below the interstitial dimer. The attractive potential due to these charges could be the driving force attracting the interstitial dimer. Each atom forming an interstitial dimer is 6-fold coordinated, as is achieved in the high pressure phase. The energy of an interstitial dimer is rather high, but it remains stable until up to 5 psec.

A recent calculation by Roberts and Needs shows that the energy of a double missing dimer, also observed in the STM images [8], is higher than that of two single missing dimers [9]. An interstitial dimer and a dimer vacancy pair is an alternative model for a double missing dimer, since the energy of this pair could be lower than that of a double missing dimer due to compensation of compressive and tensile stress in the second layer (Fig. 5).

In conclusion, the Si(001) surface structures calculated using *ab initio* molecular dynamics reproduce observed reconstruction patterns composed of asymmetric dimers. The calculated atom displacement in the second layer reveals the relaxation mechanism making observed geometries more favorable than other possible geometries. The simulation also yields a characteristic defect giving novel interpretations of missing and double missing dimers.

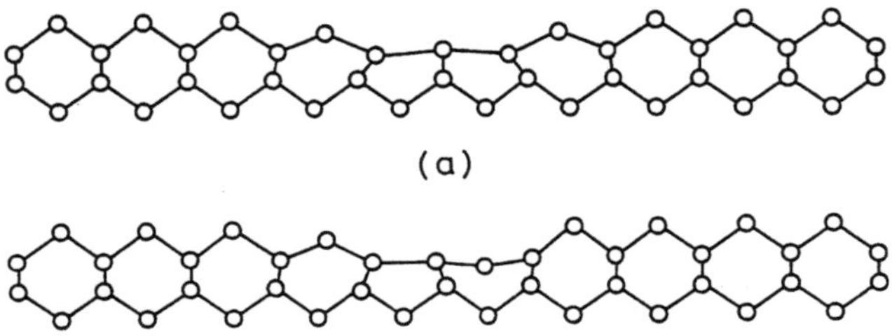

(a)

(b)

Fig. 5. Side views of the Si(001) surface with (a) double missing dimer and (b) single missing and interstitial dimer pair structures (schematic).

References

1. R. Car and M. Parrinello, Phys. Rev. Lett., **55**, 2471 (1985).
2. Z. Zhu, N. Shima and M. Tsukada, Phys. Rev. **B40**, 11868 (1989).
3. R. E. Schlier and E. Farnsworth, J. Chem. Phys. **30**, 917 (1959).
4. M. T. Yin and M. L. Cohen, Phys. Rev. **B24**, 2303 (1981).
5. J. Ihm, D. H. Lee, J. D. Joannopoulos and J. Xiang, Phys. Rev. Lett., **51** , 1872 (1983).
6. R. J. Hamers, R. M. Tromp and J. E. Demuth, Phys. Rev. **34**, 5343 (1986).
7. S. Ihara, S. L. Ho, T. Uda and M. Hirao, Phys. Rev. Lett., **65**, 1909 (1990).
8. R. J. Hamers and U. K. Kohler, J. Vac. Sci. Tecnol., **A7**, 2854 (1989).
9. N. Roberts and R. J. Needs, Surf. Sci., **236**, 112 (1990).

Traditional and Ab-Initio Molecular Dynamics Simulations – Applications to Glass Transition and Amorphous SiH

F. Yonezawa[1] *and S. Sakamoto*[2]

[1]Department of Physics, Keio University,
3-14-1 Hiyosi, Kohoku-ku, Yokohama 223, Japan
[2]Department of Applied Physics, Seikei University,
3-3-1 Kichijoji-Kitamachi, Musashino-shi, Tokyo 180, Japan

A purpose of the present article is to demonstrate the outstanding success of computer simulations in the study of non–crystalline or amorphous systems. To this end, we first apply the traditional molecular dynamics (MD) techniques to a system composed of classical particles, and study the glass transition as well as the stability of glasses constructed through the glass transition. In the second place, we develop the ab–initio MD program based on the technique proposed by Car and Parrinello, and use our program to the investigation of the Staebler–Wronski effect in amorphous SiH where the physical properties are seriously influenced by the interplay between atoms (classical particles) and electrons (quantum mechanical particles). We also show that the visualization of our results by computer graphics provides us with a great help in a profound understanding of physical properties.

1. Introduction

The central purpose of this article is first to show how successful the techniques of molecular dynamics (MD) simulations could be in the study of disordered systems, and secondly to demonstrate how helpful the computer graphics could be for a better understanding of the microscopic world.

For this purpose, we study both traditional (naturally classical) MD and ab–initio MD which is classical in nature but extended to deal with quantum mechanical problems by Car and Parrinello. Since the traditional MD technique, is widely known, we do not go into detail here. As for the ab–initio MD, the paper due to Car in these proceedings gives the explanations of the basic idea, and therefore we start with the applications of the method.

In § 2, we show the application of the traditional MD to the investigation of the glass transition achieved by a rapid quenching of a liquid into an amorphous solid. We are also interested to find out what aspects of the microscopic features are really characteristic of amorphous structures. In § 3, we make a full use of the ab–initio MD to study the structural and electronic properties of amorphous (a–)SiH.

Before we start the discussion of each subject, it would be interesting and instructive to explain why we choose amorphous systems in applying several MD methods. The answer to this question is found from the following description.

In the field of condensed matter physics, the investigations of crystalline materials has been by far advanced when compared with those of non–crystalline materials. It was as early as 1910's when the X-ray diffraction technique for the analysis of crystalline structures were first proposed, and the *atomic structures* of crystals have been fully analyzed since then. After the establishment of quantum mechanics in 1920's, the fundamental theories for the *electronic properties* in crystals were established essentially in the form of the Bloch's theorem.

In these experimental and theoretical investigations of crystals, a full use has been made of *periodicity*, or in other words, *translational symmetry*.

In the class of condensed matter, there still exists another category of materials which includes non–crystalline solids and liquids in general. The atomic configurations of the materials in this category are *non–crystalline*, *non–periodic*, *liquid*, *disordered*, *random*, *amorphous*, and *glassy*. All these adjectives are normally used synonymously.

Experiments of these amorphous materials give only limited information about the atomic configurations in there materials, the limited information being such as the absence of the long–range order (hereafter abbreviated to LRO) in the sense that there remains no vestige of *periodicity* or of *translational symmetry*, and such as the existence of the short–range order (hereafter abbreviated to SRO) in the sense that the local environment of each atom in a disordered structure is similar to that in a crystal.

With this situation in mind, we can easily predict that it is these disordered systems in which the methods of computer simulations, in particular MD techniques, give full play to prove their worth. As a matter of fact, since the introduction of the MD methods in 1950's, the most widely studied subject has been liquids. As mentioned, this is one of the reasons why we apply our MD program to amorphous materials.

2. Glass Transition

By making use of the constant–pressure MD, we carry out simulations of quenching to construct glasses, then we observe the physical properties of the obtained glasses and investigate the microscopic configurations characteristic of amorphous structures.[1,2]

We study a system composed of atoms interacting via the Lennard–Jones potentials. We use the periodic boundary condition. The number of atoms is 864, and the time step for integration is 0.5×10^{-14}s.

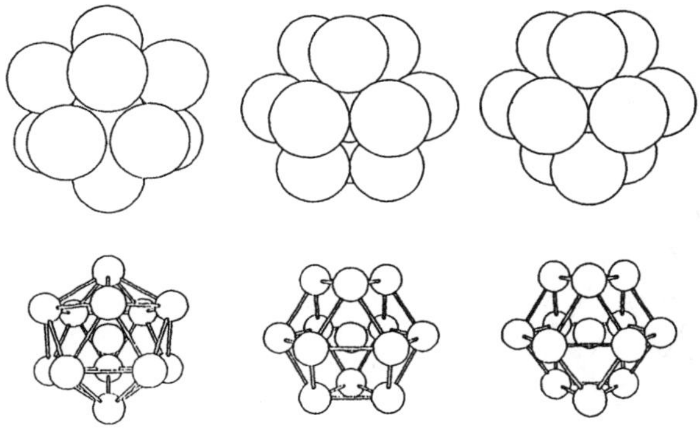

(a) (b) (c)

Fig. 1. Clusters in which 12 atoms surround a central atom (the number in the bracket indicates the number of the nearest–neighbour bonds in the luster): (a) an icosahedral cluster (42); (b) an fcc cluster (36); and (c) an hcp cluster (36).

We have made a movie titled "*Glass Transition*" in which some essential results are included. At the Nishinomiya Symposium, we presented the movie.[3]

In particular, we investigate the quench–rate dependence of the stability of glasses for the purpose of clarifying the atomic configurations which stabilize amorphous structures.

It turns out that when the glasses are annealed at some temperature below the glass transition temperature, glasses prepared with extremely high quench rates tend to relax into crystalline structures more easily than do glasses prepared with the lower quench rates.

This result is explained in the following way. When the quench rate is very high, atoms in the system are not allowed to spend enough time to adjust their positions so as to realize a structure with energy as low as possible under the given condition. Accordingly, the redistribution of atoms over a wide range is nearly impossible, and the least energetic configuration is found by adjusting only the immediate neighbour of each atom, thus resulting in a configuration containing many icosahedra.

Icosahedra constructed in this process of high–speed quenching are of the kind shown in Fig.1(a), in which 12 atoms in the first shell surround a core atom and the number of the nearest–neighbour bonds is 42. In the figure, the other clusters composed of 13 atoms are also illustrated for the sake of comparison, the clusters being fcc and hcp clusters, either of which has 36 nearest–neighbour bonds. Since each bond gains binding energy comparable with the minimum value of the pairwise interatomic interaction such as the

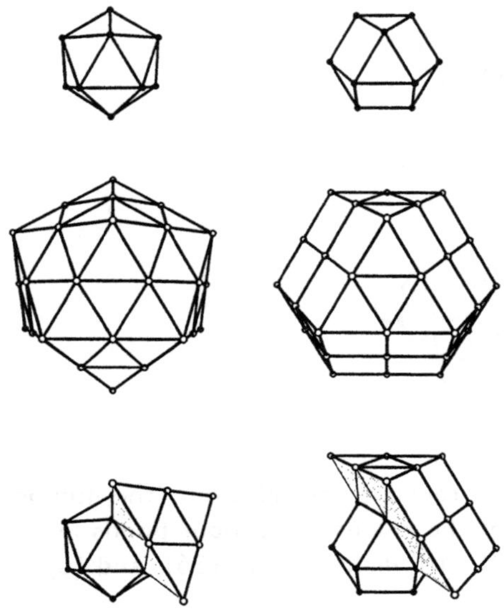

Fig. 2. (a) Illustration of a double–shell icosahedral cluster composed of 55 atoms with 234 nearest–neighbour bonds: (top) an icosahedral cluster in which a central atom is surrounded by 12 atoms belonging to the first shell; (center) 42 atoms in the second shell; (bottom) configurations of atoms in the second shell relative to atoms in the first shell. (b) Illustration of a double–shell fcc cluster compose of 55 atoms with 216 nearest–neighbour bonds; (top) an fcc cluster in which a central atom is surrounded by 12 atoms belonging to the first shell; (center) 42 atoms in the second shell; (bottom) configurations of atoms in the second shell relative to atoms in the first shell.

Lennard–Jones potential, the larger the number of the nearest–neighbour bonds in a cluster, the lower is the energy of the cluster. This is the reason why amorphous solids of isotropic atoms have local icosahedral symmetry.

In high–speed quenching, there exist many icosahedra of this kind, but the extent of each icosahedral symmetry is limited in space, and, consequently, the atoms in the icosahedral clusters easily reconstruct themselves into crystalline configurations.

When the quench rate is not high enough, the atoms are allowed to spend more time in the readjustment of configurations, and therefore the redistribution of atoms beyond the immediate neighbour is not completely excluded. As a result, there may appear clusters whose sizes are larger than a simple icosahedron, as shown in Fig.1(a), and yet whose symmetry is icosahedral. One candidate for such a cluster is the double–shell icosahedron, as depicted in Fig.2(a), which consists of 55 atoms and contains 234 nearest–neighbour bonds. An fcc cluster of the same size is shown in Fig.2(b) for comparison, which has 216 nearest–neighbour bonds. Here again, the icosahedral cluster has lower energy.

The large icosahedral clusters of this kind are stable once they are constructed because the coherence of the symmetry is longer in space. Another important point about this double–shell icosahedron is that the relative positions of the atoms in the core and the atoms in the first and second shells satisfy the fcc crystalline order. In the process of annealing, the double–shell icosahedron can grow by gathering atoms on the surface so that the third shell, the fourth shell, and so on, are formed. These outer shells stack so that they fulfill the fcc order. In this way, the coexistence of the icosahedral and fcc symmetry becomes possible, which is expected to be stable because quite an amount of energy is required to dissolve such multishell icosahedra.

3. Ab–initio molecular dynamics simulations applied to a–SiH

The purpose of our work presented in this section is to elucidate the microscopic mechanism of light–induced defect formation in the Staebler–Wronski (SW) effect.

With a view to applying the ab–initio MD method[4,5] to the study of a–SiH, let us first describe the situation of this material in connection with the exploitation of solar cells.

Today, amorphous SiH is one of the most promising devices for solar cells which could save the energy problem in the 21st century. Before 1975, however, the photovoltaic efficiency of a–Si was so low that the application to solar cells was practically out of question. The low efficiency was entirely due to dangling bonds which exist abundantly in a–Si.

In 1975, a–Si solar cells became realistic when the preparation technique by plasma decomposition of SiH_4 was introduced. Because a number of H atoms are present in the process of this preparation method, these H atoms are included in Si network and fill almost all dangling bonds, thus preventing the generation of gap states.

Since then, the photovoltaic efficiency of a–SiH has been improved remarkably. In this sense, H was the salvation.

In 1977, Staebler and Wronski[6] found that, after a sample of a–SiH prepared from SiH_4 by plasma decomposition is exposed to intensive light, both photo conductivity and dark conductivity decrease by a few orders of magnitude while the number of defects (mostly dangling bonds) increases. The effect is perfectly reversible by thermal annealing at 150°C. This effect is named the *Staebler–Wronski* effect.

What we know from experiments is that H atoms exist only in the form of Si–H bonds, and the mechanism of this light–induced defect formation is predicted to reflect the process that photons break Si–H bonds, thus creating dangling bonds. Hydrogen atoms then diffuse out and break weak Si–Si bonds, thus creating further dangling bonds.

The created dangling bonds degrade the photovoltaic efficiency very seriously. In this sense, H seems to be the cause of trouble.

It is therefore important and rather urgent to clarify, beyond predictions, the microscopic mechanism of the SW effect.

As we described, we study the fundamental processes of the SW effect by means of the ab–initio MD simulations. In this method, atoms and electrons are treated on the same basis, which is particularly appropriate for amorphous covalent systems since in these systems the distribution of atoms dominates the electron density at each point, which defines the actual forms of interatomic potentials, which in turn determine the distribution of atoms. This means that the electronic configuration must be calculated for the atomic arrangement at each MD step, which normally requires the diagonalization of a large–size matrix. An efficient way to avoid the diagonalization was proposed by Car and Parrinello, which we use here.

Since the fundamental processes of the SW effect are regarded as being local events and independent of the whole structure, we produce atomic and electronic configurations whose local environments simulate the configurations realized in a–SiH. We do this either by introducing disorders in ordered structures or by constructing amorphous structures via quenching.

In practice, we study the following three problems.[7,8]

1. We show that weak Si–Si bonds do exist in amorphous network.
2. We see what happens if there exists a hydrogen atom in the vicinity of a weak Si–Si bond.
3. We calculate the bond–breaking energy as well as the bond–switching energy of an Si–H bond.

In our simulations, the time step for integration is 1.7×10^{-16}s, the number of atoms in a simulation cell is 8, 16, 32, or 64, the maximum energy for the plane waves is 13.5 Ry. We use the norm–conserving pseudo potential due to Bachelet, Hamann and Schlüter.[9]

4. Structural and Electronic Properties of a–SiH

4.1. Weak Si–Si bond

From the analysis of the results obtained by the ab–initio MD, we find that weak Si–Si bonds do exist in a–SiH. Examples of forms of weak Si–Si bonds are as follows.

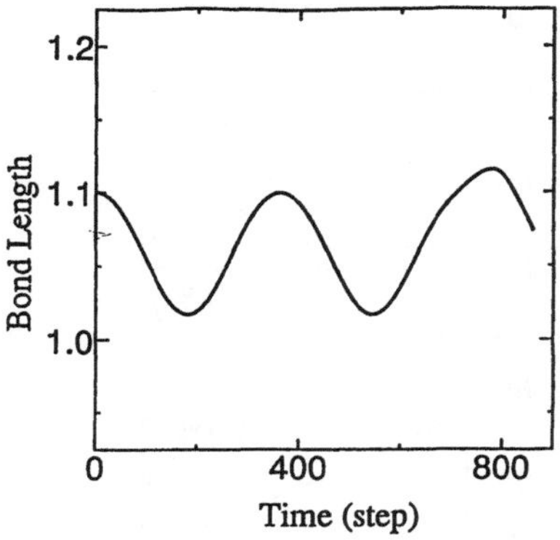

Fig. 3. Length of an Si–Si bond vs timesteps.

1. We quench liquid into amorphous solid and the obtained a–structure contains weak Si–Si bonds mainly because of geometrical requirements.
2. We show that an Si–Si bond adjacent to Si–H bond tends to be weakened.

We also find that two Si atoms in a weak Si–Si bond show optical vibration around a distance larger than the bond length in c–Si as shown in Fig.3, where the length of the Si–Si bond versus timesteps is presented. In this article, the length is scaled by 2.35 Åwhich is the length of an Si–Si bond in c–Si. This result guarantees that weak Si–Si bonds can exist in a more or less stable manner.

In principle, the method we use can be applied only to the ground state. The application of the method to excited states may fail to give accurate electron levels, but is regarded to give an approximate picture as far as the atomic configurations are concerned. For the sake of reference, we also study an excited state which simulates the situation that electrons are excited by illumination of light. The results we achieve indicate that the length of a weak Si–Si bond is increased. In other words, weak bonds are weakened further as a result of light illumination.

4.2. H near weak Si–Si bond

As for the behaviour of an H atom located in the neighbourhood of a weak Si–Si bond, we find the following results.

When the distance between two Si atoms in the weak Si–Si bond under consideration is not very long, there exists an *"off- bond"* metastable site for H, from which H diffuses out relatively easily, or to which H diffuses in relatively easily as well.

When the distance between two Si atoms in the weak Si–Si bond is quite long, one (Si–H) bond and one (Si–Dangling bond) are created. This micro-

Fig. 4

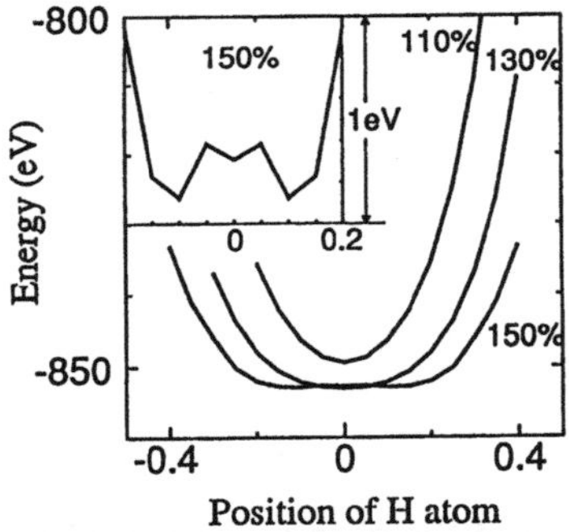

Fig. 5

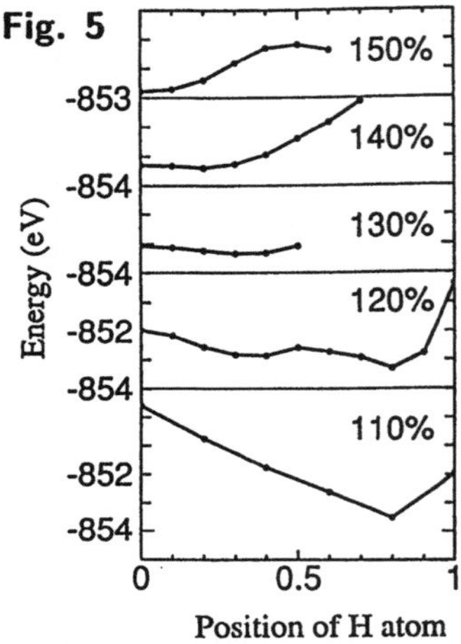

Fig. 4. Total energy vs position of H on Si–Si bond, the position being measured from the center of the bond in reduced scale. The percentage denotes the degree by which the Si–Si bond is stretched

Fig. 5. Total energy vs position of H on a line vertical to the Si–Si bond, the position being measured from the center of the bond in reduced scale. The percentage denotes the degree by which the Si–Si bond is stretched

scopic configuration is relatively stable at room temperature, while H diffuses away when the temperature is higher.

We can achieve these results first by calculating the total energy of the system for the configuration in which H is located on Si–Si bond. The results are shown in Fig.4, where the energy is given as a function of the position of H atom measured from the center of the Si–Si bond in a scaled length in which 2.35 Å is taken to be unity. The percentage in the figure denotes the Si–Si bond length as measured by taking this length as 100 %.

When the length of the Si–Si bond is 1.1, then the center of the bond is the most stable site for H along the bond, When the length of the Si–Si bond is 1.3, the center of the bond is still most stable along the bond. but the energy cost for H displacement is much lower. When the length of the Si–Si bond is 150%, there exist double minima along the bond.

The most stable site along the bond is not always the local minimum because the site could be unstable off the bond.

For the purpose of finding the site of H at which the total energy becomes a local minimum, we also calculate the energy for the configuration in which H atom is off Si–Si bond as shown in Fig.5, where the energy is given as a function of the position of H atom measured from the center of the Si–Si bond in a scaled unit. The percentage in the figure denote the length of the Si–Si bond.

When the length of the Si–Si bond is 110 %, there exists no stable site near the bond. When the length of the Si–Si bond is between 120 % and 140%, there is a stable site off the bond and it is ascertained that the site is stable in any direction, thus the site being confirmed to be the local minimum. The important point indicated by our results is that the stable position for H is not necessarily on the Si–Si bond, but it is often found off the bond.

4.3. Contours of the electron density

The situation can be seen more clearly when we look at the contours of the electron density. For the sake of comparison, the contour along the Si network in c–Si is shown in Fig.6, in which the electron density is highest at the red part, and decreases in order to yellow, green, blue and white. In the figure, the black circles denote Si atoms.

In Fig.7 is given the contour for a weak Si–Si bond. We can see from the figure that, when we say a bond is *"weak"*, that means that the bond is longer as well as that the density of electrons is lower in the bonding orbital.

It is also possible to see that a Si–Si bond adjacent to Si–H bond is weak. In Fig.8 is shown the contour along H–Si–Si–Si bonds. the electron density around H atom is high because the electronegativity of H is higher, and accordingly the Si–Si bond adjacent to Si–H bond is weak because electron transfer to H atom takes place. The Si–Si bond adjacent to Si–H bond is weak in both senses that the electron density is low on that bond and that the length of the bond is increased by a few percent.

In Fig.9 is given the contour for the *off–bond* metastable configuration. The electrons are shared not only by the two Si atoms under consideration but also by the Si atom in the nearest–neighbour position, thus the total energy being lowered.

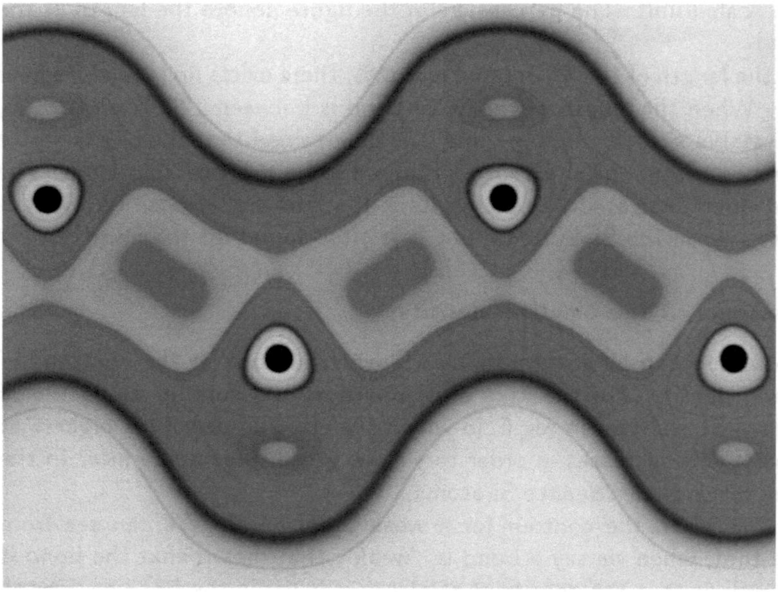

Fig. 6. Contour of the electron density along the bonds of Si atoms in c–Si.

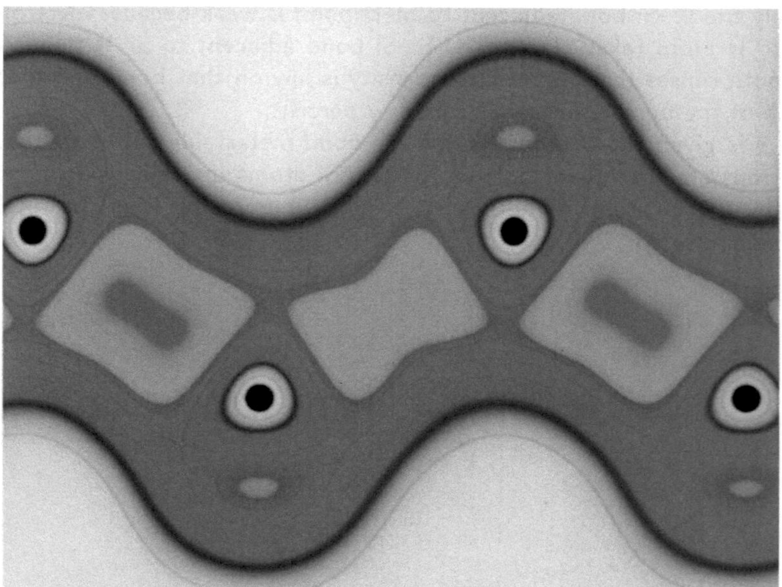

Fig. 7. Contour of the electron density along the bonds of Si atoms in which the length of a Si–Si bond is 110 % as long as that in c–Si.

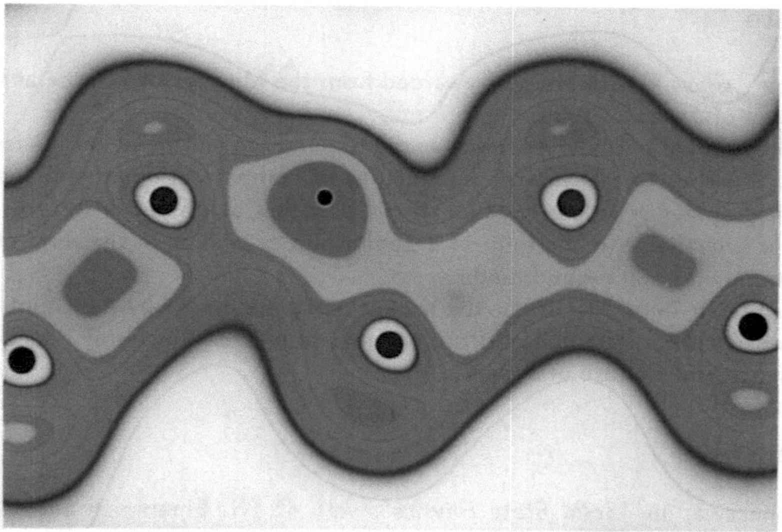

Fig. 8. Contour of the electron density in which H is placed at the distance 0.4 on a line vertical to the Si–Si bond, 0.4 being measured from the center of the Si–Si bond.

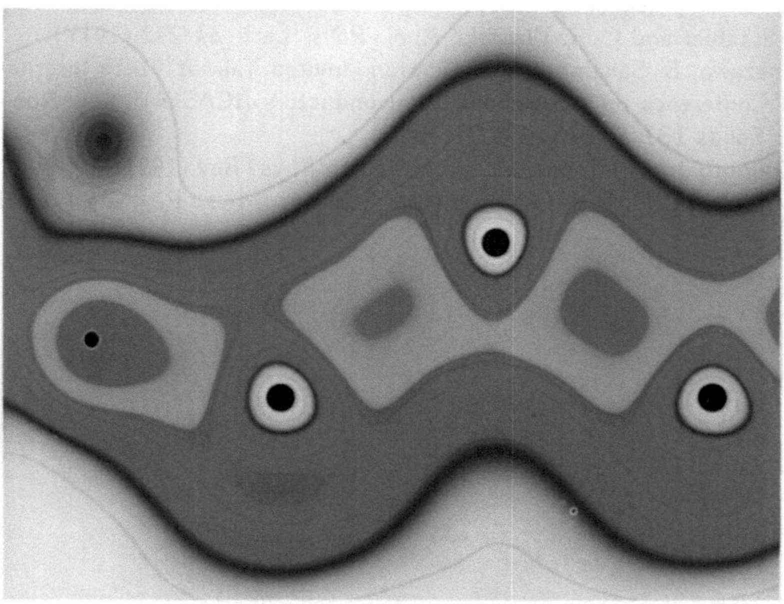

Fig. 9. Contour of the electron density along the bonds of Si–Si–Si–H.

5. Note

Our movie *"Glass Transition"* was awarded from the Minister of Science and Technology of Japan in 1989.

We like to mention that, if anyone wants to purchase our movie on *"Glass Transition"* either to present the possibilities of computer simulations and computer graphics or to use for educational purposes at classes or lectures, the movie is available in the form of a 16mm film as well as in the form of a video tape with any desired mode.

As for the ab–initio MD study, the work is partly supported by the Grant–in–Aid for Scientific Research on Priority Areas from the Ministry of Education, Science and Culture of Japan.

References

1. F. Yonezawa: in *"Solid State Physics"*, vol. 45 (H. Ehrenreich and D. Turnbull, eds.), Academic Press, 1991.
2. M. Kimura and F. Yonezawa: in *"Topological Disorder in Condensed Matter"* (F. Yonezawa, ed.), Springer, 1983.
3. F. Yonezawa, S. Nosé and S. Sakamoto: Neue Folge **156**, 77 (1988).
4. R. Car and M. Parrinello, Phys. Rev. Lett. **55** 2471 (1985).
5. R. Car, these proceedings (1992).
6. D. L. Staebler and C. R. Wronski, Appl. Phys. Lett. **31** 292 (1977).
7. F. Yonezawa, S. Sakamoto, and M. Hori Invited Talk at "14th International Conference on Amorphous Semiconductors (ICAS14)" : J. Non–Cryst. Solids **137 & 138** (1991) 135.
8. S. Sakamoto and F. Yonezawa, submitted to Phys. Rev.(1991).
9. G. B. Bachelet, D. R. Hamann and M. Schlüter, Phys. Rev. **26** (1982) 4199.

Part II

Strongly Correlated Electron Systems

Projection Monte Carlo Method and Choice of the Optimal Trial Wavefunction

S. Sorella

SISSA, Via Beirut 2, I-34014 Trieste, Italy

We present here a brief and schematic description of a quantum Monte Carlo method for fermions. The ground state of a many-body fermion system is achieved by propagating in imaginary time a properly chosen trial wavefunction. It is shown that use of a trial wavefunction of definite total spin can enormously improve the convergence in imaginary time in the half filled Hubbard model. Fluctuations of the staggered magnetization become very small in a subspace of definite spin and large scale simulations are possible at a little expense of computer time. The antiferromagnetic broken symmetry state which occurs at half filling is the main reason of this and the physical consequences on the interpretation of the Mott insulator are discussed. Application of the methodology just away from half filling is also possible because good convergence in imaginary time is reached before the fermion sign problem becomes relevant.

Recently a remarkable improvement[1-3] of a Monte Carlo technique[4] for fermions has made possible large scale simulations at low temperatures. The method was then particularly efficient for the half filled Hubbard model where no fermion sign problem exist. The problem is that when this occurs the sign of the Monte Carlo weight can vanish exponentially leading to inefficient sampling[5,6]. In this report we show that use of a good trial wavefunction can enormously improve the convergence in temperature so that we can achieve ground state properties before fermion sign problem becomes a relevant problem.

We use the standard Monte Carlo algorithm[1,3] to evaluate ground state properties of the positive U Hubbard model using the so -called projection Monte Carlo technique.

We filter out the ground state of the Hubbard model by propagating in imaginary time a trial wavefunction ψ_T, assumed to be a free particle Slater determinant state:

$$|\psi_G > \propto \lim_{\beta \to \infty} e^{-H\beta} |\psi_T > + O(e^{-\Delta\beta}) \quad , \tag{1}$$

where Δ is the finite size gap to the first excited state. In this method the choice of the trial wavefunction is extremely important to improve the efficiency of the algorithm and will be discussed in some detail in the present report.

All the correlation functions of the quantum many body system -expectation value of an operator $\hat{O}$- can be evaluated using the following relation:

$$< \psi_G|\hat{O}|\psi_G >= \lim_{\beta\to\infty} \frac{< \psi_T|e^{-t'H}\hat{O}e^{-tH}|\psi_T >}{< \psi_T|e^{-\beta H}|\psi_T >} \quad , \tag{2}$$

where $t + t' = \beta$ and in the previous expression we can choose $t \sim t' = \beta/2$. In actual calculations it is convenient to average in the interval $1/4\beta < t < 3/4\beta$ in order to reduce the statistical fluctuations.

For the statistical evaluation of the expression (2), we use the Trotter formula and discretize the imaginary time interval $(0, \beta)$ in P Trotter time slices:

$$e^{-\beta H} = \left(e^{-\frac{\Delta\tau}{2}K} e^{-U\Delta\tau\sum_r n_{\uparrow,r}n_{\downarrow,r}} e^{-\frac{\Delta\tau}{2}K}\right)^P + o(\Delta\tau^2) \quad , \tag{3}$$

where K is the kinetic one body term in the Hubbard hamiltonian and $\Delta\tau = \frac{\beta}{P}$. The many body part of the short time propagator at the lth time slice can be then decomposed in a sum of 2^{N_a} one body propagators, using the discrete Hubbard Stratonovich transformation:

$$e^{-U\Delta\tau\sum_r n_{\uparrow,r}n_{\downarrow,r}} = C \sum_{\sigma_{r,l}=\pm1} e^{\lambda\sum_r \sigma_{r,l}(n_{\uparrow,r}-n_{\downarrow,r})} \quad , \tag{4}$$

where $\lambda = \cosh^{-1}(e^{\frac{U\Delta\tau}{2}})$ and the overall constant $C = 2^{-N_a}e^{-\frac{U\Delta\tau}{2}N}$ and, we have introduced the discrete fields $\sigma_{r,l}$ for each of the different P time slices.

After substitution of expression (3) and (4) in Eq.(2), the ground state expectation value of any operator becomes a ratio of two functional depending on the discrete Hubbard Stratonovich fields $\sigma_{r,l}$. This field due to the discreteness of the HST can assume 2^{N_aP} different configurations. For each configuration since ψ_T is a Slater determinant the problem consists in a time dependent propagation of a Slater determinant in a fluctuating discrete magnetic field:

$$e^{-H\beta}|\psi_T >= C^P \sum_{\sigma_{r,l}=\pm1} U_\sigma(\beta, 0)|\psi_T > \quad , \tag{5}$$

where

$$U_\sigma(\beta, 0) = \prod_{l=1}^{P} e^{-\frac{\Delta\tau}{2}K} e^{\lambda\sum_r \sigma_{r,l}(n_{\uparrow,r}-n_{\downarrow,r})} e^{-\frac{\Delta\tau}{2}K} \quad . \tag{6}$$

This propagation can be easily and efficiently implemented in a numerically stable way[2,3,7]. The sum over the 2^{N_aP} configurations can be only achieved using a statistical method. We have in fact followed the standard procedure to generate configurations of the fields according to the weight:

$$w_\sigma =< \psi_T|U_\sigma(\beta, 0)|\psi_T > \tag{7}$$

using the single spin flip update of Hirsch[1]. After that the expectation value of the operator $\hat{O}$ (2) becomes a classical average $\frac{\sum_\sigma E^t_O(\sigma) w_\sigma}{\sum_\sigma w_\sigma}$ of a corresponding estimator $E^t_O(\sigma)$ at the imaginary time slice t:

$$E^t_O(\sigma) = \frac{< \psi_T | U_\sigma(\beta, t) \hat{O} U_\sigma(t, 0) | \psi_T >}{< \psi_T | U_\sigma(\beta, 0) | \psi_T >} \tag{8}$$

For each configuration of the fields the evaluation of the estimator can be performed very efficiently using Wick's theorem. In fact (8) represents a matrix element of the operator $\hat{O}$ between two free particle states $U_\sigma(t,0)|\psi_T>$ and $<\psi_T|U_\sigma(\beta,t)|$.

A further complication arises since the weight (7) is not always positive definite. This problem can be easily circumvented by taking into account the average sign of the weigh $< S >_\beta = \frac{\sum_\sigma w_\sigma}{\sum_\sigma |w_\sigma|}$ which leads to straightforward generalization of previous expressions[4]. However, from the computational point of view, the average sign can became very small for large imaginary time, since it has been shown that $< S >_\beta$ may decrease exponentially with β [5,6]. This is one of the most important limitation of fermion Monte Carlo schemes.

In the present QMC approach we can try to optimize the trial wavefunction to improve the efficiency of the algorithm. This can be done for two main reasons The first is that we can directly improve the average sign by choosing a good wavefunction ψ_T[8] and the second is that if the trial wavefunction has all the symmetries of the ground state one can significantly increase the gap Δ to the first excited state. In fact the full many body propagator e^{-Ht} conserves by definition all the symmetries of the chosen trial function. In this way the propagated state is automatically orthogonal to the excitations of different symmetries and we obtain the considerable advantage to achieve a good convergence to the ground state with considerably less imaginary time β. From (1) a good convergence is in fact expected for $\beta \sim \frac{1}{\Delta}$.

In order to show the efficiency of this simple idea we have plotted in Fig.1 the result of two different simulation done in a 4×4 Hubbard model at half filling. Here the spin spin correlation function at the maximum distance is shown as a function of the imaginary time for an AF trial wavefunction with no definite spin and a closed shell trial wavefunction with total spin $S = 3$. In a perfect singlet , like the half filled ground state, the transverse part of the spin spin- correlation function

$$C_{xy}(r) = < \psi_G | 2 S^+_R S^-_{R+r} | \psi_G > \tag{9}$$

is equal to the usual z-axis correlation

$$C_z(r) = 4 < \psi_G | S^z_R S^z_{R+r} | \psi_G > - 4 < S^z_R >^2 \tag{10}$$

because there is no preferred direction in spin space. Here the factors 2 and 4 are chosen

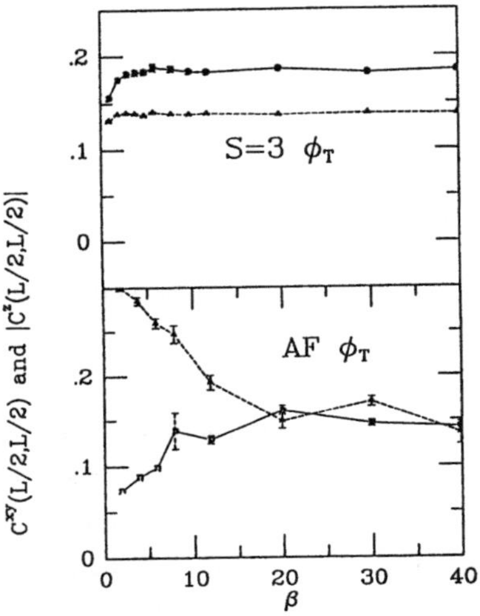

Fig. 1 x-y component (continuous lines) and z-component (dashed lines) of the spin-spin correlation function as a function of the imaginary time β in a 4×4 square lattice for $U = 4t$, $\Delta\tau t = 0.1$. The lower curves are obtained using an antiferromagnetic trial wavefunction, where only electrons at the outermost shell are combined with maximum antiferromagnetism (i.e. with $c^\dagger_{k,\sigma} \pm c^\dagger_{k+Q,\sigma}$ orbitals). The remaining 10 electrons fill the inner "closed shell". The upper curves refer to the trial wavefunction described in the text which has a definite spin. $C^z(L/2, L/2)$ is negative in this case.

for convenience and we have subtracted the constant $4 < S^z_R >^2$ in (10) which is non vanishing only if the total spin projection is finite.

When the trial wavefunction has not a definite spin as in the first case (lower part of Fig.1) the convergence in imaginary time is very poor and the condition $C_{xy} = C_z$ that should be fulfilled in the singlet ground-state is reached only for $\beta \sim 20$. Instead in the second case, when we have set the spin of the trial wavefunction to $S = 3$ and $< S^z_R >= 3/16$ has a non vanishing value , we actually sample an excited state, i.e. the ground state in this spin sector, and the convergence to this state is reached very soon. This also show that with this method we can sample also some of the excited state in a very stable way. This is in principle not trivial because the Hubbard Stratonovich field does not conserve the total spin. The data in Fig.1 were obtained using the same amount of computer time ($\sim$ 100 seconds of Cray YMP for each point), however the fluctuation of the staggered magnetization are much smaller using a trial wavefunction of definite spin.

We believe that the use of a good trial wavefunction may lead to a great advantage even for the larger sizes, and in general the most useful trial wavefunction is the one which is most symmetric and with lowest possible energy.

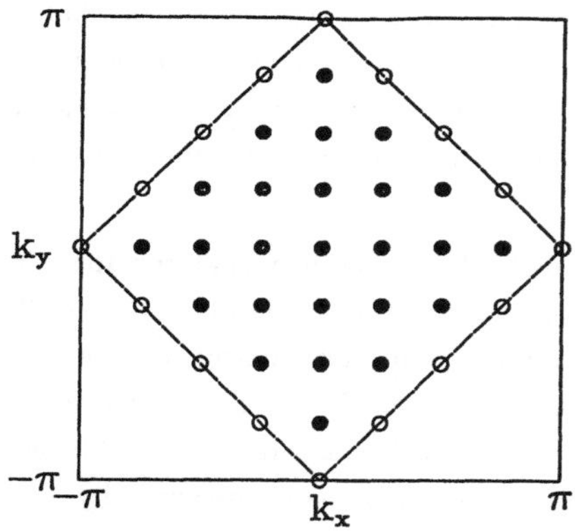

Fig.2 Nested Fermi surface (dashed line) for the half filled hubbard model at $U = 0$. The dots are the allowed k-points for an 8×8 lattice with periodic boundary conditions. Black dots are inside the Fermi surface while the empty dots are allowed momenta lying just in the $U = 0$ Fermi surface.

In a finite system, we have always chosen square lattices $L \times L$ or rotated by 45 degrees $\sqrt{2}L \times \sqrt{2}L$ with even L. It is convenient to choose the trial wavefunction as a simple product of its spin-up and spin down part: $\psi_T = \psi_T^{\uparrow} \times \psi_T^{\downarrow}$ (i.e. all the orbitals have definite spin $\uparrow$ or $\downarrow$). It has been shown[1,5] that the sign of the weight (7) is always positive at half filling if the down spin orbitals are related to the up spin ones by the charge conjugation symmetry. Thus in the following we describe how we have defined the orbitals $\phi_l(r)$ in the spin up trial state $\psi_T^{\uparrow} = \det \phi_l(r_i)$ for the half filled case.

Since we want to work with definite momentum for the full many body wavefunction $\psi_T^{\uparrow}$, we have to use plane wave orbitals, i.e. the tight binding eigenstate of the non -interacting one body problem. Despite the momentum restriction there is a huge degeneracy of the non interacting problem at half filling (and with one hole) since the last shell which contains $k-$points at the non interacting Fermi surfaces (see Fig.2) can be only partially occupied. In fact exactly half of the allowed non-equivalent k points ($2L-2$ for an $L \times L$ lattice and $4L-2$ for $L\sqrt{2} \times L\sqrt{2}$ lattice) can be filled in the subspace with vanishing spin projection S_z. On the other hand it is extremely important to select trial states with definite total spin S, since due to broken symmetry the different spin subspace are very close in energy. The ground state of the half filled Hubbard model is a singlet[9]. Unfortunately it is quite difficult to build up a single Slater determinant which is a singlet and has a good energy, that is with not too many doubly occupied sites. A simple way to satisfy this rule (without destroying charge conjugation symmetry) is to depart form the condition of vanishing S_z. If we take all the down spin electrons in the outermost shell and bring them in the up spin one, we fulfil the closed shell condition in the up and down spin band and satisfy in some sense the Hund's rule.

We have therefore a trial wavefunction of definite spin with only one Slater determinant. This can be easily shown by direct calculation of the square of the total spin $< S^2 >$ on this kind of wavefunction. As a result such an expectation value is $< S^2 >= S_z(S_Z + 1)$. Then the spin $S \geq S_z$ has to coincide with S_z, the minimum value.

In principle we are interested in the spin 0 sector but for large enough size there is not a macroscopic difference in spin (it is much smaller than the maximum $S_{max} = L^2/2$) and physical correlation functions are not affected by this small change of spin. As it will be discussed later the restriction to this "closed shell" sequence is equivalent to the simulation of the Hubbard model in presence of a Zeeman magnetic field that vanishes as $\frac{1}{L}$. In fact a finite magnetic field (assuming finite pauli-susceptibility χ_{st}, e.q. $\chi_{st} = \frac{S(S+1)}{8J}$ in the classical limit) has to increase the total S_z macroscopically in the size and linear in H.

The chosen sequence of closed shell sizes is also convenient to minimize the finite size effect because the total momentum can be kept fixed, say at $p = 0$, for any size.

As we have mentioned we are not working in the singlet ground state sector but in a sector that contain a representation of the "thermodynamic ground state" with broken symmetry and no definite spin. As well known from the classical theory of antiferromagnets a state with broken symmetry, i.e. with the staggered magnetization in a definite direction, has no definite spin. The fluctuation of the total spin in this state are small compared to the volume of the system,$\sim L$. Therefore in a finite system we have a chance to peak up a component of the "broken symmetry state" even in a spin sector $S \sim L$. That is just what we have done. In order to confirm that this semiclassical hypothesis is correct we have just to show[10] that spin rotational invariant quantities are not affected by the chosen representation of the broken symmetry state. In fact spin rotational invariant quantities should not depend on possible directions of the antiferromagnetic order parameter.

In the following we give a numerical test of this interpretation of the broken symmetry phenomena. To this end we have plotted in Fig.3 numerical simulation of the spin-spin correlation function following the sequence of sizes $L \times L$ from 4×4 to 16×16 for the half filled case. In Fig.4 it is clearly evident that the broken symmetry occurs only in the $x - y$ plane. This can be easily understood using classical arguments, since our choice of the $S_z = S$ component may be considered as an effective uniform magnetic field along the z axis (vanishing in the thermodynamic limit).

Due to the mentioned anisotropy the staggered magnetization can be calculated by:

$$m^2 = \lim_{L \to \infty} 2C_{xy}(L/2, L/2) \quad .$$

Here the factor 2 ,instead of the usual factor 3, is because only the $x - y$ components

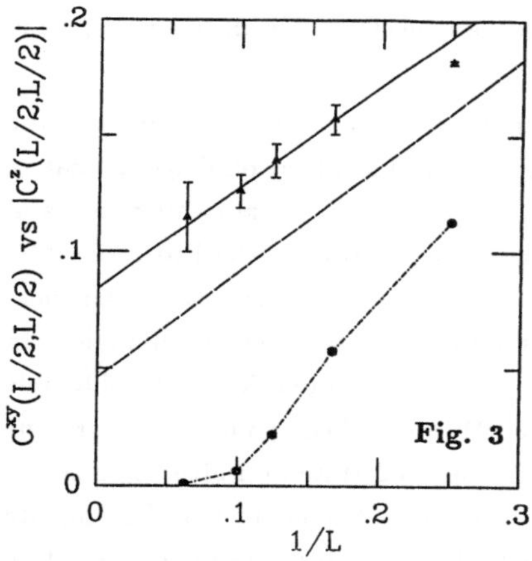

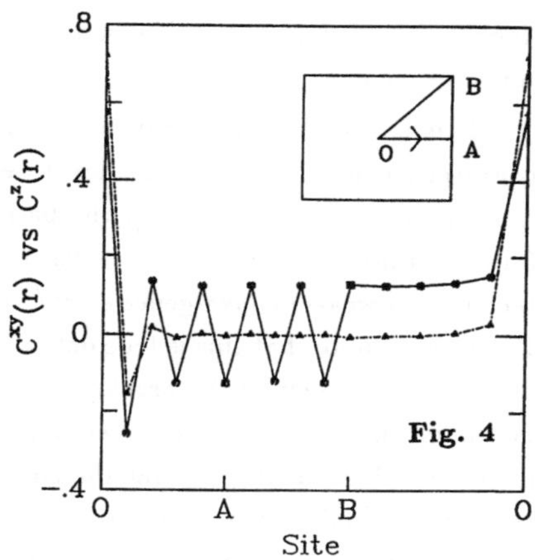

Fig. 3 Size scaling of the spin-spin correlation function at the maximum distance at half filling. The negative z-component (triangles) $|C^z(L/2, L/2)|$ and the planar component $C^{xy}(L/2, L/2)$ (circles) are plotted vs. $1/L$. The continuous line is a least square fit for C^{xy} for the largest sizes while the dashed one is the corresponding fit for $C_z(L/2, L/2)$ in the singlet case[7]. The dotted line is only a guide to the eyes.

Fig. 4 Spin spin correlation function, z-component (triangles) and planar component (circles) as a function of distance for a 10×10 half filled Hubbard model at $U = 4t$, $\beta = 20$ and $\Delta\tau = 0.1$. The path in the finite lattice is shown in the inset. The lines are guide to the eyes.

of the spin contribute to the staggered magnetic oscillations. The extrapolated order parameter shown in Fig.3 is then in quite good agreement with the analogous calculation done on the same sequence of sites in the singlet sector[7]. Indeed in Fig.3 the extrapolation at infinite size of $C^z(L/2, L/2)$ in the singlet case is roughly a factor $\frac{2}{3}$ smaller (as it should) than our extrapolated $C^{xy}(L/2, L/2)$. It is also interesting that the correspondence of the order parameter (once we neglect the z component) is approximately correct even for finite size for $L \geq 6$ (see Fig.3), since the two curves are approximately parallel. Moreover in our simulations the z-component of the staggered magnetization seems to vanish in an exponential way.

This data are also an independent verification that the thermodynamic limit for the half filled Hubbard model is well defined and that a non–macroscopic fluctuation of the spin does not affect thermodynamic properties, due to broken symmetry phenomena.

We can apply the same strategy for the study of the properties of one hole in the Hubbard model where sign problem is particularly serious. We have performed an exact diagonalization study in a $2\sqrt{2} \times 2\sqrt{2}$ system. For $U = 4$ the ground state has spin 1/2, and the gap to the first spin 3/2 excitation is only $\Delta = 0.06t$. However if we select a trial wavefunction of definite spin, following the Hund's rule as before, the gap is enormously improved by a factor 25. The same factor that we expect to gain in imaginary time convergence in (1) and (2).

It is usually believed that use of the standard trace algorithm[1,7] can improve the convergence in β by a factor 2 since if we use the trace in Eq.(2) the off diagonal matrix elements of the operator O -in the basis which diagonalize H- do not contribute and 0 temperature is reached with $e-\Delta\beta$ convergence rate. Instead in the Projection MC technique a weaker convergence $e^{-\Delta\beta/2}$ is expected due to off-diagonal matrix elements. However we note that this is true only if we use the worst possible trial wavefunction. In fact as we have seen in the previous example we can easily improve Δ by a considerable factor much larger than the irrelevant factor 2. When broken symmetry occurs one expect by classical theory that the gap to the first excited state in a given spin sector scales as $\frac{1}{L}$, while much more dramatically as $\frac{1}{L^2}$ without restriction of spin. Taking into account that the average sign can vanish exponentially with β and that the computation is roughly proportional to $< S >^{-2}$, we expect significant advantages in the use of the projection Monte carlo scheme, at least for the study of ground state properties.

Acknowledgements

I would like to thank M. Rice A. Parola and E. Tosatti for stimulating discussion throughout this work. I also acknowledge P.A. Bares, M. Luchini and B. Puttika for useful comments. In particular I am in debt to W.P. Petersen for optimization of the Monte Carlo code. Use of Cray-YMP has been supported by and ISI (Institute for Scientific Interchange) in Torino.

References

1) S. Sorella, S. Baroni, R. Car and M. Parrinello, *Europhys. Lett.* **8**, 663 (1989).
2) M. Imada and Y. Hatsugai *J. Phys. Soc. Jpn.* **58**, 3752 (1989).
3) J.E. Hirsch, *Phys. Rev.* **B 31**, 4403 (1985).
4) R. Blankenbecker, D.J. Scalapino and R.L. Sugar *Phys. Rev. D* **24**, 2278 (1981).
5) S.Sorella *Int. J. Mod. Phys.* **B 5** 937 (1991).
6) E.Y. Loh, J.E. Gubernatis, R.T. Scalettar, S.R. White, D.J. Scalapino, and R.J. Sugar *Phys. Rev.* **B 41**, 9391 (1990).
7) S.R. White, D. J. Scalapino, R.L. Sugar, E.Y. Loh, J.E. Gubernatis and R.T. Scalettar *Phys. Rev.* **B 40**, 506 (1989).
8) N. Furukawa and M. Imada *J. Phys. Soc. Jpn.* **60,** 3669 (1991).
9) E.H. Lieb , Phys. Rev. Lett. **62**, 1201 (1989).
10 M. Gros, E. Sánches-Velasco, and E. Siggia,*Phys. Rev.* **B39** 2484 (1989).

Simulation Method for Strongly Correlated Systems and Its Applications

M. Imada and N. Furukawa

Institute for Solid State Physics, University of Tokyo,
Roppongi 7-22-1, Minato-ku, Tokyo 106, Japan

Abstract. Recent progress in quantum Monte Carlo algorithm of the Hubbard model is summarized. The ground state properties of the square-lattice Hubbard model near the half-filling are discussed. Several aspects of spin, charge and superconducting correlations clarified in recent simulations are reviewed.

1. Introduction

Properties of strongly correlated fermion systems are key subjects of recent intensive studies of condesed matter phenomena such as oxide superconductors and heavy fermion systems. Various lattice fermion models have been proposed to describe essential physics of the ground state and low energy excitations of realistic systems. Simplified lattice models provide a way to extract essence of phenomena in real complicated systems. Although the reduction of the degrees of freedom in simplified models such as the Hubbard model from cuprate superconductors is drastic, even these lattice models do not allow us to solve them exactly except few examples. Many approximations to the lattice models have been proposed based on various physical concept and they sometimes contradict each other. On one hand, to understand whether these simplifications and reductions in constructing lattice fermions may correctly reproduce essential properties of real systems, and on the other hand, to check the reliability of various approximations, numerical analysis of the models has been widely used.

Basically two approaches are known as methods free of uncontrolled approximations. One is the exact diagonalization technique of model hamiltonians. Because all the physical properties including dynamical aspects are obtainable from diagonalization study, it is often powerful to discuss overall features of the models. Because the exact diagonalization gives very accurate numerical values for physical quantities of small sized cluster, asymptotic properties in the thermodynamic limit can sometimes be discussed with the combination of the finite size scaling analysis. Apparently, a serious limitation of the numerical diagonalization study is that tractable system size is small. If one wishes to overcome the limitation, the only available to date is a stochastic sampling method called simulations. Quantum simulation method of lattice fermions have been very rapidly developed in the last decade. A major obstacle in simulations of lattice fermions is the notorious sign problem. In this paper, we will discuss the present status of this difficulty in an example of the Hubbard model and show how it is reduced in the actual Monte Carlo simulation. We will also discuss the present understanding of the Hubbard model on a square lattice derived from various diagonalization studies and Monte Carlo simulations.

2. Sign Problems in Quantum Monte Carlo Method

In quantum simulation algorithms, the path integral formalism is used by introducing an additional real or imaginary time dimension to d-dimensions of real space. The path integral is defined as an integral over the Hilbert space on this $d+1$ dimensional space. To introduce a sampling of this integral, we have to define a measure of the integral. The sign problem is the difficulty that we do not know a way to keep this measure positive-definite in certain classes of lattice fermions and also in frustrated quantum spin models.

To be more specific, we introduce the Hubbard model, one of the simplest models of correlated lattice fermions, on a square lattice by the following hamiltonian :

$$H = -t \sum_{\substack{<ij> \\ \sigma}} (c_{i\sigma}^{\dagger} c_{j\sigma} + h.c.) + U \sum_{i} n_{i\uparrow} n_{i\downarrow}, \tag{2.1}$$

where the nearest-neighbor transfer term proportional to t and the interaction term proportional to U operate to fermions with spin σ. The partition function is calculated from

$$Z = \sum_{\{\psi\}} \langle\psi| \exp(-\beta H) |\psi\rangle, \tag{2.2}$$

where the trace summation over ψ is taken at the inverse temperature β. In this paper, we concentrate on the ground state properties. If $|\psi\rangle$ is not orthogonal to the ground state, the ground state properties are obtained from a single initial state $|\psi\rangle$ without the trace summation at sufficiently large β. In the path integral, the density matrix

$$\rho(\tau : \psi) = \langle\psi| \exp(-\tau H) |\psi\rangle \tag{2.3}$$

is represented by the summation over the Hilbert space of intermediate states as

$$\rho(\tau : \psi) = \sum_{\{s\}} W(\{s\}, \tau : \psi), \tag{2.4}$$

where W is the fermion determinant. The summation over a set of auxiliary variables s are introduced by the Stratonovich-Hubbard transformation to the Hubbard model and the interacting fermion problem is transformed to noninteracting fermions under τ dependent auxiliary variables $\{s\}$, with a choice of single particle state for ψ. Because W is not necessarily positive definite, we take $|W|$ as the weight of the importance Monte Carlo sampling and the average of a physical quantity A is obtained from the identity

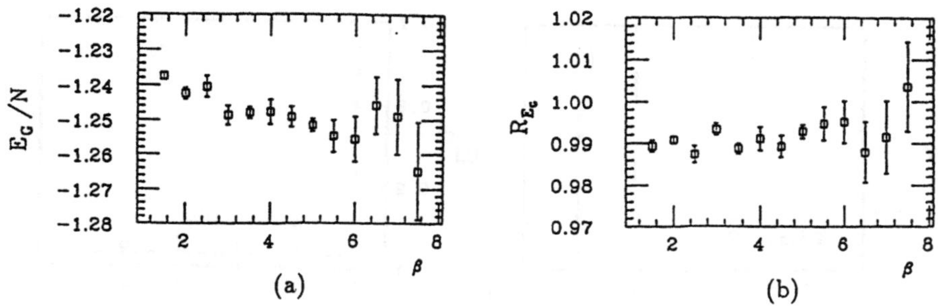

Fig.1.a. τ-dependence of $\langle E_G \rangle$ computed from direct measurement. See text.
1.b. τ-dependence of R_{E_G}.

$$\langle A \rangle = \sum_i A_i |W_i| \mathrm{sign} W_i \Big/ \sum_i |W_i| \mathrm{sign} W_i, \tag{2.5}$$

where i denotes a configuration of $\{s\}$. It has been shown numerically [1-3] that the average of the sign defined by

$$\langle S \rangle \equiv \frac{W_+ - W_-}{W_+ + W_-}, \tag{2.6}$$

$$W_+ = \frac{1}{2} \sum_i (|W_i| + W_i) \quad \text{and} \quad W_- = \frac{1}{2} \sum_i (|W_i| - W_i) \tag{2.7}$$

follows $\langle S \rangle \propto \exp[-\tau\Delta]$ with a positive constant Δ in usual cases of the Hubbard model away from half-filling. Because the average sign decreases exponentially with the increase of τ, estimates of both the numerator and the denominator of eq. (2.5) and hence the ratio become extremely difficult due to the statistical error. Although the form (2.7) is strictly speaking correct for asymptotically large τ, it has been shown that the exponential dependence (2.7) follows even from very small imaginary time τ in many cases [1-3]. Then using the identity

$$\langle A \rangle = \frac{Q_A}{\langle S \rangle} A^*$$

with $\langle A \rangle^* = \sum_i A_i |W_i| / \sum_i |W_i|$ and $Q_A = \sum_i A_i W_i / \sum_i A_i |W_i|$, the average $\langle A \rangle$ can be calculated from the product of the sign neglected average $\langle A \rangle^*$ and its correction $R_A = Q_A / \langle S \rangle$. This R-correction method [3] is more efficient than the direct measurement of $\langle A \rangle$, if $\langle A \rangle$ and $\langle A \rangle^*$ converge slowly to the ground state value while R_A converges faster, because the statiscal error due to the sign problem gets larger for larger τ not in $\langle A \rangle^*$ but in R_A. An example showing efficiency of R-correction method as compared to the direct measurement is given in Fig.1 for the

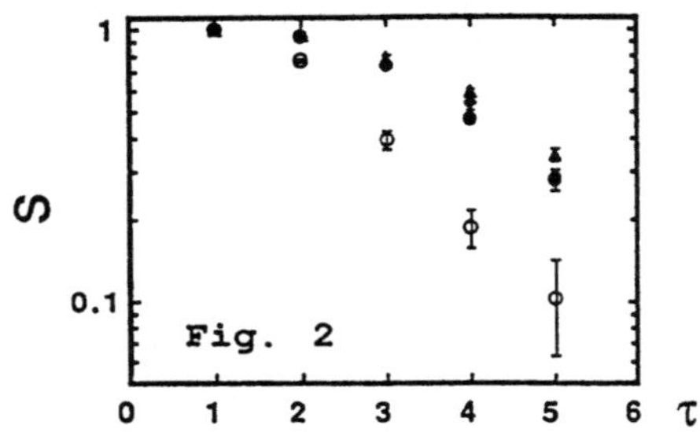

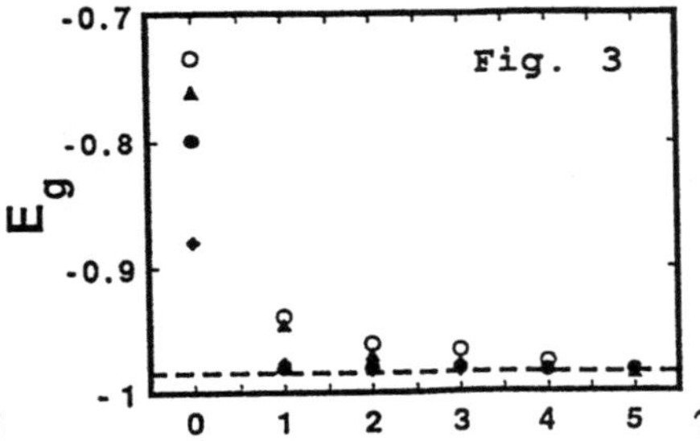

Fig.2. τ-dependence of the sign ratio for the paramagnetic trial state (open circle), unrestricted Hartree-Fock trial state at $U_{\text{init}} = 0.1$ (filled circle), $U_{\text{init}} = 4.0$ (diamond) and $U_{\text{init}} = 8.0$ (triangle).
Fig.3. τ-dependence of the energy with the same notation as Fig.2.

case of the ground state energy $\langle E_{\text{G}} \rangle$ of the 3×2 square lattice Hubbard model at $U = 8$, where the convergence of the direct measurement $\langle E_{\text{G}} \rangle$ and $\langle E_{\text{G}} \rangle^*$ are seen only for $\tau \geq 4$ but $R_{E_{\text{G}}}$ is already convergent at very small $\tau \sim 1$. Because we can observe the convergence of $\langle E_{\text{G}} \rangle^*$ without the sign difficulty, A faster convergence of $R_{E_{\text{G}}}$ than $\langle E_{\text{G}} \rangle$ shows that the R-correction method is a better way than the direct measurement. In this case, the R-correction method gives -1.240 ± 0.003 while the exact value is -1.2442.

Another and more efficient prescription [4] to the sign problem is to take better initial trial state ψ. Within the framework of a pure state in single particle basis with conserving number of fermions, the solution of an unrestricted Hartree Fock equation is the optimal state as ψ. The eigenvalue problem for the Hartree Fock hamiltonian

$$H_{\text{UHF}} = -t \sum_{<ij>,\sigma} (c_{i,\sigma}^\dagger c_{j,\sigma} + h.c.) + U_{\text{init}} \sum_i \left[\langle n_{i\downarrow} \rangle \hat{n}_{i\uparrow} + \langle n_{i\uparrow} \rangle \hat{n}_{i\downarrow} - \langle n_{i\uparrow} \rangle \langle n_{i\downarrow} \rangle \right]$$

is solved for an appropriate choice of U_{init} and the M-particle state is obtained by filling the lowest M eigenstates. Using the unrestricted Hartree Fock state, the sign problem is reduced in two points. First the negative sign ratio decreases as compared to other choices such as the paramagnetic free fermion ground state. An example of the reduction is seen in Fig.2 for the case of 7 up and down fermions each on a 4×4 lattice with $U = 4$. The reduction appears to be due to the increase of the prefactor S_0 in the exponential dependence of the form

$$\langle S \rangle = S_0 \exp[-\tau \Delta].$$

The second point is that the convergence to the ground state value becomes faster for the optimal choice. An example is seen in Fig.3 for the energy in the same case as Fig.2. These two effects combined together make it possible to obtain converged ground state properties of the Hubbard model for various choices of lattice sizes. In particular, several ground state properties near half-filling with configurations of open-shell filling at the fermi level have been reliably obtained through this technique. The optimization of the initial trial state is only possible in the canonical ensemble approach and has proven to be advantageous over the approach from the grand canonical ensemble if one wishes to obtain the ground state properties. In the next section some results obtained by this technique will be summarized.

3. Two-Dimensional Hubbard Model in the Ground State

There is a lot of numerical studies on the Hubbard model on a square lattice especially after the discovery of high T_c oxides [5-7]. In particular, by implementing the algorithmic progress described in §2, the sign difficulty, which is serious near the half filling, is reduced to a substantial extent. Through this progress, charge, spin and superconducting correlations have been calculated extensively [8-10] and the ground state properties of the Hubbard model is then more clearly understood than before.

The Hubbard model with $U \neq 0$ has a charge gap at the half-filling. At $U = 4$, the gap is estimated to be 0.58 ± 0.08. This value is substantially smaller than the Hartree Fock estimate 1.38. Upon doping to the half-filled system, the chemical potential μ shifts immediately with the amount of the charge gap. Upon further doping, an interesting feature emerges. The charge compressibility defined by

$$\kappa = \left(\frac{\partial \mu}{\partial n}\right)^{-1} \quad \text{and} \quad \mu = \frac{\partial E}{\partial n},$$

where n is the fillingness and E is the ground state energy, appears to follow

$$\kappa \propto (1-n)^{-1} \tag{3.1}$$

near the half-filling. In Fig.4 the chemical potential vs. the fillingness is plotted.

This singular feature is similar to the one-dimensional Hubbard model where the exact solution [11] shows the same behavior as Eq.(3.1). Therefore the divergence of the charge compressibility near the half-filling seems to be a common feature of the Mott insulator both in one- and two-dimensions. This excludes the possibility that the singularity is a consequence of the singularity in the density of states inherently found in one-dimensional systems. The charge compressibility seems to suggest that the mass itself is diverging as the system approaches the Mott insulator.

The spin correlation of the square-lattice Hubbard model shows another interesting feature of a strongly correlated system. It is widely accepted that the antiferromagnetic long range order exists in the half-filled case. The incommensurate spin correlation replaces the commensurate long range order upon doping [5]. The peak value of the spin structure

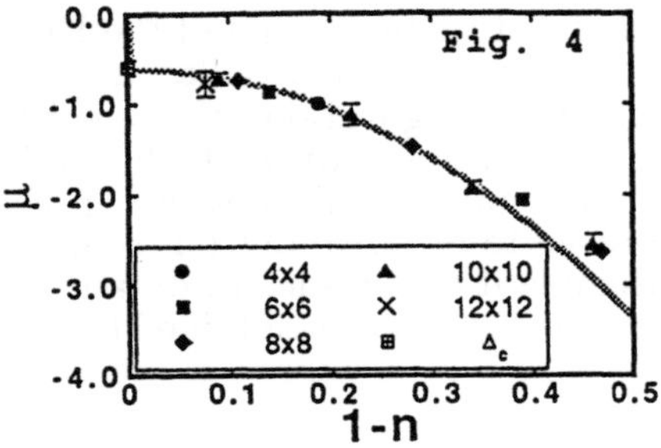

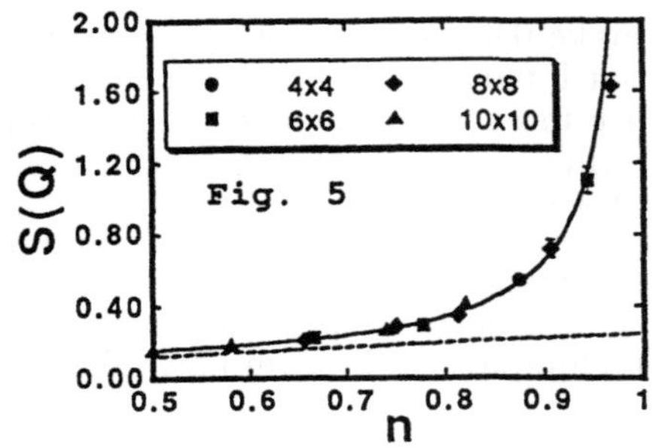

Fig.4. Chemical potential μ vs. fillingness n.
Fig.5. *S(Q)* vs. fillingness n.

factor $S(Q)$ at the incommensurate wave vector Q is illustrated in Fig.5 [9]. It is well fitted by the form

$$S(Q) \propto \frac{1}{N} \sum_{r} \mathrm{e}^{-r/\xi}$$

with the correlation length

$$\xi \simeq 0.4/\sqrt{(1-n)}.$$

This indicates that the correlation length is proportional to the mean hole-hole distance and only the short ranged correlation is seen in the finite doping regime.

Many of the features in spin correlations and the charge responses of the two-dimensional Hubbard model described above are in good agreements with experimental properties of high-T_c oxides observed by neutron scattering, X-ray absorption experiment, optical conductivity and so on.

The superconducting correlation of the square lattice Hubbard mode in the ground state is also obtained by the improved Monte Carlo method Measurements for various choices of pairing correlations show that th superconducting correlation is very short ranged both for s and d wav symmetries [8,10].

Concluding Remarks

We have seen that improvements of the Monte Carlo simulation tech nique make it possible to perform simulations of the two-dimensiona Hubbard model in the ground state at arbitrary fillingness. This al gorithmic progress has clarified several features of the Hubbard mode near the half-filling. The divergent behavior of the charge susceptibility as $n \to 1$ may be a fundamental property of the Mott transition and th present data have shown clearer evidence of it consistently with a previou finite temperature result [12]. Generally speaking, the spin correlation and the charge degrees of freedom on the square lattice Hubbard mode bear many resemblances to the experimental indications in the high T_c oxide superconductors. However, crucially important property, namely the superconductivity is lacking in the Hubbard model. This contras between these resemblances and the absence of superconducting order in numerical simulations of the Hubbard model, one of the most puzzling questions, seems to be the key point for the understanding of the mech anism of high temperature superconductivity. Combined with furthe efforts to improve the simulation algorithm, the quantum simulation wil play crucial roles in clarifying various fundamental aspects of strongly correlated systems.

References

1. E. Y. Loh, Jr., J. E. Gubernatis, R. T. Scalettar, S. R. White, D. J Scalapino and R. L. Sugar, Phys. Rev. B**41**, 9301(1990)
2. D. R. Hamann and S. B. Fahy, Phys. Rev. B**41**, 11352(1990)
3. N. Furukawa and M. Imada, J. Phys. Soc. Jpn. **60**, 810(1991)
4. N. Furukawa and M. Imada, J. Phys. Soc. Jpn. **60**, 3669(1991)
5. J. E. Hirsch, Phys. Rev. B**31**, 4403(1985)
6. M. Imada and Y. Hatsugai, J. Phys. Soc. Jpn. **58**, 3752(1989)
7. D. J. Scalapino, *High Temperature Superconductivity Proceedings*, ed K. S. Bedell et al. (Addison-Wesley, 1990)p.314
8. M. Imada, J. Phys. Soc. Jpn. **60**, 2740(1991)
9. N. Furukawa and M. Imada, J. Phys. Soc. Jpn. **60**, 3604(1991)
10. N. Furukawa and M. Imada, in preparation
11. N. Kawakami and A. Okiji, Phys. Rev. B**40**, 7066(1989)
12. H. Otsuka, J. Phys. Soc. Jpn.**59**, 2916(1990)

Recent Progress in the Auxiliary-Field Many-Fermion Simulation Method

D.R. Hamann

AT&T Bell Laboratories, Murray Hill, NJ 07974, USA

Abstract. This paper discusses three topics pertaining to quantum Monte Carlo simulations of the ground states of many-fermion systems based on the Hubbard-Stratonovich transformation. These are the recently introduced positive-projection approximation, the extension of these methods to lattice models with long-range interactions, and the passage from lattice models to the continuum limit.

1. Introduction

There is considerable current interest in achieving accuracy in the calculation of the electronic ground state properties of solids beyond that afforded by the local density approximation. Quantum Monte Carlo methods provide a promising way to approach this problem. Among several classes of such methods, the Hubbard-Stratonovich (HS) auxiliary field method seems most promising for situations when the correlations in the ground state are strong and complex. Other methods are more dependent on the possibility of guessing a closed-form expression for an approximate wave function that embodies the most important aspects of the correlations. In addition, it is compatible with the familiar tools of local density functional theory, such as basis functions and non-local pseudopotentials. This method has been widely applied to the Hubbard model, but no application to a realistic hamiltonian for a solid has been attempted. The motivation for my recent work in this area, all of which has been carried out in collaboration with S. B. Fahy, has been to explore this possibility.

HS simulations are often used in the context of finite temperature quantum statistical mechanics simulations. A well known problem of extending HS simulations to ground state properties is divergent statistical noise at low temperatures, otherwise known as the "sign problem." This will be briefly reviewed in Sec. 2.

Since dealing with the sign problem is a prerequisite for the applications in which we are interested, we have concentrated on an analysis of this problem. This analysis has led to a new approximation scheme which we call "positive projection" (PP) and which appears to yield very accurate results in situations in which we have tested it so far. These developments will be summarized, and some recent test results reported in Sec. 3

The initial analysis and development of PP was carried out within the familiar confines of the simple Hubbard model. We have recently studied several formulations of the HS transformation for lattice models with long range

Springer Proceedings in Physics, Vol. 70 Computational Approaches in Condensed-Matter Physics
Editors: S. Miyashita · M. Imada · H. Takayama

interactions. One of these has been tested extensively on small models at this point, and its derivation and testing in conjunction with PP will be discussed in Sec. 4. Another step in moving towards real materials is to explore the behavior of our methods in the limit in which the lattice model is allowed to become a representation of continuous space. Recent results on this topic will be discussed in Sec. 5.

2. Hubbard-Stratonovich Ground State Simulations

We are interested in applying the Hubbard-Stratonovich transformation [1] in the context of projection of the ground state from a trial state, as explored originally by Sugiyama and Koonin [2] and more recently by Sorella et al.[3] In this context, we consider the formal solution of the imaginary time Schrödinger equation applied to the trial state $|\psi_t>$,

$$|\Psi(\tau)> = e^{-\tau H}|\psi_t> = e^{-\Delta\tau H} \cdots e^{-\Delta\tau H} e^{-\Delta\tau H}|\psi_t> \underset{\tau\to\infty}{\to} |\Phi_0> , \qquad (1)$$

which projects out the ground state in the large τ limit as indicated above. We suppose that we apply the Trotter-Suzuki approximation [4] in the usual way to factor the kinetic and interaction terms of the Hubbard hamiltonian H in each time slice propagator in Eq. (1), and carry out the HS transformation [1] on the interaction term. We then reassemble the factors to obtain

$$|\Psi(\ell\Delta\tau)> = \int \prod_{i=1}^{N} \prod_{j=1}^{\ell} dx_{ij} e^{-\frac{1}{2}\sum_{ij} x_{ij}^2} e^{-\Delta\tau H_\ell} \cdots e^{-\Delta\tau H_2} e^{-\Delta\tau H_1}|\psi_t> , \qquad (2)$$

where we introduce the "time slice hamiltonians"

$$H_j = \sum_{ik\sigma} t_{ik} a_{i\sigma}^{\dagger} a_{k\sigma} + \sqrt{U/\Delta\tau} \sum_i x_{ij}(n_{i\uparrow} - n_{i\downarrow}) . \qquad (3)$$

We have found the reassembly of the Trotter-Suzuki breakup of the time slice propagators to be advantageous, as discussed below.[5] Since each propagator is a one-electron operator,

$$|\psi(X)> = e^{-\Delta\tau H_\ell} \cdots e^{-\Delta\tau H_2} e^{-\Delta\tau H_1}|\psi_t> = U_\ell \cdots U_2 U_1 |\psi_t> \qquad (4)$$

is a single determinant and a functional of the set of auxiliary fields $\{x_{ij}\} \equiv X$ which we can calculate explicitly for each X.

Averages of physical quantities such as the energy can be calculated as

$$E = \frac{\int \delta X G(X) <\psi_t|H|\psi(X)>}{\int \delta X G(X) <\psi_t|\psi(X)>} = \frac{\int \delta X G D E}{\int \delta X G D} , \qquad (5)$$

where G is the Gaussian from Eq. (2), D is the determinant of the overlap matrix between the one-electron states forming $|\psi_t>$ and those forming $|\psi(X)>$, and $E(X)$ is an energy functional.[5] Since $D(X)$ can change sign, a Monte Carlo evaluation of the δX integral requires us to use $G|D|$ as the probability functional of X, and introduce the sign $s = D/|D|$. Eq. (5) is then re-written

$$E = \int \delta X G|D| s E \,/ \int \delta X G |D| s . \qquad (6)$$

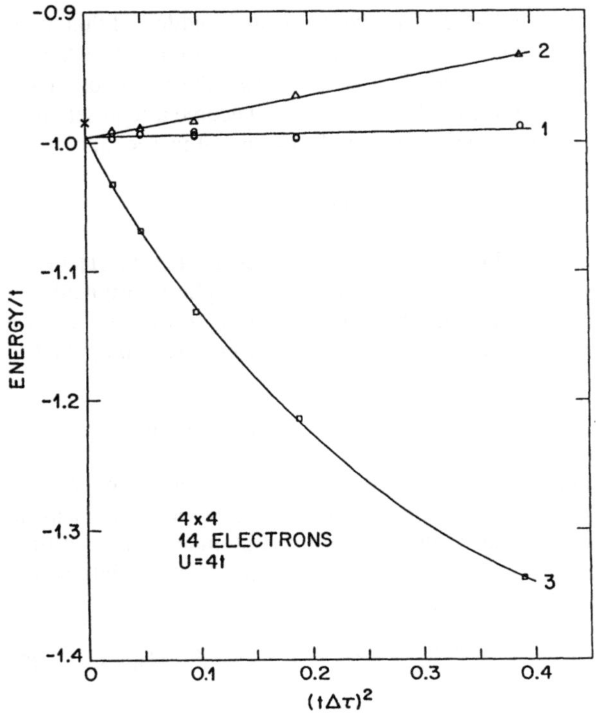

Fig. 1. $\Delta\tau$ convergence of the energy per site for simulations of a 2-dimensional Hubbard model. The different curves represent different approximations described in the text. (From reference 5.)

The conventional statement of the sign problem is that the average sign $<s>$ (defined in terms of the probability $G|D|$) vanishes exponentially at large τ through the cancellation of the positive and negative contributions. Therefore, when the numerator and denominator are evaluated by any statistical sampling technique, the "signal to noise ratio" also vanishes at large τ.

Sorella et al. found that surprisingly good results could be obtained simply by setting $s=1$ in Eq. (6).[3] In Fig. 1, we show some results of ours to illustrate this and another point. These are plots of energy as a function of $(\Delta\tau)^2$ for a 4x4 Hubbard model with 14 electrons, a particularly bad sign-problem case, calculated within what we will refer to as the "Sorella approximation." Curve 3 is based on the standard Trotter-Suzuki evaluation of the time-slice propagators, and has a $(\Delta\tau)^2$convergence rate comparable to the (rather rarely) published results for this rate.[6] Curve 2, based on measurement of the energy at a different place relative to the Trotter-Suzuki breakup, has an order of magnitude faster convergence. Finally, curve 1, based on treating the "reassembled" $U_j = \exp(-\Delta\tau H_j)$ exactly, appears to have yet another order of magnitude better convergence.[5] This is why we prefer to use the forms given in Eqs. (2) and (3). All three curves extrapolate to the same

$\Delta\tau=0$ limit, which is only ~0.01 below the exact result shown by the x. This serves as another example of the surprising accuracy of the Sorella approximation,[3] which we will come to understand better in the next section.

3. Positive Projection Approximation

It is clear from Eqs. (3) and (4) above that $|\Psi(\tau)\rangle$ is being represented through the HS transformation as a distribution of determinants,

$$|\Psi(\tau)\rangle = \int \delta X G(X) |\psi(X)\rangle \ . \tag{7}$$

We can write an alternative expression to make this observation explicit,

$$|\Psi(\tau)\rangle = \int d\psi f(\psi,\tau) |\psi\rangle \ , \tag{8}$$

where this integral is over the manifold of Slater determinants, and the relation between Eqs. (7) and (8) serves to define the distribution $f(\psi,\tau)$. As suggested by numerical experiments and confirmed by an arduous analysis of the change of integration variables between these equations,[7] f obeys a diffusion equation,

$$-\frac{\partial f}{\partial \tau} = -\frac{1}{2}\mathbf{D}(\psi) f - [\nabla_\psi V_1(\psi)]\cdot\nabla_\psi f + V_2(\psi) f \ . \tag{9}$$

The second-order differential diffusion operator $\mathbf{D}$, the drift potential V_1 and the branching potential V_2 can all be calculated explicitly in terms of the original hamiltonian and the HS transform. Solving Eq. (9) subject to the initial condition $f(\psi,0)=\delta(\psi-\psi_t)$ and substituting the result in Eq. (8) will yield $|\Psi(\tau)\rangle$. We note that Eq. (8) is a highly overcomplete representation, and that while $|\Psi(\tau)\rangle$ is unique, a range of HS transformations may be chosen for the original many-body hamiltonian, resulting in different $\mathbf{D}$'s and V's in Eq. (9) and different solutions f.

The solution of Eq. (9) can, in principle, be expanded in terms of eigenfunctions f^i and eigenvalues ε^i of its right-hand side, as

$$f(\psi,\tau) = \sum_i c_i \, e^{-\varepsilon^i\tau} f^i(\psi) \ , \tag{10}$$

We have shown that Eq. (9) is invariant under the parity operator $\psi\rightarrow-\psi$, and it follows from the general properties of diffusion equations that the dominant eigenfunction at large τ in Eq. (10) must be an even-parity solution f^+. But an even parity solution substituted into Eq. (8) gives $|\Psi(\tau)\rangle\equiv0$, since $|\psi\rangle$ and $|-\psi\rangle$ cancel exactly. Thus, while $f^+(\psi)$ dominates the magnitude of $f(\psi,\tau)$ at large τ, the leading odd-parity eigenfuntion $f^-(\psi)$ dominates all integrals involving f.[7] The rate of exponential decay of the "sign" $\langle s\rangle$ is easily shown to be given by the difference between the eigenvalues associated with these two solutions.

This analysis suggests a new approximate method for controlling the sign problem. Distributions $f(\psi)$ tend to be peaked near mean-field solutions of the many-body hamiltonian. We can schematically represent this situation by a parity-symmetric double-well branching potential $V_2(\psi)$ defined on a hypothetical 1-dimensional ψ manifold. Such a potential and the corresponding f^+ and f^- are sketched in Fig. 2(a). It is clear that if we supplement V_2 by an infinite barrier at the

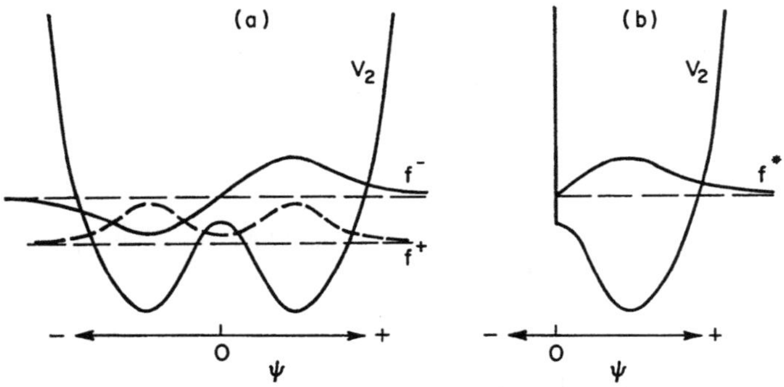

Fig. 2. Schematic representation of the growth potential for a 1-dimensional diffusion equation for $f(\psi,\tau)$, showing the lowest even-parity (dashed) and odd-parity (solid) eigenfunctions in (a), and the lowest eigenfunction for the well with an added infinite barrier in (b). (From reference 8.)

location of the node of f^-, Fig. 2(b), the leading solution of the modified problem f^* is equal to f^- in the allowed region. Thus, if we place a single nodal surface cutting the manifold of determinants, and constrain $f(\psi,\tau)$ to remain on one side of it, we expect to suppress the sign problem. The expectation value in Eq. (5) can be rewritten in an symmetric fashion, and expressed in terms of the representation in Eq. (8),

$$<E>_\tau = \frac{\int f(\psi,\tau)\, f(\psi',\tau) <\psi|H|\psi'>\, d\psi\, d\psi'}{\int f(\psi,\tau)\, f(\psi',\tau) <\psi|\psi'>\, d\psi\, d\psi'} \,. \tag{11}$$

Written in this form, it is clear that while an exactly chosen nodal surface will give the exact result for this and other expectation values, an approximately chosen surface, resulting in $f^* \sim f^-$, will give a variational upper bound for the energy. Imposing an approximate nodal surface is what we term the positive projection approximation.[8]

We observe that there is a formal similarity between the PP approximation and the fixed node approximation of diffusion Monte Carlo theory.[9] However, the physical content is entirely different. The diffusion in that theory is produced by the kinetic energy operator in real space. Here, the diffusion on the determinant manifold is caused strictly by the interaction operator of the original hamiltonian, with the original kinetic energy contributing only to V_1 and V_2. There, the unconstrained dominant state is the boson ground state, and the rate of sign decay is the difference between the boson and fermion ground state energies. Here, fermion antisymmetry of the wave functions is built in analytically by our representation as a distribution of determinants, and the sign decay rate, measured by the difference between the f^+ and f^- eigenvalues, is much smaller. The energy error of the Sorella approximation in Fig. 1 is a measure of this rate,[5] and is obviously vastly smaller than the boson - fermion difference for this problem. There, the detailed nodal structure of the wave function in space is constrained, whereas here we have an extreme simple constraint, albeit one that is rather abstract.

The present results also give us insight into the apparent accuracy of the Sorella approximation.[3] In our terms, this approximation amounts to setting $f^-(\psi) \approx f^+(\psi)\,\mathrm{sign}(<\psi|\psi_t>)$. Since a double (or multiple) well situation is expected to be physically typical, with small amplitude for both f^+ and f^- near the f^- node, this could be a good approximation, pending the choice of ψ_t.

Practical application of PP must be carried out in terms of the δX functional integral, because the diffusion equation on ψ is too complicated to work with directly. The only workable form of the constraint is a plane cutting the $|\psi>$ manifold, defined by a "constraint" state perpendicular to the plane, $|\Phi_c>$. Physically, using a single mean field determinant or a symmetrized combination of a few such determinants should keep the node out of high probability regions of f^-, and is practical for computation. If we consider a total of P $\Delta\tau$ slices, and the symmetric form of the expectation value in Eq. (11), we must impose the P separate constraints

$$<\Phi_c|U_l U_{l-1} \cdots U_2 U_1|\psi_t> \; >0, \quad l=1,...,P/2, \tag{12}$$

$$<\psi_t|U_P U_{P-1} \cdots U_{l+1} U_l|\Phi_c> \; >0, \quad l=P/2+1,...,P,$$

on the sampling process by which we perform the δX integration.

Since the Eq. (12) constraints are highly non-local in τ, a local-update sampling strategy would be extremely inefficient. Instead, we have used the hybrid Monte Carlo approach,[10] in which a "harmless" gaussian integration of velocity variables $\{v_{ij}\}$ is added to the δX integration, and the resulting integrand is interpreted as the exponential of a "classical energy,"

$$\int \delta X G|D| \rightarrow \int \delta V \delta X e^{-E_{cl}(V,X)} \,, \tag{13}$$

$$E_{CL} = \frac{1}{2}\sum_{ij} v_{ij}^2 + \frac{1}{2}\sum_{ij} x_{ij}^2 - \ln|D\{x_{ij}\}| \,. \tag{14}$$

The averages we wish to perform now appear as averages over a classical partition function. We then use a slightly modified molecular dynamics approach to update all our fields simultaneously, motivated by the ergodic hypothesis. At each update step, we pick a gaussian distribution of $\{v_{ij}\}$ with unit variance (note we have unit classical "temperature"), and integrate the equations of motion n time steps of length Δt_{MD}. We use an inexact integration algorithm for efficiency, and to deliberately let the system pass through the logarithmic barriers caused by the zeroes of $|D|$(which do occur even with positive projection). to correct the algorithm's errors, we accept each move with probability $p=\min[1,\exp(E_{CL}^{final} - E_{CL}^{initial})]$, which makes our averaging procedure rigorously correct as long as our approximate algorithm is reversible and phase-space conserving.[10] At the end of all this, we test our updated $\{x_{ij}\}$ using Eq. (12), and reject the move if it fails to satisfy the PP constraints. Our tests on small models indicate that the $|\Phi_c>$ barrier repels the X random walk well enough that we only increase our rejection rate a modest amount in this last step.

PP has proven effective in defeating the sign problem, with the PP sign saturating near 1 at large τ values. Our tests have also shown that PP gives accurate estimates of energies and correlation functions for several small Hubbard models.[8] The same considerations that motivated PP also explain why the Sorella approximation can give good results. A logical question is whether the difficulties of PP are worth the trouble compared to the simplicity and better statistics of the Sorella approximation. We have been aided in attempting to answer this question by a recent study of Furukawa and Imada, which identified a simple model in which the Sorella approximation gives qualitatively incorrect numbers for certain quantities.[11] This is the 3x2 2-dimensional Hubbard model with 4 electrons, which has a relatively bad sign problem. We have carried out Sorella approximation, PP, and exact diagonalization calculations for this model for $U=4$ and $U=8$, and report the results in Table 1, including energy per site E, kinetic energy KE, occupancy of kinetic energy eigenstates $n(k_x, k_y)$, and pair correlation functions for parallel (ppc) and antiparallel (apc) spins.

This model has an unusual anisotropy induced by the periodic boundary conditions. In the 2 (y) direction, the effective hopping is $2t$ instead of t as in the 3 (x) direction. The kinetic energy eigenvalues are { -4, -1, -1, 0, 3, 3}, with the lower three all corresponding to the y-symmetric states, and the fermi level at the {-1, -1} pair for the non-interacting system. The constraint state $|\Phi_c>$ for these calculations was chosen to be essentially a non-interacting state. A small potential was imposed to break the degeneracy of the pair of states at the fermi level, and put spin-up and spin-down electrons in opposite members of the pair. The good acceptance rate obtained indicated that efforts to devise a more elaborate state were not warranted.

Most of the simulation results are close to each other and to the exact results. However $n(0,\pi)$, which is the occupancy of the lowest y-antisymmetric state, is overestimated by a factor of 4 by the Sorella approximation. If this system were confined to the hilbert space spanned by the lowest 3 states, it would be effectively 1-dimensional. For electrons to "get by" each other using the second dimension, the upper set of states is involved. PP, which includes the fluctuating sign in its averaging, gets this right. The fact that these numbers have larger relative error bars than the rest is an indication that cancellations are more important. (These calculations were performed for $\tau=20$, so a conventional simulation would have hopelessly large error bars for all quantities.) The other quantity for which the Sorella approximation is qualitatively wrong is the first-neighbor parallel correlation function in the y direction, ppc(1y), which undoubtedly involves related effects.

In the PP approach as described, the quantity being averaged may only be measured in the center of the overall τ range, between the left and right constraint regions of Eq. (12). It is possible to terminate the constraint before getting to the innermost time slices, and leave an unconstrained region in the center. Of course, the sign will decay over this region, but if its total length is kept short (say, $\tau_{UC}=2$), the improved statistics from the ability to average the calculation of the measured quantity over all the time slices in this region may make this "release node" procedure worthwhile. In addition, the unconstrained system may be able to cancel some residual errors imposed by the constraint within this region. This generalization is used for results discussed in the subsequent sections.

Table 1. Comparison of Sorella approximation (SOR), positive projection (PP), and exact (DIAG) results for a 3x2 Hubbard model with 4 electrons. The remaining notation is described in the text.

	SOR	PP	DIAG
U=4			
E	-1.3905(12)	-1.3995(16)	-1.4009
KE	-1.5807(14)	-1.5943(19)	-1.6044
$n(0,0)$	0.9727(12)	0.9690(6)	0.9738
$n(2\pi/3,0)$	0.4608(64)	0.4916(51)	0.4924
$n(0,\pi)$	0.0818(25)	0.0222(33)	0.0191
$n(2\pi/3,\pi)$	0.0118(10)	0.0127(3)	0.0111
ppc(1x)	0.0787(11)	0.0848(11)	0.0847
ppc(1y)	0.0048(3)	0.0006(3)	0.0002
ppc(2)	0.0855(14)	0.0816(7)	0.0819
apc(0)	0.0476(3)	0.0487(2)	0.0509
apc(1x)	0.1226(15)	0.1214(9)	0.1219
apc(1y)	0.1335(17)	0.1173(10)	0.1157
apc(2)	0.1209(15)	0.1289(8)	0.1282
U=8			
E	-1.2489(28)	-1.2437(36)	-1.2442
KE	-1.4367(37)	-1.4572(43)	-1.4798
$n(0,0)$	0.9221(29)	0.9105(25)	0.9200
$n(2\pi/3,0)$	0.4069(127)	0.4780(125)	0.4778
$n(0,\pi)$	0.1994(80)	0.0581(110)	0.0592
$n(2\pi/3,\pi)$	0.0321(15)	0.0377(12)	0.0327
ppc(1x)	0.0725(18)	0.0877(23)	0.0874
ppc(1y)	0.0170(11)	0.0003(16)	0.0008
ppc(2)	0.0856(21)	0.0788(24)	0.0790
apc(0)	0.0235(3)	0.0266(2)	0.0295
apc(1x)	0.1270(26)	0.1187(31)	0.1193
apc(1y)	0.1465(49)	0.1421(46)	0.1367
apc(2)	0.1213(23)	0.1302(26)	0.1310

4. Long Range Interactions

As a step towards applications to real materials, we wish to treat long-range interactions. We consider an extended Hubbard-type model,

$$H = \sum_{ij\sigma} t_{ij} a^{\dagger}_{i\sigma} a_{j\sigma} + \frac{1}{2} \sum_{ij\sigma\sigma'} V_{ij} n_{i\sigma} n_{j\sigma'} , \tag{15}$$

where we might in general expect the 2-particle interaction V_{ij} to have a homogeneous form $V(|\vec{x}_i - \vec{x}_j|)$.

A generalization of the familiar gaussian integral identity used to derive the HS transformation [1] can be obtained by manipulating the integral

$$\int_{-\infty}^{\infty} d\zeta_1 \cdots d\zeta_n \exp(-\sum_{i,j=1}^{n} K_{ij}\zeta_i\zeta_j) = \pi^{n/2}(\mathrm{Det}[K_{ij}])^{-1/2} , \tag{16}$$

in which K_{ij} is a positive-definite real symmetric matrix.[12] Through appropriate changes of variables, this can be maneuvered into the form

$$e^{\sum_{i,j=1}^{n} K_{ij}\phi_i\phi_j} = (2\pi)^{-n/2} \int_{-\infty}^{\infty} d\eta_1 \cdots d\eta_n \, e^{-\frac{1}{2}\sum_{i=1}^{n}\eta_i\eta_i - \sum_{i,j=1}^{n}(2K)^{1/2}_{ij}\eta_i\phi_j} , \tag{17}$$

which appears to be what is needed to carry out a generalized HS transformation of time-slice propagators involving the interaction in Eq. (15). Unfortunately, because of the minus sign in $e^{-\Delta\tau H}$ and the requirement that K_{ij} be positive definite, this will only work for strictly attractive potentials V_{ij}, which rules out most cases of interest.

While there are several strategies one might follow to circumvent this problem, we found that one which most closely follows the "magnetic auxiliary field" approach usually used with the simple Hubbard model works best.[13] Considering the general case, we first determine the eigenvalues and eigenvectors of the interaction potential matrix,

$$\sum_{j=1}^{n} V_{ij}\xi_{jk} = v_k \, \xi_{ik} . \tag{18}$$

We can now easily split V into positive-definite and negative-definite components,

$$V_{ij} = V^+_{ij} + V^-_{ij} = \sum_{k,v_k>0} v_k \, \xi_{ik}\xi_{jk} + \sum_{k,v_k<0} v_k \, \xi_{ik}\xi_{jk} . \tag{19}$$

The matrix HS transformation can be applied to V^- with no further difficulties. If we manipulate the V^+ term into a negative $S_{zi}S_{zj}$ form, we find we must add compensating positive terms representing *independent* up-up and down-down interactions. These terms are similar to spinless fermion interactions, and can be shifted by a constant times the unit matrix without adding any non-physical interaction.[2] The resulting expression is

$$\sum_{i,j=1}^{n} V_{ij}(n_{i\uparrow}+n_{i\downarrow})(n_{j\uparrow}+n_{j\downarrow}) = \sum_{i,j=1}^{n} V^-_{ij}(n_{i\uparrow}+n_{i\downarrow})(n_{j\uparrow}+n_{j\downarrow}) \tag{20}$$

$$- \sum_{i,j=1}^{n} V^+_{ij}(n_{i\uparrow}-n_{i\downarrow})(n_{j\uparrow}-n_{j\downarrow}) + 2\sum_{\sigma}\sum_{i,j=1}^{n} (V^+_{ij} - v_M\delta_{ij}) \, n_{i\sigma}n_{j\sigma} ,$$

where we have used the minimum possible shift v_M, the maximum eigenvalue of V.

A generalized HS transformation of each time-slice propagator can now be carried out using Eq. (16). The transformed time slice hamiltonian is

$$H^\ell = \sum_{ij\sigma} t_{ij}a^\dagger_{i\sigma}a_{j\sigma} + \sum_{i,j=1}^{n} x^\ell_i (V^-/\Delta\tau)^{1/2}_{ij}(n_{j\uparrow}+n_{j\downarrow}) + \sum_{i,j=1}^{n} x^\ell_i (V^+/\Delta\tau)^{1/2}_{ij}(n_{j\uparrow}-n_{j\downarrow})$$

$$+ \sum_{i,j=1}^{n} y^\ell_i [2(V^+-v_M I)/\Delta\tau]^{1/2}_{ij} n_{j\uparrow} + \sum_{i,j=1}^{n} z^\ell_i [2(V^+-v_M I)/\Delta\tau]^{1/2}_{ij} n_{j\downarrow} , \tag{21}$$

where I is the unit matrix, and the time slice index is now the superscript ℓ. An independent transformation of each of the 4 terms in Eq. (20) was necessary. However, the construction of V^+ and V^- in Eq. (19) causes each to be rank-deficient. The resulting terms in Eq. (20) can thus share x_i^ℓ , since the matrices act on two orthogonal groups of linear combinations of its components. We see from Eq. (21) that the 3 auxiliary fields associated with each site act not just on electrons occupying that site, but as sources of long-range potentials which act on all the sites within the same time slice. This, in fact, adds negligibly little computational effort to the simulation.

Results of several tests on small models are reported in the Tables 2 to 4 below, which compare 3-field positive projection (3F-PP) with exact (DIAG) results, and display their difference (Δ). Tables 2 and 3 are for 4 site ring models with 4 electrons and interactions $V = 4, 1.5, 0$ and $V = 3, 2, 0$, respectively, for zeroth, first, and second neighbors. The first is positive definite, and has a spin density wave Hartree-Fock ground state, while the second has a negative eigenvalue and a charge density wave HF state. The long range interaction breaks the electron-hole

Table 2. Comparison of simulation and exact results for a 4-electron, 4-site ring with $V = 4, 1.5, 0$. Notation follows Tab. 1. (From reference 13.)

	3F-PP	DIAG	Δ
E	0.8621(48)	0.8594	0.0027
KE	-0.8837(39)	-0.8838	0.0001
ppc(1)	0.0991	0.1000	-0.0009
ppc(2)	0.3016	0.3000	0.0016
apc(0)	0.0947	0.0935	0.0012
apc(1)	0.3563	0.3564	-0.0001
apc(2)	0.1924	0.1937	-0.0013

Table 3. Comparison of simulation and exact results for a 4-electron, 4-site ring with $V = 3, 2, 0$. Notation follows Tab. 1. (From reference 13.)

	3F-PP	DIAG	Δ
E	0.9676(52)	0.9684	-0.0008
KE	-0.8416	-0.8423	0.0007
ppc(1)	0.0609	0.0636	-0.0027
ppc(2)	0.3782	0.3729	0.0053
apc(0)	0.3873	0.3793	0.0080
apc(1)	0.1010	0.1047	-0.0037
apc(2)	0.4108	0.4115	-0.0007

Table 4. Comparison of simulation and exact results for a 8-electron, 8-site cube with $V = 2, 1, 0.5, 0$. Notation follows Tab. 1. (From reference 13.)

	3F-PP	DIAG	Δ
E	1.0217(40)	1.0218	0.0001
KE	-1.4668	-1.4668	0.0
ppc(1)	0.1771	0.1774	-0.0003
ppc(2)	0.2597	0.2607	-0.0010
ppc(3)	0.1896	0.1859	0.0037
apc(0)	0.2175	0.2161	0.0014
apc(1)	0.2531	0.2534	-0.0003
apc(2)	0.2490	0.2488	0.0002
apc(3)	0.2764	0.2774	-0.0010

symmetry which, for the simple Hubbard model, removed the sign problem at half filling.[14] Symmetrized combinations of 2 HF-like states were used for $|\Phi_c>$, and a $\tau_{UC}=2$ unconstrained region was allowed.

The third long-range interaction model tested was cube of sites with an interaction constructed to be positive definite. The HF state in this case is uniform, but antiferromagnetic correlations tend to polarize spins dynamically on two tetrahedral sublattices. The non-interacting ground state, which is a filled-shell state, was used for $|\Phi_c>$. Results are given in Table 4. The lowest one-electron states of this system have s and p character, and it can be considered a toy model for the correlations of the valence electrons in a rare gas atom.

We have only given error bars for the energy in the above tables, because the analysis is difficult, and is incomplete for the remaining quantities. A rough estimate is 0.001 to 0.002 for the pair correlation functions. This suggests that the 3-field PP results are essentially exact within statistical error, which may be a premature conclusion. We studied $\Delta\tau$ convergence extensively in the course of these tests, and found an unexpected result. Over the studied range of $\Delta\tau$ from 0.5 to 0.05, the kinetic energy and pair correlation functions were best fit by $a + b\Delta\tau + c(\Delta\tau)^3$. The energies were fit approximately equally well by this expression and by the expected $a + b(\Delta\tau)^2$. The former fit was used for all the extrapolated results in the tables. The multiple fields and large v_M terms added to the interactions make the convergence rate of the energy (interpreteted by the quadratic fit) worse by factors of 20 to 40 compared to the results in Fig. 1, even though the exact (non-Trotter-Suzuki) treatment of the time slice propagators is used. Despite the advantage of of this treatment apparent in the earlier tests,[5] the HS transformation introduces a $\sqrt{\Delta\tau}$ factor, and the reassembly of the Trotter-Suzuki decomposition might then conceivably lead to a linear error.

5. Continuum Limit

The next step towards real materials is to go back to real hamiltonians. A systematic approach based on our previous developments is to consider that the lattice models with long range interactions represent a grid approximating continuous space, and that the hopping operator approximates ∇^2. We have conducted tests on 1 and 2 dimensional models where we kept the number of electrons fixed at 4, and introduced a gaussian interaction $V_{ij} = \exp(-\gamma|\vec{x_i}-\vec{x_j}|^2)$, with γ set to give essentially the familiar $U=4$ simple Hubbard model for 4 sites.

We subsequently increased the number of sites, decreasing the lattice spacing a proportionally. The hopping t scales as a^{-2}, and v_M scales as a^{-D}, where D is the dimension. Thus the physical energy scale, measured by the average kinetic and interaction energies of the electrons, remained essentially constant, while the energy scale entering auxiliary field simulation diverged rapidly. Our first goal was to find which scale determined $\Delta\tau$ convergence. Results are shown in Fig. 3 for constant $\Delta\tau=0.25$ and 1-dimensional rings of 4, 8, 12, and 16 sites. While the error bars (for comparable samples) grow, no systematically increasing deviation from the exact results (fit by the solid line) is apparent. This also implies that high kinetic energy states play only a minor role, which we confirmed by measuring their occupation in the simulation. This should permit the introduction of a truncated basis set.

The less favorable aspects of these calculations concern the sampling, and are shown in Table 5. As we decreased a, we had to cut the length of the molecular dynamics run between PP constraint tests drastically to get even marginally useful acceptance rates. This implies that this constraint is becoming much more difficult to satisfy. Such effects were worse in 2 dimensions (4x4 and 6x6 models). It became much more difficult to generate statistically independent samples because of this.

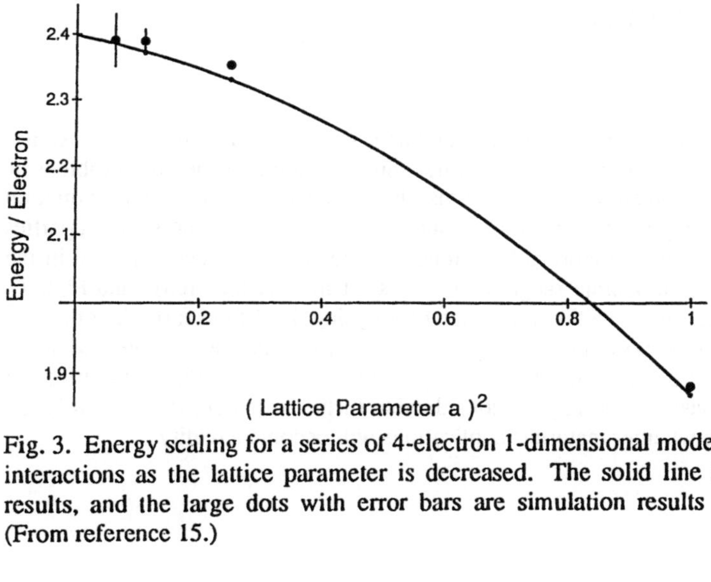

Fig. 3. Energy scaling for a series of 4-electron 1-dimensional models with gaussian interactions as the lattice parameter is decreased. The solid line is a fit to exact results, and the large dots with error bars are simulation results at constant $\Delta\tau$. (From reference 15.)

Table 5. Continuum limit test for 4-electron 1 and 2 dimensional models giving molecular dynamics time, move acceptance rate, unconstrained τ region, and average sign. (From reference 15.)

Sites	t_{MD}	<acc>	τ_{UC}	$<s>$
4	1.8	0.77	2.0	0.97
8	0.6	0.56	2.0	0.36(4)
12	0.6	0.53	2.0	0.32(3)
16	0.6	0.46	2.0	0.32(4)
2x2	1.8	0.75	2.0	0.99(1)
4x4	0.4	0.27	2.0	0.02(2)
4X4	0.4	0.24	0.5	0.16(7)
6X6	0.4	0.24	0.5	0.18(9)
6X6	0.4	0.20	0	0.51(5)

We used an unconstrained region, as discussed at the end of Sec. 3, initially of length $\tau_{UC}=2$. Our average sign got smaller with decreasing a, but appeared to saturate at a usable level in 1 dimension. In 2 dimensions, we found our sign vanishing too rapidly, but recovered usable values by cutting τ_{UC} to 0.5, and finally by eliminating the unconstrained region altogether. The more rapid unconstrained sign decay and the greater difficulty in constrained sampling are undoubtedly manifestations of the same effect. A better sampling strategy in which the constraint is used to guide the molecular dynamics evolution of the fields rather than just to reject trial moves is clearly called for, but finding a practical scheme of this sort remains an unsolved problem.

6. Conclusions

This paper constitutes a progress report, and as such reaches few firm conclusions. We believe we have firmly established a new understanding of the sign problem in auxiliary field fermion simulations. We also have shown that a new approximation scheme, positive projection, is useful for calculating accurate ground state properties within the realm in which Hubbard-Stratonovich methods have been applied in the past. Extensions to moderate range interactions, of interest for multi-band high T_c models, also appear to be well within reach of PP using the 3-field formulation.

As to the much more ambitious goal we have set ourselves of applications to real materials, there are some encouraging results. However, the difficulties we discussed in the last section require considerable further research. It is certainly too early to speculate whether materials applications will become a reality.

References

1. J. Hubbard, Phys. Rev. Lett. **3**, 77 (1959); R. L. Stratonovich, Dokl. Akad. Nauk. SSSR **115**, 1097 (1957) [Soviet Phys. Doklady **2**, 416 (1957)].
2. G. Sugiyama and S. E. Koonin, Annals of Phys. **168**, 1 (1986).
3. S. Sorella, E. Tosatti, S. Baroni, R. Car, and M. Parrinello, Int. J. Mod. Phys. B, **1**, 993 (1988); S. Sorella, S. Baroni, R. Car, and M. Parrinello, Europhys. Lett. **8**, 663 (1989).
4. H. F. Trotter, Proc. Am. Math. Soc. **10**, 545 (1959); M. Suzuki, Commun. Math. Phys. **51**, 183 (1976).
5. D. R. Hamann and S. B. Fahy, Phys. Rev. B **41**, 11352 (1990).
6. S. R. White, D. J. Scalapino, R. L. Sugar, E. Y. Loh, J. E. Gubernatis, and R. T. Scalettar, Phys. Rev. B **40**, 506 (1989).
7. S. B. Fahy and D. R. Hamann, Phys. Rev. B **43**, 765 (1991).
8. S. B. Fahy and D. R. Hamann, Phys. Rev. Lett. **65**, 3437 (1990).
9. J. B. Anderson, J. Chem. Phys. **65**, 4121 (1976).
10. S. Duane, A. D. Kennedy, B. J. Pendleton, and D. Roweth, Phys. Lett. B **195**, 216 (1987); R. T. Scalettar, D. J. Scalapino, and R. L. Sugar, Phys. Rev. B **34**, 7911 (1986).
11. N. Furukawa and M. Imada. J. Phys. Soc. Japan **60**, 810 (1991).
12. J. W. Negele and H. Orland, *Quantum Many-Particle Systems* (Addison-Wesley, New York, 1988), p. 33.
13. D. R. Hamann and S. B. Fahy, unpublished.
14. J. E. Hirsch, Phys. Rev. B **31**, 4403 (1985).
15. D. R. Hamann and S. B. Fahy, unpublished.

S-Wave Superconductivity in the $t - J$ Model with Two Planes

E. Dagotto

Institute for Theoretical Physics, University of California
at Santa Barbara, Santa Barbara, CA 93106, USA

ABSTRACT: We discuss an electronic model consisting of two planes, each described by a $t - J$ model, coupled by an antiferromagnetic interaction J' between planes. For $J' \geq J$ we show the presence of a spin-gap in the spectrum and hole-pair formation upon doping. The model superconducts away from half-filling. We support our claims by numerical studies of binding energy of holes and pairing correlations on finite clusters. The pairing operator is an $s - wave$ spin-singlet with one member of the pair in each plane. Although all interactions are explicitly *repulsive*, our model belongs to the universality class of the $U < 0$ Hubbard model with $|U| \gg t$. We argue that this model can be physically realized in the two Cu-O layers of the unit cell of $YBa_2Cu_3O_{6+x}$ and also in the orthorhombic compound $(VO)_2P_2O_7$.

The discovery of high-T_c superconductors[1] has induced considerable theoretical work in strongly correlated electronic systems. Hubbard-like models have been proposed as simple two-dimensional (2D) Hamiltonians that may describe the physics of the new cuprates. Computer simulations and other techniques have shown that many properties of the normal and insulating states of the new materials are qualitatively well described by these models. For example, the optical conductivity and the density of states of the Hubbard and $t - J$ models are in good agreement with experiments in an intermediate region of parameter space.[2] However, superconductivity still remains elusive. Some nonperturbative techniques like diagrammatic, variational or various mean-field approaches have predicted the existence of superconductivity in the 2D Hubbard model. However, extensive numerical studies of the one and three band models have not supported these claims.[3] Although it is difficult to show that the numerical studies have been performed for large enough clusters and low enough temperatures, it is clear that superconductivity in these models, if it exists, will not be easily accessible to computer simulations.[4] Then, there is still no robust unbiased calculation supporting the exciting conjecture that purely electronic models, like the 2D Hubbard, can explain the behavior of the high-T_c materials. It may be suspected that only electron-phonon interactions can induce superconductivity (or, if the phononic degrees of freedom are integrated out, explicitly *attractive* electronic interactions). Then, a natural question arises: is there a purely electronic model with *repulsive* interactions in the Hamiltonian that superconducts?[5] It is the purpose of this paper to present a variation of the $t - J$ model which is indeed superconducting without the need of phonons and having only repulsive forces in the Hamiltonian. This work has been done in collaboration with J. Riera and D. Scalapino.[6]

Springer Proceedings in Physics, Vol. 70 Computational Approaches in Condensed-Matter Physics
Editors: S. Miyashita · M. Imada · H. Takayama

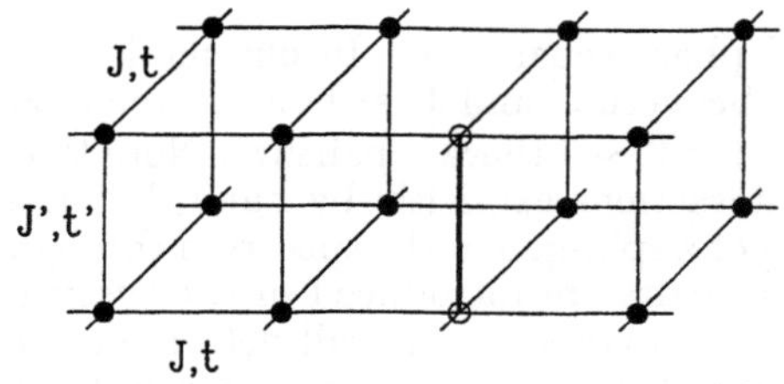

1. Schematic definition of our model. Dots represent sites of the 2D $t-J$ planes which are in interaction through vertical links with couplings J', t'. Full dots are occupied sites while open dots denote holes. A tight bound state of two holes is shown.

Consider the model implicitly defined in Fig.1. It consists of two adjacent square lattices, each one described by the $t-J$ model, and having $t'-J'$ couplings between planes. More specifically, our model is defined by the Hamiltonian

$$H = J\sum_{\mathbf{i},\hat{l}} \mathbf{S}_{\mathbf{i}}.\mathbf{S}_{\mathbf{i}+\hat{\mathbf{l}}} + J'\sum_{\mathbf{i},\hat{\mathbf{z}}} \mathbf{S}_{\mathbf{i}}.\mathbf{S}_{\mathbf{i}+\hat{\mathbf{z}}} - t\sum_{\langle\mathbf{ij}\rangle,s}(\bar{c}^{\dagger}_{\mathbf{i},s}\bar{c}_{\mathbf{j},s} + h.c.) - t'\sum_{\mathbf{i},s}(\bar{c}^{\dagger}_{\mathbf{i},s}\bar{c}_{\mathbf{i}+\hat{\mathbf{z}},s} + h.c.), \tag{1}$$

where J is the coupling of the spin-spin interaction and t corresponds to the hopping parameter of the kinetic energy of holes, both in the planes. The same parameters but between planes are denoted by J' and t'. $\bar{c}_{\mathbf{i},s}$ are *hole* destruction operators with spin s at sites $\mathbf{i}$ of the lattice shown in Fig.1. We use periodic boundary conditions in the planes. N denotes the number of sites *per* plane. $\hat{\mathbf{l}} = \hat{\mathbf{x}}, \hat{\mathbf{y}}$ are unit vectors along the crystal axes in the planes while $\hat{\mathbf{z}}$ is a unit vector between planes. The rest of the notation is standard.

To gain insight about the physical properties of this model let us consider the limit where J' is much larger than any other scale in the problem. At half-filling (one particle per site) the model decouples into N vertical links and in each link the lowest energy state is a spin singlet where one electron is in the upper plane and the other in the lower one. Naturally, there is a spin-gap in the spectrum of order J' which corresponds to having a triplet state in one of the links. When the system is doped with holes, it is energetically favorable to break as few of the singlets that form the ground state as possible. Then, a pair of holes added to the system will minimize its energy by breaking the same spin-singlet and thus it will form a tight bound state having a hole in each plane along the vertical links, as schematically shown in Fig.1. Then, in this scheme it is relatively easy to form pairs by a "dimerization" of the lattice in the vertical direction which produces a spin-gap.[7] The pair is in a spin singlet and it is invariant under a rotation of the planes in $\pi/2$ and, thus, we label it as a "local" s-wave bound state whose related pairing operator[8] is $\Delta_{\mathbf{i}} = \frac{1}{\sqrt{2}}(\bar{c}_{\mathbf{i},\uparrow}\bar{c}_{\mathbf{i}+\hat{\mathbf{z}},\downarrow} - \bar{c}_{\mathbf{i},\downarrow}\bar{c}_{\mathbf{i}+\hat{\mathbf{z}},\uparrow})$. The existence of hole pairs in Hamiltonian Eq.1 is a nontrivial feature that is difficult to obtain in purely two-dimensional systems[9] without the complication of phase separation. In our model the pairs are formed due to the anisotropy between J' and J (although numerical studies presented below show that even for $J' \sim J$ there is pairing). Note that we have not introduce "attractive" interactions explicitly by hand, but the hole attraction comes from the opening of a spin-gap in the spectrum through the dimerization of the lattice. If more pairs are introduced in the system, the residual interaction between them and between spins will determine the various phases of the model and the type of long-range order. We would like to remark that if a Hubbard model with couplings t',t and U is used instead of our $t - J - t' - J'$ model, then there is no obvious limit where pairs are formed. For example, when $t'/t \to \infty$ the model decouples into N vertical links described only by a hopping term (like a two site $U = 0$ Hubbard model) and there is no pair formation. Then, the use of the $t - J$ Hamiltonian seems crucial for the existence of superconductivity.

The mobility of the hole-pairs depends on the parameter t^2/J'. In the subspace of pairs it can be easily shown that the effective Hamiltonian is equal to a negative-U model[7] with $|U| \sim J'$. If the density of holes corresponds to quarter-filling, then, the ground state at small t will consist of a checkerboard of hole singlets and spin singlets. Replacing hole singlets by doubly occupied sites and spin singlets by empty sites, this resembles the ground state of the negative-U Hubbard model at half-filling which we know is superconducting (although degenerate with a CDW).[10,11] From this analogy it is clear that model Eq.1 will have a superconducting s-wave ground state at T=0 in an appropriate region of parameter space. Actually, the vertical links in Fig.1 that induce hole binding when J' is large can be thought of as "negative-U centers"[12,7] that produce "on-site" attraction in the s-channel.[13] If J' is large and as a function of temperature, the model will have a critical temperature where superconductivity disappears with $T_c \sim t^2/J'$. At higher temperatures $T \sim J'$ pairs will be dissociated. In this respect this model resembles a Bose condensation of small size pairs. This limit is just the opposite of the standard BCS theory where the size of the pairs is larger than the distance between them. In our case the size of the pairs is of order one lattice spacing in the vertical direction (local pairs with respect to the planes) while the distance between pairs is larger and depends on the hole concentration, the ratio J/t and other possible in-plane interactions.[10] What is the influence of the perpendicular hopping t'? Consider the case of two holes on a 2×2 cluster with t' and J' being the dominant interactions. If t' is larger than $\frac{3}{8}J'$ the holes are *not* localized in one link. Then, for our scenario to work we will take t' fixed at a small number compared with J'.

Although in the limit where J' is the largest scale we have shown that our model superconducts, it is important to find if this assumption can be relaxed. For that purpose we studied our model numerically. We analyzed two planes with 8 or 10 sites each and we also replaced planes by chains and consider the particular case of a *ladder* i.e. two 1D chains coupled by J' rungs.[14] First, let us study the existence of a spin-gap and also the issue of binding of holes. We have analyzed the case of zero and two holes in model Eq.1. We measured the spin-gaps $\Delta S_0 = E(0,1) - E(0,0)$, $\Delta S_2 = E(2,1) - E(2,0)$ in the subspace of zero (half-filling) and two holes, respectively, where $E(n,S)$ is the ground

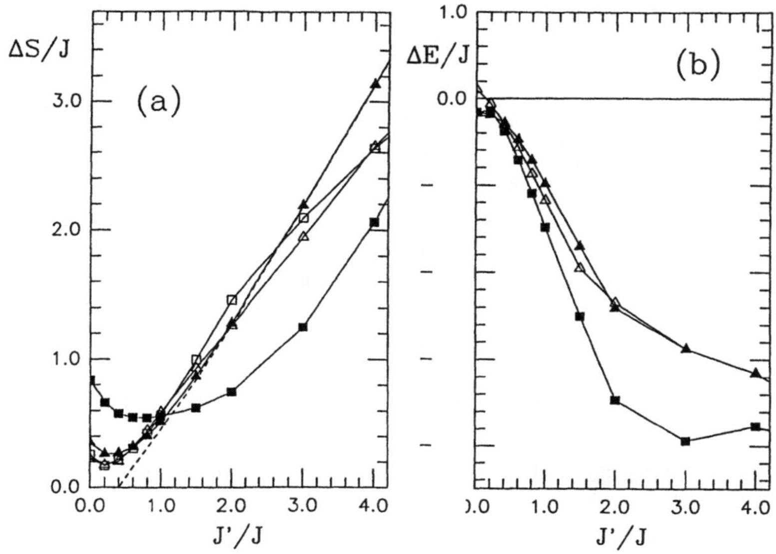

2.a Spin-gap as defined in the text fo ifferent lattices as a function of J'/J. ▲ denote results at half-filling foı × 12 ladder while ■ represent results for two planes of 10 sites each. Tl lashed line is the extrapolation to the bulk limit of the ladder gaps at h -filling. △ and □ indicate results for the case of two holes and $J = 0.4$ $= 1$ for the 2×8 ladder and planes, respectively.

2.b Binding energy as defined in the :t as a function of J'/J. △, ▲ and ■ denote results for the 2×6, 2×8 a coupled planes clusters, respectively.

state energy of the model with n hol ınd total spin S. We also studied the binding energy defined as $\Delta E = E(2$ $+ E(0,0) - 2E(1,1/2)$. The results are shown in Fig.2. Studying lattice: $2 \times N$ sites with $N = 4,6,8,10$ and 12, we found that our results have sm finite size effects. Extrapolating the spin-gap of the undoped ladder versı ./N we can safely conclude that the critical ratio for the opening of the g is $J'/J|_c \sim 0.4 - 0.5$ or smaller and thus the assumption $J' \gg J$ is not ne sary for the success of our scenario.

We obtain similar results for the l ler doped with two holes and for two *planes* in interaction (in this case w ınnot make a finite size scaling but the shape of the curve in Fig.2.a sugg that the gap also opens for $J' \sim J$). Our numerical studies are not accur enough to show if indeed a *finite* ratio is necessary to open the gap or i his occurs immediately for a nonzero J'/J. It would be very interesting to form a Quantum Monte Carlo study of the undoped model (where they aı .easible) to find this critical ratio. In the limit where $J'/J \gg 1$, the binding energy converges to the energy of an isolated singlet ($E_s \sim -\frac{3}{4}J'$) showing that the pair is strongly localized in one vertical link as expected. However, note that $\Delta E < 0$ even for $J' \sim J$ or smaller. The results for the coupled planes do not differ qualitatively from the ladder. Then, the presence of a spin-gap and pairing of holes seems to

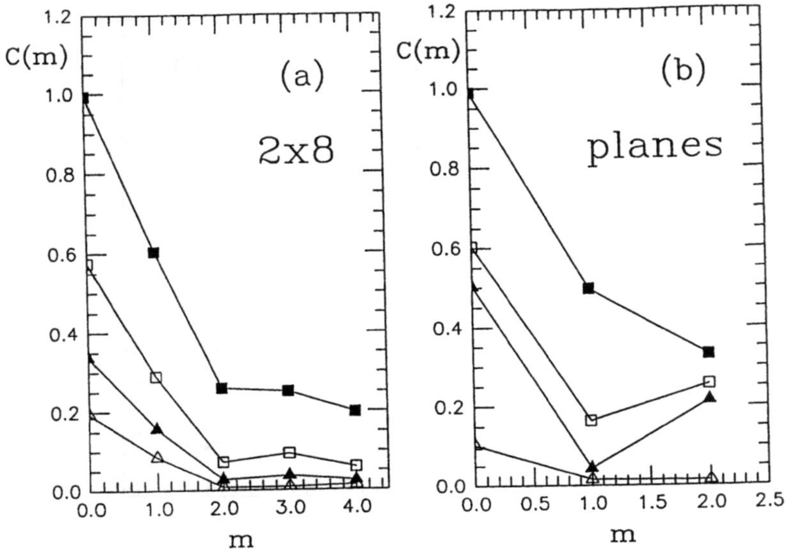

3.a Pairing correlation function C(m), as defined in the text, as a function of distance for different values of J'/J on a 2×8 ladder. The couplings are $J = 0.4$, $t = 1.0$ and $t' = 0.1$. △,▲,□, and ■ denote results for $J' =$ 0.01,0.4,1.6 and 40, respectively. The number of particles corresponds to quarter-filling i.e. there are 8 electrons in the figure.

3.b Same as Fig.3.a but for two planes of 8 sites each in interaction.

be a robust property of this model and both seem to be correlated. We have checked explicitly that there is no phase separation in our model by adding one more pair of holes to the clusters.

After checking that pairs are properly formed we need to analyze the existence of long-range order in the plane. Consider e.g. the case of quarter-filling with $J = 0.4$, $t = 1$, $t' = 0.1$ and varying J'. Let us define the equal-time pairing correlation function as

$$C(m) = \frac{1}{N} \sum_{i} \langle \Delta_i^{\dagger} \Delta_{i+m} \rangle, \tag{2}$$

where Δ_i is a spin singlet, s-wave operator that was previously defined and $\langle\rangle$ denotes expectation value in the ground state. In Fig.3 we show our results for the ladder and also for planes in interaction. It is clear that at the maximum distance allowed in our small clusters the correlation function is enhanced by increasing J' showing that a tendency towards long-range order is being developed. This correlation at $J' \sim 0$ is negligible. The physical properties of two coupled chains or two coupled planes seem very similar.

We again observed that results for different ladders $2 \times N$ show small finite size-effects. A tentative schematic phase diagram of our model is shown

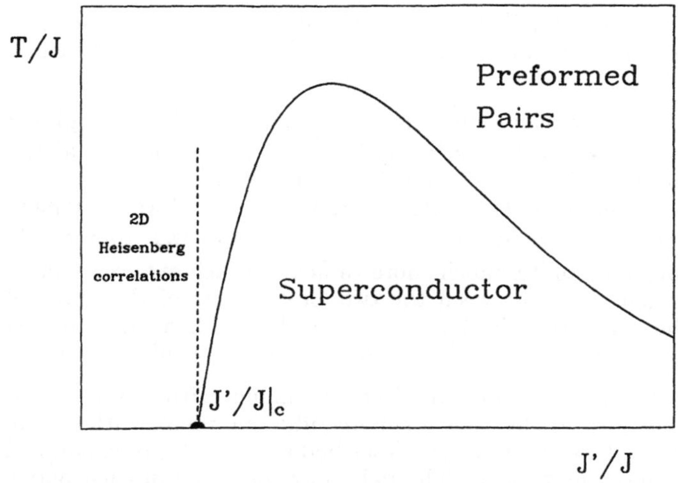

4. Schematic phase diagram of our model in the plane temperature T/J - coupling J′/J. The doping is assumed to be low enough such that there is antiferromagnetism for $J' = 0$.

in Fig.4 for *low* doping of holes. Below the critical ratio $J'/J|_c$ there are antiferromagnetic correlations similar to those of the $t - J$ model near half-filling. This is different from the results obtained for the $U < 0$ Hubbard model where there is *no* region of stability for a spin density wave. With respect to the superconducting phase and based on the analogy with the negative-U model in 2D, we believe that T_c should present a Kosterlitz-Thouless transition[10] with the approximate shape shown in Fig.4 and having preformed pairs above T_c at least as large J'/J.

Can our model be physically realized? There are materials like $(VO)_2$ P_2O_7 whose structure is that of a spin-$\frac{1}{2}$ Heisenberg model on a ladder with super-exchange interactions in both directions of equal magnitude[15] $J' \sim J \sim 0.05eV$. From our analysis we predict that this material should become superconducting upon doping. Another interesting scenario is the following: consider the cuprate $YBa_2Cu_3O_{6+x}$. In this case there are two CuO_2 planes in the unit cell.[16] Constructing superlattices of this material separated by insulating layers of $PrBa_2Cu_3O_{6+x}$, it has been claimed[17] that a single unit cell of $YBa_2Cu_3O_{6+x}$ is still superconducting with a T_c of $\sim$ 19K. Then, the Cu-O double layer of the unit cell seems enough to induce superconductivity. Recent experiments[18] have found that the distance between Cu atoms in the Cu-O plane is of $\sim$ 3.8Å while between layers in the unit cell is of $\sim$ 3.4Å. However, the direct exchange interaction between Cu(2) of these layers was found to be 40 times smaller than the in-plane superexchange[19] i.e. $J'/J \sim 1/40$. Naively, it seems that this value is outside the range of validity of our approximations. However, note that for the case of the spin-gap at half-filling, the critical ratio J'/J is much smaller than expected and thus it may occur that our model still qualitatively describes superconductivity in $YBa_2Cu_3O_{6+x}$.

To further increase J' we can reduce the distance between layers applying pressure. This has been studied experimentally[18] and it was found that the critical temperature barely changed replacing Y by other rare-earth atoms (chemical pressure). However, the distance between Cu-O layers changed only from ~ 3.3Å to ~ 3.45Å and thus it is not clear if J'/J was appreciably affected.[18] A second possibility, much more difficult to achieve experimentally, is to add oxygen atoms between the two Cu-O layers of the unit cell creating links in the c-direction similar to those in-plane. Then, the superexchange constants would be similar ($J' \sim J$) and our model would be valid.

Summarizing, we have presented an electronic model with purely repulsive interactions that nevertheless has a superconducting phase with s-wave symmetry. It consists of two planes each described by a $t-J$ model coupled by an antiferromagnetic interaction. The pairs have one hole in each plane. This model may describe superconductivity in $YBa_2Cu_3O_{6+x}$.

We specially thank M. Imada for his kind invitation to this conference. We also thank useful conversations with D. Coffey, J. Guimpel, A. Moreo, M. Imada and D. Clougherty. This project was supported by grant from the NSF PHY89–04035. The computer calculations were done at NCSA, Urbana, Illinois.

REFERENCES

1. J. G. Bednorz and K. A. Müller, Z. Phys. **B 64**, 188 (1986).

2. E. Dagotto, A. Moreo, F. Ortolani, J. Riera and D. Scalapino, *Phys. Rev. Lett.* **67**, 1918 (1991); E. Dagotto *et al.*, NSF-ITP-91-58 preprint.

3. See for example, M. Imada, J. Phys. Soc. Jpn. **57**, 42 (1988); J.Hirsch and S. Tang, *Phys. Rev. Lett.* **62**, 591 (1989); G. Dopf *et al.*, *Phys. Rev. B* **41**, 9264 (1990); M. Frick et al., *Phys. Rev. B* **42**, 2665 (1990); A. Moreo and D. Scalapino, *Phys. Rev. B* **43**, 8211 (1991); M. Imada, ISSP preprint, Tokyo; E. Dagotto, *et al.*, NSF-ITP-91-54; A. Moreo, UCSBTH-91-33.

4. Superconductivity in the 2D $t-J$ model is not excluded for large values of J/t. See A. Moreo, UCSBTH-91-32.

5. W. Kohn and J. Luttinger, *Phys. Rev. Lett.* **15**, 524 (1965) showed that a *dilute* system of fermions superconducts with nonzero orbital momentum and a low T_c. However, we are interested in finding s-wave superconductivity near half-filling.

6. E. Dagotto *et al.* NSF-ITP-91-121 preprint.

7. Similar ideas have been previously proposed in the context of the dimerized $t - J$ model by M. Imada, J. Phys. Soc. Jpn. **60**, 1877 (1991). See also M. Ogata, M. Luchini and T. M. Rice, to appear in Phys. Rev. **B**.

8. Similar ideas have been discussed recently in the context of the Hubbard model by N. Bulut, D. Scalapino and R. Scalettar, preprint.

9. E. Dagotto *et al.*, *Phys. Rev. B* **41**, 811 (1990); G. Fano *et al.*, *Phys. Rev. B* **42**, 6877 (1990); J. Riera and A. P. Young, *Phys. Rev. B* **39**, 9697 (1989).

10. A. Moreo and D. Scalapino, *Phys. Rev. Lett.* **66**, 946 (1991); R. Scalettar, *et al.*, *Phys. Rev. Lett.* **62**, 1407 (1989).

11. In our approach a strong singlet is formed because the length of the lattice in the c-direction is of only one lattice spacing. If more layers (L) were added and keeping J' as the dominant interaction, the model would reduce to a bundle of Heisenberg chains when $L \to \infty$ which are known to be gapless. Then, our model Eq.1 superconducts when the number of layers is small enough such that a spin-gap exists.

12. P. W. Anderson, *Phys. Rev. Lett.* **34**, 953 (1975).

13. Other models have been presented in the literature where tunneling of hole *pairs* between the Cu-O layers plays an essential role. See J. M. Wheatley *et al.*, *Phys. Rev. B* **37**, 5897 (1988); J. C. Phillips, AT&T preprint. However, in these approaches the pairs are already *preformed* in each plane while in our model each hole of the pair belongs to a different plane.

14. The physics of both geometries (planes and chains) should be similar. However, note that the ladder system can be mapped into a 1D model with two different species of fermions per site. Such a model does not have long-range order since quantum fluctuations are strong even at T=0.

15. D. C. Johnston *et al.*, *Phys. Rev. B* **35**, 219 (1987); J. Bonner *et al.*, *Phys. Rev. B* **27**, 248 (1983). See also E. Dagotto and A. Moreo, *Phys. Rev. B* **38**, 5087 (1988); *ibid* 44, 5396 (1991)(E); L. Hubert and A. Caillé, *Phys. Rev. B* **43**, 13187 (1991); K. Hida, J. Phys. Soc. Jpn. **60**, 1347 (1991).

16. J. Jorgensen, Physics Today **44**, no. 6, 35, June 1991 and references therein.

17. T. Terashima *et al.*, *Phys. Rev. Lett.* **67**, 1362 (1991); See also D. Lowndes *et al.*, *Phys. Rev. Lett.* **65**, 1160 (1990); Q. Li *et al.*, *Phys. Rev. Lett.* **64**, 3086 (1990).

18. A. Fernandes, J. Santamaria, S. Bud'ko, O. Nakamura, J. Guimpel and I. Schuller, UCSD preprint; Y. Le Page *et al.*, *Phys. Rev. B* **36**, 3617 (1987).

19. J. M. Tranquada *et al.*, *Phys. Rev. B* **40**, 4503 (1989). The coupling J' between the Cu-O layers of the unit cell should not be confused with the coupling between double-layers which governs the Néel temperature and is much smaller.

Calculation of Excitation Spectra Based on the Numerical Renormalization Group Method

Y. Shimizu and *O. Sakai*

Department of Physics, Tohoku University, Sendai 980, Japan

Abstract. A method to calculate the excitation spectra of impurity Anderson model, which is based on the numerical renormalization group(NRG) technique is reviewed. The Kondo effect in systems with complex internal structures, such as the Sm-ion-like and the two magnetic impurities is studied.

In the NRG method [1,2], the Hamiltonian is transformed to an expression represented by the shell orbits:

$$H = H_f + \sum_m V(f_m^+ s_{0m} + h.c.) + \sum_{\ell=0}^{\infty} \sum_m t_\ell (s_{\ell m}^+ s_{\ell+1m} + h.c.), \qquad (1)$$

where f_m denotes the electron in the atomic f-orbit with symmetry m, and $s_{\ell m}$ the ℓ-th shell orbit, and V the strength of the c-f hybridization. The hopping matrix, t_ℓ, decreases as $\Lambda^{-\ell/2}$ when ℓ increases, where $\Lambda(> 1)$ is the discretization parameter. The index m runs over from $-j$ to j, and the degeneracy factor, N, is given as $2j+1$. We classify the atomic states by their electron number(n) and the total angular momentum(J) in H_f, and denote their energy by $E(f^n J)$. If we consider the usual Anderson model with constant f-f interaction, U, $E(f^n J)$ is given as $\varepsilon_f n + Un(n-1)/2$, where ε_f is the energy level of f-orbit.

Starting from the H_f term, the total Hamiltonian is diagonalized step by step by adding a shell orbit from $\ell = 0$ successively. We denote by H_N the Hamiltonian including up to the N-th shell orbit. In the NRG method, only a small number of low energy states are kept to next iteration step because of the computational capacity. At step N, low energy states may be good approximate eigen states of H_N. When we proceed the step, the very low energy states are affected by the added shell orbit, while intermediately low energy states do not show any essential change. We calculate transition probabilities by using states in this energy region.

Springer Proceedings in Physics, Vol. 70 Computational Approaches in Condensed-Matter Physics
Editors: S. Miyashita · M. Imada · H. Takayama

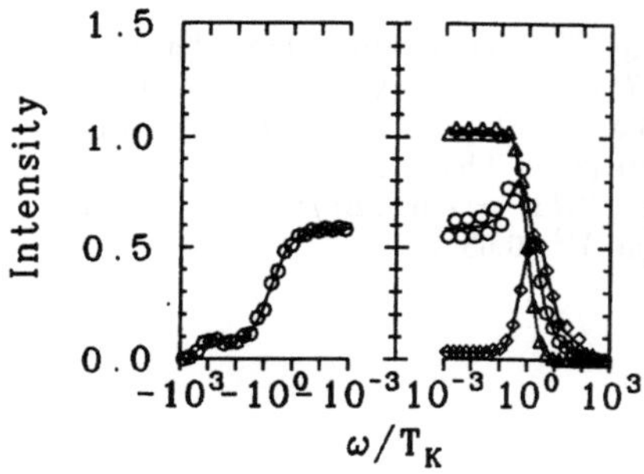

Fig.1. Spectral intensities calculated by NRG method for $N = 3$. The symbol o gives the data points of $(\pi\Delta\rho_f(\omega)/N)$, $\triangle\,(3T_K^2 N\chi''_m(\omega)/\pi\omega j(j+1))$ and $\diamond\,(\chi''_c(\omega)/2\pi\omega N)$, where Δ is the hybridization width$(\pi V^2/2D)$, D(half of the band width) $= 1$ and $V^2 = 0.04$.

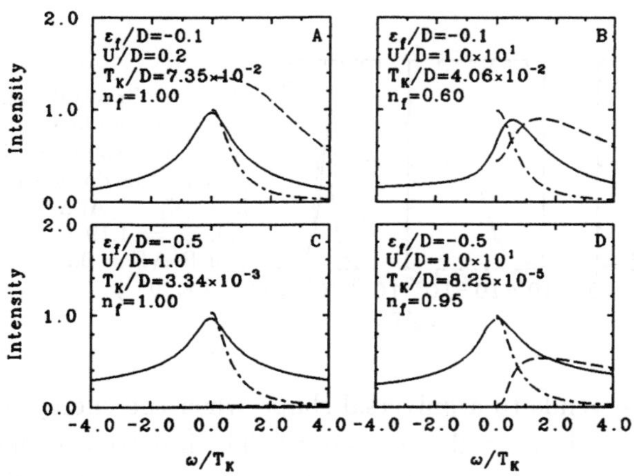

Fig. 2. Spectral intensities of N=2 model. The solid line gives $\pi\Delta\rho_f(\omega)/N$, the dot-dashed line $3T_K^2 N\chi''_m(\omega)/\pi\omega j(j+1)$ and the dashed line $\chi''_c(\omega)/2\pi\omega N$. D=1 and V^2=0.06.

The calculated excitation spectra are smoothed by the Gaussian shape function in the logarithmic energy scale. In Fig. 1, the symbol o gives the data points of the single particle excitation(SPE) spectrum of f-electron(ρ_f) and $\triangle$ and $\diamond$ are those of the imaginary part of the magnetic(χ''_m) and the charge(χ''_c) susceptibilities [3]. Smooth curves are obtained when we join series of data points in only odd, or only even N steps. But they do not agree with each other in general. The lines in the figure are obtained by averaging the two curves. The averaged line is not so sensitive to the choice of smoothing parameters. Other averaging methods of data have been proposed by Brazilian group [4].

In Fig. 2, we show the excitation spectra for usual Anderson model with $N = 2$ for various parameter cases of ε_f and U. The solid line gives $\rho_f(\omega)$, the dashed line $\chi''_c(\omega)/\omega$ and the dot-dashed line $\chi''_m(\omega)/\omega$, and the Kondo temperature T_K is defined through the relation $\chi(T \to 0) = j(j+1)/3T_K$. The column A and C corresponds to the electron-hole symmetric case, and the Coulomb interaction increases in the lower

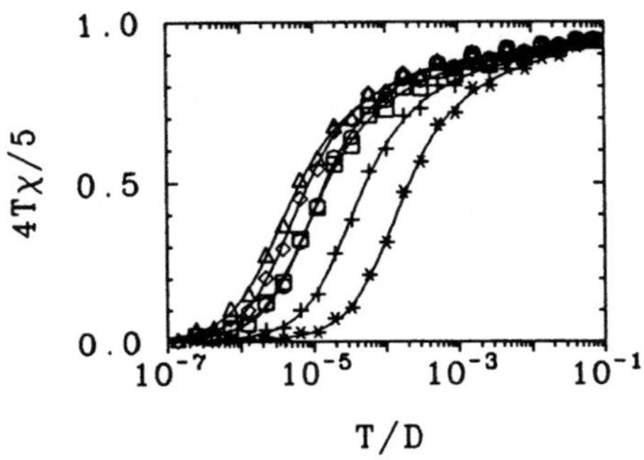

Fig. 3. Temperature dependence of χT for Sm-like system. ΔE=10(○), 2(◇), 0.8(△), 0.4(□), 0.02(+) and 0.04(∗). The other parameters are $E(f^1 3/2) = -0.3, E(f^2 2) = 0$, D=1 and V^2=0.03.

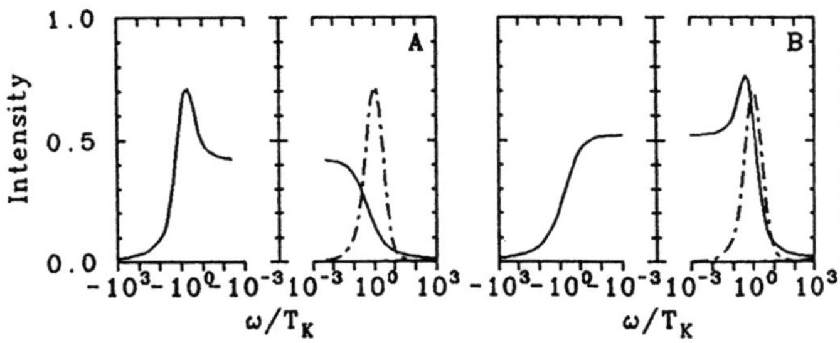

Fig. 4. Spectral densities of Sm-like system. The solid line is ρ_f and the dot-dashed line χ''_{m}. A(ΔE = 2.0, T_{K} = 1.19 × 10^{-5}) and B(ΔE = 0.2, T_{K} = 7.25 × 10^{-5}).

row. When U is large, ρ_f tends to a universal shape when the energy is scaled by T_{K} and the occupation number n_f is fixed. The intensity of the tail part is large compared with the Lorentzian shape, especially the BIS part is large when n_f is less than unity. The quantity $\chi''_{\mathrm{m}}/\omega$ has almost Lorentzian shape with width about $0.63T_{\mathrm{K}}$(HWHM) in the $n_f \sim 1$ case.

The valence of Sm ion fluctuates between $Sm^{3+}(^6H_{5/2})$ and Sm^{2+} in which the excited state($^7F_1$) is located very near the ground state($^7F_0$). It has not been known how the Kondo effect appears in such system.

We consider a model with $j = 3/2$, and simulate the multiplet states of Sm ion by $f^1(J = 3/2), f^2(J = 0)$ and $f^2(J = 2)$. Other atomic states are excluded in the calculation. In Fig. 3 [5], we show the temperature dependence of χT for various values of the energy difference, $\Delta E = E(f^2 2) - E(f^2 0)$, in primarily f^1 stable case. The ground state becomes always singlet. The temperature at which χT decreases to a small value does not show monotonic change as ΔE changes.

An electron is bound similarly to the Kondo effect in Yb impurity case when ΔE is large, while a hole is bound similarly to Ce when ΔE is small. This change of the nature is seen from the SPE spectrum shown in Fig. 4. The peak of the Kondo resonance appears in the PES side when ΔE is large, while it appears in the BIS side same as Ce case

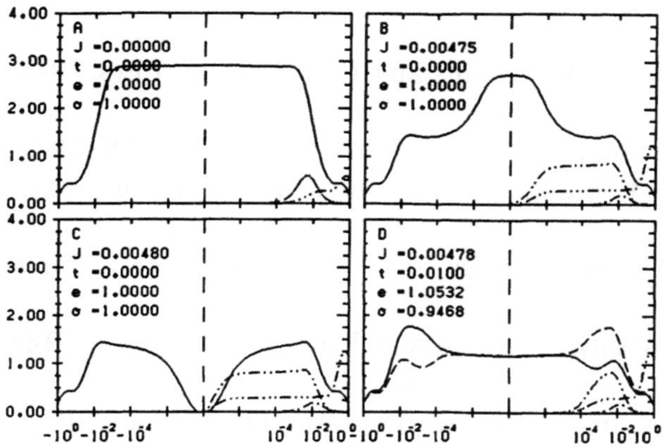

Fig. 5. Spectral intensities of two impurity model. Hybridization matrix is constant (V^2=0.06) common to two channels, D = 1, T_K = 7.68 × 10^{-3}. $e(o)$ is the f-electron occupation number of even (odd) parity, t the transfer matrix. The solid (dashed) line is SPE of even (odd) component, the dot-dashed line (χ'',uniform susceptibility), the two-dot-dashed line (χ''_a) and the three-dot-dashed line (χ''_s).

when ΔE is small. The Kondo temperature becomes lowest in the cross over region between two types of states. The hole-binding-type state is expected in actual Sm ion, and this may explain the experimental results of a thermopower, which show positive sign similar to the Ce impurity case [6].

The problem of the two impurities in metal has been extensively restudied recently. JONES *et al.* [7] have pointed that the γ-coefficient and the antiferromagnetic(AF) susceptibility diverges when the AF coupling between local spins ($J > 0$) has a critical value which separates the Kondo singlet and the local-spin-singlet like states. The authors have shown that the critical transition is suppressed when the f-electron occupation number shows parity splitting [8].

In Fig. 5, the excitation spectra are given. In A ~ C, terms causing parity splitting are neglected as JONES *et al.* did. When J increases, the low energy edge of the AF excitation (χ''_a) shifts to low energy side at first, and approaches unlimitedly to zero as J goes to the critical value $J_c \cong 4.78 \times 10^{-3}$. The low energy edge begin to return to high energy side when J increases further. The SPE shows a peak structure at the Fermi energy with energy scale of the low energy edge of χ''_a, and suddenly changes to valley structure as J increases beyond J_c. The susceptibility of the singlet non-uniform superconductivity(χ''_s) shows similar behavior to that of χ''_a, but the intensity is about 1/10 of the latter. When we consider the f-f transfer term, the SPE spectra show parity splitting as seen from D. The softening of χ''_a to zero energy is bounded by an energy scale approximately given as $T_K \times (t/T_K)^2$. Details are given in ref. [8].

References

1. H.R. Krishna-murthy, J.W. Wilkins and K.G. Wilson: Phys. Rev. **B21** (1980) 1003, **B21** (1980) 1044.
2. O. Sakai, Y. Shimizu and T. Kasuya: J. Phys. Soc. Jpn. **58** (1989) 3666, O. Sakai and Y. Shimizu: to appear in Prog. Theor. Phys. Supple. (1992).
3. Y. Shimizu, O. Sakai and T. Kasuya: to be submitted to J. Phys. Soc. Jpn.
4. M. Yoshida, M.A. Whitaker and L.N. Oliveira: Phys. Rev. **B41** (1990) 9403.
5. Y. Shimizu, O. Sakai and T. Kasuya: Physica. **B163** (1990) 401.
6. E. Umlauf, P. Sütsch and E. Hess: *Crystalline Electric Field and Structural Effect in f-Electron Systems* eds. J.E. Craw, R.P. Guertin and T.W. Mihalisin (Plenum, 1980) p.341.
7. B.A. Jones, C.M. Varma and J.W. Wilkins: Phys. Rev. Lett. **61** (1988) 125.
8. O. Sakai, Y. Shimizu and T. Kasuya: Solid State Commun. **75** (1990) 81, O. Sakai and Y. Shimizu: submitted to J. Phys. Soc. Jpn.

Real Space Numerical Renormalization Groups for Hubbard Models

S.R. White

Department of Physics, University of California, Irvine, CA 92717, USA

Abstract. We discuss developments in real-space numerical renormalization groups. In particular, we discuss a treatment of boundary conditions on blocks for one-dimensional *single-particle* systems which makes the renormalization group exact. For the many-particle, interacting case we discuss a technique for producing basis states for a block from wavefunctions on larger systems of which the block is a subset.

1. Introduction

Currently the most interesting and difficult challenge in the area of numerical studies of Hubbard and related models involves understanding very low temperature behavior in systems with competition between different types of order. Away from half-filling, does the 2D Hubbard model exhibit superconductivity, Fermi-liquid behavior, or perhaps some other type of order? The answer is still not clear. Quantum Monte Carlo[1] is able to answer many questions about this system at higher temperatures, but is prevented from reaching low temperatures by the sign problem[2]. From exact diagonalization[3] we can extract information about the ground state and some excited states on small lattices (up to about 4×4), but it is not clear that one can resolve the more difficult questions regarding competing types of order with lattices this small.

The renormalization group (RG) was designed to resolve just this kind of question. Given the dramatic success of Wilson's numerical RG treatment of the Kondo problem[4], one might expect similar numerical RG approaches to be much more widely used than they are. I believe there are two reasons for the current lack of interest in such methods. First, early attempts at real-space numerical RGs were largely unsuccessful[5], for reasons discussed below, and the technique gained a poor reputation. Second, it was thought that Wilson's Kondo technique could only be applied to impurity problems (where only the radial coordinate was important), and was not useful for higher dimensions. Recently, I have developed a new method of implementing Wilson's technique as a momentum-space technique for finite Hubbard lattices[6], but in this paper I will discuss only real-space techniques. (A completely different approach to the methods considered here uses finite-temperature functional-integrals[7]. Here we consider only Hamiltonian-based methods.)

This paper presents several new ideas directly related to real-space numerical RGs. However, the implementation of these ideas in a full RG algorithm is still underway and no results will be presented here.

2. Single-particle problem

The standard real-space blocking approach for quantum problems can be described most simply by considering a 1-D single-particle tight-binding chain. This problem is equivalent to a 1-D particle in a box. We take as the Hamiltonian matrix H_{ij} (i, j are sites on the chain)

$$H_{ij} = \begin{cases} 2 & i = j \\ -1 & |i-j| = 1 \\ 0 & otherwise. \end{cases} \tag{1}$$

The standard approach consists of considering a group of sites to be a "block", and diagonalizing that block to find a set of eigenstates. One then truncates the set of eigenstates, keeping only the lowest m states (ordered by energy), and uses those states to construct an approximate Hamiltonian for a new, larger block composed of two of the old blocks. At each iteration s we can write the Hamiltonian of the infinite chain in terms of diagonal blocks H^s and off-diagonal blocks T^s as

$$H = \begin{pmatrix} H^s & T^s & 0 & 0 & 0 & \dots \\ T^{s\dagger} & H^s & T^s & 0 & 0 & \dots \\ 0 & T^{s\dagger} & H^s & T^s & 0 & \dots \\ 0 & 0 & T^{s\dagger} & H^s & T^s & \dots \\ \vdots & \vdots & \vdots & \vdots & \vdots & \end{pmatrix} \tag{2}$$

Initially, the block size is 1 and H^1 and T^1 are 1×1 matrices equal to 2 and -1, respectively. We start iteration s by forming the Hamiltonian matrix for a block composed of two blocks from the previous iteration

$$\bar{H}^s = \begin{pmatrix} H^{s-1} & T^{s-1} \\ (T^{s-1})^\dagger & H^{s-1} \end{pmatrix} \tag{3}$$

and

$$\bar{T}^s = \begin{pmatrix} 0 & 0 \\ T^{s-1} & 0 \end{pmatrix} \tag{4}$$

We diagonalize $\bar{H}^s$ and take the lowest m eigenvalues E^s_ℓ and eigenstates, $V^s_\ell, \ell = 1, \dots m$, discarding the rest. We then perform a change of basis to the eigenstates via

$$H^s_{\ell\ell'} = E^s_\ell \delta_{\ell\ell'} \tag{5}$$

and

$$T^s_{\ell\ell'} = \sum_{i,j} V^s_{\ell i} \bar{T}^s_{ij} V^s_{\ell' j}. \tag{6}$$

We then proceed on to the next iteration, starting with Eq. (3) and (4). The idea is that the higher energy states which are discarded at the current iteration are unimportant in making up the low energy states at a later iteration. We can increase the accuracy by keeping more states, i.e., increasing m.

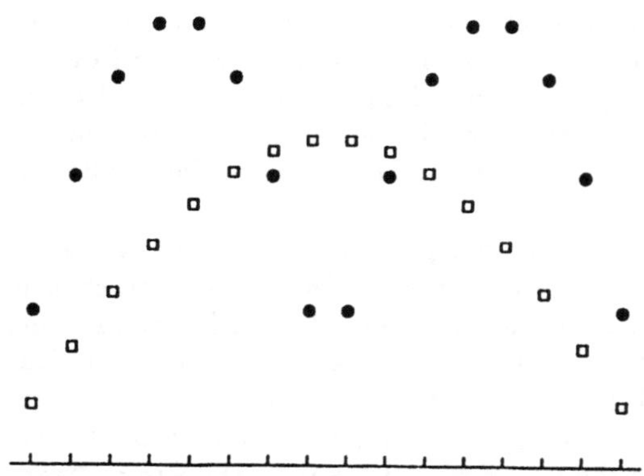

Figure 1: Lowest eigenstates of two 8-site blocks (solid circles) and a 16-site block (open circles).

It is easy to see in this simple example, however, that this procedure is quite *poor* in describing large scale, low energy behavior. The Hamiltonian in this example is just a finite difference discretization of the kinetic energy of a 1-D particle, and in the limit of large block size, the eigenstates are just particle-in-a-box eigenstates. The boundary condition of ignoring the connections T to neighboring blocks corresponds to setting the wavefunction to 0 at the sites just outside the block. Figure 1 illustrates the difficulty. Any state made only of low-lying states from the previous iteration must have a "kink" in the middle. In order to accurately represent states in the larger block, one must make use of nearly all the states in the smaller block: any truncation leads to large errors.

Of course, the procedure in the many-particle case is more complicated. In particular, instead of getting twice as many states when we combine two blocks, we *square* the number of states. Nevertheless, the effect of the edges is similar. It is just more obvious in the single-particle case.

The above example is due to K. G. Wilson, who discussed it during an informal group talk in 1987. He pointed out that unless an approach could deal with this simple, noninteracting problem, it could hardly be trusted in the much more complicated interacting case. He suggested that the problem might be cured by integrating out the higher energy states via a unitary transformation, taking into account the interactions with other blocks perturbatively rather than ignoring them. (This approach is closely related to standard degenerate perturbation theory.) K. Ingersent and I recently tried a variety of ways of implementing this, but were unable to find a satisfactory approach. I now believe that the set of low-lying states kept in the standard approach is too incomplete for the perturbative unitary transformation to help.

The key to fixing this simple example lies in the treatment of boundary conditions. The standard approach (in the large block limit) causes the eigenstates to vanish at the edges of a block, causing "kinks". Several alternative choices of boundary conditions do not work either: for example, periodic boundary conditions would result in the low-lying eigenstates being identical at the edges, which would prevent the representation of low-lying excited states on larger blocks. Free boundary conditions (obtained by changing H_{ii} from 2 to 1 on the edges) result in the slope of the eigenstate vanishing at the edges, and again we cannot represent low-lying excited states on larger blocks. In order to obtain a working approach, we must *combine* eigenstates obtained from different boundary conditions.

The following procedure solves the simple problem exactly, in the sense that a finite number of the lowest energies of large blocks are obtained exactly even after many iterations, keeping only a few states at each iteration. One performs the method outlined above using eigenstates from each of the four possible combinations of free and fixed boundary conditions at the two edges simultaneously. In other words, we keep track of four H matrices at each iteration, $\bar{H}^s_{bb'}$, where b represents the left edge and takes on the values "free" or "fixed", and b' similarly represents the right edge. For example, for an initial two-site block one would have

$$H^1_{\text{free,fixed}} = \begin{pmatrix} 1 & -1 \\ -1 & 2 \end{pmatrix}. \tag{7}$$

We diagonalize $\bar{H}^s_{\text{free,fixed}}$ and extract the lowest $m/4$ eigenstates. Similarly, we extract the lowest $m/4$ eigenstates for $\bar{H}^s_{\text{free,free}}$, $\bar{H}^s_{\text{fixed,free}}$, and $\bar{H}^s_{\text{fixed,fixed}}$.

This set of m states is not orthogonal, so we next orthonormalize them. This set of states forms the new basis, taking the place of the $V^s_\ell, \ell = 1, \cdots m$. (The orthogonal complement to these states are the states that are discarded.) We next perform a change of basis on the matrices $\bar{T}^s$ and $\bar{H}^s_{\text{free,fixed}}$, etc., using V^s to obtain $T^s, H^s_{\text{free,fixed}}$, *etc.*, as in Eq. (6) [but we may not use Eq. (5)].

Finally, for the next iteration, we replace Eq. (3) by

$$\bar{H}^s_{bb'} = \begin{pmatrix} H^{s-1}_{b,\text{fixed}} & T^{s-1} \\ (T^{s-1})^\dagger & H^{s-1}_{\text{fixed,b}'} \end{pmatrix}. \tag{8}$$

(One must always use the original fixed boundary conditions for the inner edges when joining two blocks together).

Table 1 compares the results of this procedure with the standard one. The eigenvalues of $\bar{H}^s_{\text{fixed,fixed}}$ are shown for the new procedure. A total of 8 states ($m = 8$) were kept for both procedures. We find that the new procedure yields exact energies for the first four states even after 10 (or more) iterations, whereas the results from the standard procedure bear little relationship to the exact results. I have only numerical evidence that this procedure gives exact results; nevertheless, the results are quite convincing.

The lesson to be learned from this example is that one set of boundary conditions is not sufficient, and it is difficult to correct this insufficiency by keeping many states or by applying perturbative corrections. Instead, one should apply several different boundary conditions, keeping a few low-lying states from each. The influence of the surrounding blocks not taken into account in the current iteration is to effectively apply different boundary conditions to the current block.

Table 1. Lowest energies after 10 blocking transformations for the noninteracting single particle on a 1-D chain.

	Exact	Standard	New procedure
E_0	2.3508×10^{-6}	1.9207×10^{-2}	2.3508×10^{-6}
E_1	9.4032×10^{-6}	1.9209×10^{-2}	9.4032×10^{-6}
E_2	2.1157×10^{-5}	1.9214×10^{-2}	2.1157×10^{-5}
E_3	3.7613×10^{-5}	1.9217×10^{-2}	3.7613×10^{-5}

3. Many-particle systems

How do we apply this to interacting, many-particle systems? It is not clear how to impose "free" boundary conditions in this case. For example, a characteristic of the free boundary condition case in the single-particle problem is that the ground state wavefunction is constant (independent of the site), a characteristic shared by the ground state of the infinite system. In the case of a 1-D Hubbard model with Hamiltonian

$$H = -t\sum_{i,\sigma}(c^\dagger_{i,\sigma}c_{i+1,\sigma} + c^\dagger_{i+1,\sigma}c_{i\sigma}) + U\sum_i n_{i\uparrow}n_{i\downarrow} - \mu\sum_i(n_{i\uparrow}+n_{i\downarrow}), \tag{9}$$

one might try to adjust a local chemical potential on the edge sites of a block in order to make the *density* $\langle n_{i\uparrow}+n_{i\downarrow}\rangle$ constant. However, one finds that in the half-filled case ($\mu = U/2$) the density is *already* constant, but the *double-occupancy* $\langle n_{i\uparrow}n_{i\downarrow}\rangle$ varies from site to site. (It is a minimum at the edges.) One can then try to adjust t, U, and μ for the last site in order to make both the density and double occupancy constant, but this becomes a rather unsatisfactory fitting process. If one has adjusted the local parameters at the edges, then it is tempting to *exclude* these special edge sites from the block to avoid any hidden effects from them in the eigenstates, or equivalently, add extra boundary sites at each edge of the block. This leads to the idea of *projecting out* basis states for a block from an eigenstate on a "superblock" (The superblock is the block plus extra edge blocks; see Figure 2. Any adjustments to t, U, and μ would be comfined to the edge blocks.) I believe this idea of projecting out basis states is the *key idea* needed in the interacting problem.

This idea would be implemented as follows. We diagonalize the Hamiltonian on the superblock for a few of the lowest eigenstates. From each superblock

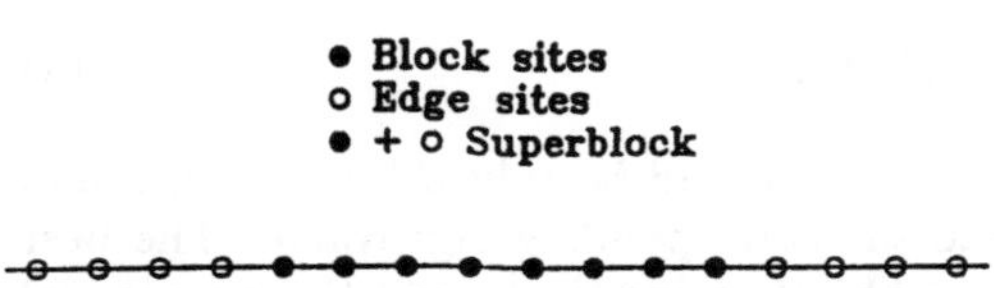

Figure 2: A superblock, containing a block and extra edge sites.

eigenstate we project out a few block basis states, orthogonalize them and then use them to represent the block. The use of several superblock eigenstates would in effect apply different boundary conditions to the block.

The need for this type of projection in a successful numerical renormalization group can be further illustrated by a simple example. Consider a Hubbard model with $\mu = U/2$, $U >> t$. In this case there is an antiferromagnetic gap between states with N particles and either $N+1$ or $N-1$. If one chose only a limited set of eigenstates with the lowest energy, *all* would have exactly N particles. However, in order for hopping from block to block to take place, we must have states with $N \pm 1$ particles on the block. The projection technique described below gives states with $N \pm 1$ particles in a natural way, occuring early in the expansion.

In the single-particle case the projection operation is trivial: given ψ_i defined on a superblock, we define its projection V_i on the block as $V_i = \psi_i$, and then adjust the normalization of V_i. In the many-particle case the situation is much more complicated. Let $|\alpha\rangle$ indicate a complete set of basis states for the block, and $|\beta\rangle$ correspond to states of the edge blocks. Then any wavefunction for the superblock can be written as

$$|\psi\rangle = \sum_{\alpha,\beta} \psi_{\alpha\beta} |\alpha\rangle|\beta\rangle. \tag{10}$$

It is clear that any truncation of the set $|\alpha\rangle$ will result in some errors in our ability to accurately reproduce $|\psi\rangle$. What we seek are the most natural bases $|V\rangle$ and $|W\rangle$ to replace $|\alpha\rangle$ and $|\beta\rangle$ so that only a few terms dominate and the remaining basis states can be dropped. We write

$$|\psi\rangle \approx \sum_{\gamma} \psi_{\gamma} |V_{\gamma}\rangle|W_{\gamma}\rangle. \tag{11}$$

Finding $|V_\gamma\rangle$ and $|W_\gamma\rangle$ is equivalent to finding the most rapid expansion of a rectangular matrix in terms of outer products of vectors.

Given a rectangular matrix M_{ij}, suppose we want an approximate representation of M_{ij} as

$$M_{ij} \approx e v_i w_j. \tag{12}$$

We minimize the squared error in this approximation

$$S = \sum_{ij} (M_{ij} - e v_i w_j)^2 \tag{13}$$

obtaining

$$Mw = ev \; ; \; v^{\dagger} M = ew. \tag{14}$$

If M were a square matrix, then Eq. (14) would be solved by setting v and w to any eigenvector of M, and w to the corresponding eigenvalue. The best approximation to M would be obtained by choosing the largest (in magnitude) eigenvalue. For an expansion involving several terms, one takes the largest several eigenvectors. For the more general rectangular case, the procedure is the same: one defines generalized left and right "eigenpairs" v and w via Eq. (14), and expands M as

$$M_{ij} \approx \sum_{\gamma} e^{\gamma} v_i^{\gamma} w_j^{\gamma} \tag{15}$$

Thus the procedure for projecting out a superblock wavefunction $|\psi\rangle$ onto the block is reduced to finding the largest few "eigenpairs" of the matrix $\psi_{\alpha\beta}$, v_{α}^{γ} and w_{β}^{γ}. Then

$$|V^{\gamma}\rangle = \sum_{\alpha} v_{\alpha}^{\gamma} |\alpha\rangle \tag{16}$$

It is straightforward to find the largest eigenvalues and eigenvectors of a matrix numerically, even for the general rectangular case. The simple iteration

$$v' \propto Mw \ ; \ w' \propto v^{\dagger} M \tag{17}$$

will project out the largest left and right eigenvectors. One can then subtract out the eigenvector approximation from M

$$M'_{ij} = M_{ij} - e v_i w_j \tag{18}$$

in order to find the next largest eigenvectors, and the process can be repeated.

I have tested this projection technique by diagonalizing to find the ground-state of an 8-site Hubbard chain and then projecting onto a 4-site subblock. The tests showed that relatively few terms dominate in Eq. (11), as one would hope. In addition, some of the dominant eigenvalues e^{γ} corresponded to states with $N \pm 1$ particles, an important feature as discussed above. The next step is an attempt to incorporate this into a blocking transformation and do several iterations, comparing to exact results for a finite system. Program development for this next test is currently underway.

4. Conclusions

Many of the most difficult problems for the numerical techniques currently available for the study of strongly interacting fermion systems would be very suitable for treatment by real-space numerical RG methods. These methods have not been pursued extensively both because of their computational complexity and because early attempts were not very successful. I hope I have made it clear that we now have a much better understanding of the reasons for the lack of success. I believe the new developments discussed here provide several possible paths for overcoming these difficulties, and further advances may bring numerical RG methods back to the forefront.

Acknowledgements

I would like to thank K. Ingersent for a number of helpful discussions and collaboration. This work was supported by the Office of Naval Research, Grant

No. N00014-91-J-1143. This research was also supported in part by the University of California through an allocation of computer time on the UC Irvine Convex.

References

1. See, for example, S.R. White, D.J. Scalapino, R.L. Sugar, E.Y. Loh, J.E. Gubernatis, and R.T. Scalettar, *Phys. Rev.* **B**40, 506 (1989).

2. E.Y. Loh, J.E. Gubernatis, R.T. Scalettar, S.R. White, D.J. Scalapino, and R.L. Sugar, *Phys. Rev.* **B**41, 9301 (1990).

3. See, for example, A. Parola, S. Sorella, M. Parrinello, E. Tosatti, *Phys. Rev.* **B**43, 6190 (1991).

4. K.G. Wilson, *Rev. Mod. Phys.* 47, 773 (1975).

5. J.W. Bray and S.T. Chui, *Phys. Rev.* **B**19, 4876 (1979); S.T. Chui and J.W. Bray, *Phys. Rev.* **B**18, 2426 (1978); J.E. Hirsch,*Phys. Rev.* **B**22, 5259 (1980); C. Dasgupta and P. Pfeuty, *J. Phys.* **C**14, 717 (1981).

6. S.R. White, to be published.

7. S.R. White, J.W. Wilkins, and K.G. Wilson, *Phys. Rev. Lett.* **56**, 412 (1986).

Bayesian Methods, Maximum Entropy, and Quantum Monte Carlo

J.E. Gubernatis[1], *R.N. Silver*[2], *and M. Jarrell*[3]

[1]Theoretical Division, Los Alamos National Laboratory, Los Alamos, NM 87545, USA
[2]Theoretical Division and Los Alamos Neutron Scattering Center, Los Alamos National Laboratory, Los Alamos, NM 87545, USA
[3]Department of Physics, University of Cincinnati, Cincinnati, OH 45221, USA

Abstract. We heuristically discuss the application of the method of maximum entropy to the extraction of dynamical information from imaginary-time, quantum Monte Carlo data. The discussion emphasizes the utility of a Bayesian approach to statistical inference and the importance of statistically well-characterized data.

1. Introduction

Recently, we proposed procedures that for the first time permit the consistent extraction of dynamical properties from imaginary-time quantum Monte Carlo data. The technical details of these procedures, as well their application to several different physical models, studied by several different quantum Monte Carlo methods, have been published [1, 2]. In these proceedings, we will summarize our current understanding of how these procedures work by discussing them in a more heuristic manner than previously possible.

Most quantum Monte Carlo simulations at finite temperatures are done in imaginary time. The Wick rotation, $it \to \tau$, converting real time t to imaginary time τ, transforms oscillatory exponentials e^{-itH} of the Hamiltonian H into diffusive ones, $e^{-\tau H}$. Monte Carlo sampling of a function that becomes highly oscillatory at large, real times is inefficient as sampling on smaller and smaller times scales becomes necessary just to achieve proper self-cancellations [3]. In most cases, however, it is a spectral density, which describes the distribution of elementary excitations associated with various one- and two-particle Green's functions, and not the real-time behavior which is generally sought. Still again, the difficulty in obtaining highly oscillatory, long-time information to insure proper low-frequency behavior is inhibiting.

In imaginary-time, a different difficulty arises. With quantum Monte Carlo estimates $\bar{G}(\tau)$ of a one- or two-particle Green's function $G(\tau)$, the generic problem is to find the associated spectral density $A(\omega)$ from

$$G(\tau) = \int_{-\infty}^{+\infty} d\omega \, \frac{e^{-\tau\omega} A(\omega)}{1 \pm e^{-\beta\omega}} \tag{1}$$

where the $\pm$ refers to Fermi and Bose statistics and $\tau > 0$. The difficulties arise at large positive and negative frequencies because of the exponential behavior of the kernel

$$K(\tau, \omega) = \frac{e^{-\tau\omega}}{1 \pm e^{-\beta\omega}}\,. \tag{2}$$

For the forward problem, i.e., "Given A, what is G?", $G(\tau)$ is insensitive the large $|\omega|$ details of $A(\omega)$. For the inverse problem, i.e., "Given G, what is A?", this insensitivity leads to an ill-posed problem, so with incomplete and inaccurate information about $G(\tau)$, an infinite number of solutions exist. The task is then to select from these solutions one that is best by some criterion.

Previously, several criteria have been proposed. None lead to a satisfactory procedure with the potential of handling the more difficult problems. The procedures we have proposed [1] succeed because they use more relevant prior information about the behavior of $A(\omega)$ and are accompanied by a closer assessment of the statistical character of the data produced by the quantum Monte Carlo method. Their structure is completely tied to the Bayesian approach to statistical data analysis and the use of the principle of maximum entropy.

2. Bayesian Inference and Maximum Entropy

Given two events a and b, Bayes's theorem says [4]

$$\Pr[a,b] = \Pr[a|b]\Pr[b] = \Pr[b|a]\Pr[a] \tag{3}$$

where $\Pr[a]$ is the probability of a and $\Pr[a|b]$ is the conditional probability of a given b, and $\Pr[a,b]$ is the joint probability function for a and b. Also from probability theory, we have the condition for the marginalization of a variable

$$\Pr[a] = \int db\, \Pr[a,b]\,. \tag{4}$$

Our events are the functions $\bar{G}(\tau)$ and $A(\omega)$, where $\bar{G}(\tau)$ is our estimated (measured) value of $G(\tau)$. Our *initial* criterion for a best solution will be the function $\hat{A}(\omega)$ that maximizes $\Pr[A|\bar{G}]$. We take

$$\Pr[A|\bar{G}] = \Pr[\bar{G}|A]\Pr[A]/\Pr[\bar{G}]\,. \tag{5}$$

$\Pr[A|\bar{G}]$ is called the posterior probability, $\Pr[\bar{G}|A]$, the likelihood function, and $\Pr[A]$, the prior probability. $\Pr[\bar{G}]$ is a normalization constant [5], called the evidence.

Since the spectral density $A(\omega)$ is non-negative and normalizable to unity, it can be interpreted as still another probability function. In the absence of data, the principle of maximum entropy [4] says that the values of a probability function $A(\omega)$ are to be assigned by maximizing the entropy expression

$$S = -\int d\omega A(\omega) \ln A(\omega)\,.$$

In the absence of data (or with the lack of information about the likelihood function), $\Pr[A|G] \propto \Pr[A]$. Hence, the A that maximizes the posterior probability is the same as the one that maximizes S when

$$\Pr[A] \propto e^{\alpha S} . \tag{6}$$

Thus, the entropy associated with the probability function fixes the prior probability up some constant α [6].

If $A(\omega)$ is not normalized, we could maximize

$$S - \lambda\Big(1 - \int d\omega A(\omega)\Big)$$

where λ is a Lagrange multiplier, or equivalently, we can assign probabilities, i.e., find $A(\omega)$, by maximizing

$$S = -\int d\omega \Big(A(\omega)\ln A(\omega) - A(\omega)\Big) .$$

In the literature [6, 7], a fuller form for S is used

$$S \equiv -\int d\omega \Big(A(\omega)\ln[A(\omega)/m(\omega)] - A(\omega) + m(\omega)\Big) . \tag{7}$$

The function $m(\omega)$ is called the default model, because in the absence of data, S is maximized when $A(\omega) = m(\omega)$. From one point of view, $m(\omega)$ sets the zero of the entropy. The importance and use of the default model, however, goes beyond this, as we will shortly discuss.

In the presence of data, i.e., the values of $\bar{G}(\tau)$ for N different values of $\tau = \tau_i$, we could assign probabilities by maximizing

$$S - \sum_{i=1}^{N} \lambda_i \Big(\bar{G}(\tau_i) - \int d\omega\, K(\tau_i,\omega)A(\omega)\Big) .$$

This procedure ignores any errors associated with the data, but these errors are important information needed to determine confidence in the results. The errors associated with the data are best incorporated into the analysis through the likelihood function.

3. Some Intermediate Observations

To motivate about the default model a bit more, we will assume that we know $A(\omega)$. Interpreting $A(\omega)$ as a probability function, we could fix τ in (1), evaluate the integral by a Monte Carlo procedure where we generate a sequence of ω_i drawn from the probability distribution $A(\omega)$, estimate $G(\tau)$ by

$$\bar{G}(\tau) = \frac{1}{M}\sum_{i=1}^{M} K(\tau,\omega_i) ,$$

and then assign an error σ to this estimate from

$$\sigma^2 = \frac{1}{M^2} \sum_{i=1}^{M} \Big(K(\tau, \omega_i) - \bar{G}(\tau) \Big)^2 . \tag{8}$$

In Monte Carlo integration, importance sampling [8] is often used to reduce variance by introducing a trial function $m(\omega)$ that is similar to the actual function, rewriting the integral as

$$G(\tau) = \int_{-\infty}^{+\infty} d\omega\, m(\omega)\, K(\tau, \omega) \frac{A(\omega)}{m(\omega)} ,$$

and drawing the random variables ω_i from $m(\omega)$. The estimate of $\bar{G}(\tau)$ becomes

$$\bar{G}(\tau) = \frac{1}{M} \sum_{i=1}^{M} K(\tau, \omega_i) \frac{A(\omega_i)}{m(\omega_i)} .$$

If $m(\omega) = K(\tau, \omega) A(\omega)/\bar{G}(\tau)$, it is easy to show that for large M the correct value of $\bar{G}(\tau)$ is produced with zero variance. Of course, this method of eliminating variance requires *a priori* knowledge of the exact answer, but clearly by using relevant, prior information about the expected answer, we can produce a more certain result. On the other hand, an unwise choice for $m(\omega)$ can be counterproductive.

Our operational point-of-view is that the default model is a "variance reduction" tool analogous to the trial function in Monte Carlo integration: the closer the default model is to answer, the better will the answer be. We are not trying to assume the answer, but to solve an ill-posed problem, we need to use as much the prior information as possible. The default model is a tool towards this end. In many cases, we may not know much before hand.

It is clear that the Monte Carlo integration scheme just described produces independent estimates of $G(\tau)$ at different τ values. At a given τ, successive values of ω_i, and hence of the estimators $K(\tau, \omega_i)$ of $G(\tau)$, can be expected to be reasonably statistically independent. The error estimate (8) assumes that these independent values are Gaussian distributed, but clearly their distribution depends on $K(\tau, \omega)$ and $A(\omega)$. This estimate σ is meaningful only if M is sufficiently large so that the central limit theorem applies [4].

In the quantum Monte Carlo applications, we are addressing the inverse problem, not the forward problem just described. As we will discuss, for different values of τ, the $\bar{G}(\tau)$ are not independent so care must be taken to achieve statistically independent estimates data. Reliable error measurements will also depend on the central limit theorem, but one has to face the fact that this theorem does not specify the practical size of M needed to know when the theorem applies. The need to consider this question to dictated by the choice of the likelihood function used. Almost always, a specific choice is made.

4. Likelihood Function

The method of maximum likelihood [9] is commonly used in parameter fitting problems. For simple probability functions, like a Gaussian, fitting two parameters to the data, the mean and standard deviation, is sufficient to determine completely the probability function. Using the data to find $A(\omega)$ is type of parameter fitting problem in as much as we will determine a set $\mathbf{A} = (A_1, A_2, \ldots, A_N)$ of values of A at a number of discrete values ω_i of ω, but the number N of parameters we determine is on the order of 200. From (1), we find that a given $\mathbf{A}$ makes a specific prediction for a set of $G(\tau)$ values $\mathbf{G} = (G_1, G_2, \ldots, G_L)$ at various discrete values τ_i of τ. In the method of maximum likelihood the objective is to vary $\mathbf{A}$ so that the resulting $\mathbf{G}$ maximizes some likelihood function.

In maximum likelihood method, the $\bar{G}(\tau)$ are regarded as random variables drawn from a probability distribution $f(\bar{\mathbf{G}}; \mathbf{G})$ specified by $\mathbf{G}$. For a single random sampling of $\bar{G}(\tau)$, a set of random variables $\bar{\mathbf{G}}^{(j)} = (\bar{G}_1^{(j)}, \bar{G}_2^{(j)}, \ldots, \bar{G}_L^{(j)})$ describing the values of $\bar{G}$ at discrete values τ_i of τ is realized. For a single $\bar{\mathbf{G}}^{(j)}$, we can assign a number $dP^{(j)} = f(\bar{\mathbf{G}}^{(j)}; \mathbf{G})d\mathbf{G}$. For M *independent* measurements, $dP = \prod_{j=1}^{M} f(\bar{\mathbf{G}}^{(j)}; \mathbf{G})\, d\mathbf{G}$. The *logarithm* of the likelihood function is $L = \sum_{j=1}^{M} f(\bar{\mathbf{G}}^{(j)}; \mathbf{G})$. The method of maximum likelihood is based on the assumption that a single peak in the space of the parameters $\mathbf{G}$ dominates this function so that the mode $\hat{\mathbf{G}}$ is a reliable estimate of the parameters to be fitted and the width of the peak provides some measure of confidence in the fit.

From a Bayesian point of view, the method of maximum likelihood takes $\Pr[A|\bar{G}] \propto \Pr[\bar{G}|A]$ so that the prior probability $\Pr[A]$ is ignored. Hence, finding the $\mathbf{A}$ that maximizes the posterior probability is equivalent to finding the A that maximizes $\Pr[A|\bar{G}] \propto e^L$. The method does not specify the function $f(\bar{\mathbf{G}}^{(j)}; \mathbf{G})$. In some applications, it might be known, but it is definitely an unknown in the quantum Monte Carlo case. When the number of measurements becomes large, by central limit theorem arguments, modes becomes means, etc., and the asymptotic behavior becomes known [9]

$$e^L \propto e^{-\frac{1}{2}\chi^2} \tag{9}$$

where

$$\chi^2 = \sum_{i,j}^{L} (\bar{G}_i - G_i)[C^{-1}]_{ij}(\bar{G}_j - G_j) \tag{10a}$$

and

$$\bar{G}_i = \frac{1}{M}\sum_{j=1}^{M} \bar{G}_i^{(j)}. \tag{10b}$$

C is the covariance matrix

$$C_{ik} = \frac{1}{M(M-1)}\sum_{j=1}^{M} (\bar{G}_i - \bar{G}_i^{(j)})(\bar{G}_k - \bar{G}_k^{(j)}). \tag{10c}$$

Maximizing the likelihood is equivalent to minimizing χ^2. A least-squares fitting procedure (10) is thus a special case of the method of maximum likelihood with $L = -\frac{1}{2}\chi^2$.

The least-squares method has several conditions for its proper application: independent measurements and a multi-variant Gaussian distribution of the data. It also makes an important assumption that the prior probability is unimportant or unknown. What is curious about least-squares determinations of the spectral density is that one of the first things done, when undesirable solutions were obtained, was to seek new solutions by constraining the fit to be positive and normalized. These conditions, however, are inherent to the entropic prior (6). The more significant constraint imposed, however, was the one of smoothness, which is *ad hoc* because there is no *a priori* reason to believe that the spectral density has such a property. When combined with the likelihood function, the entropic prior has a smoothing effect on the result that is principally dictated by the data and not by the user of the data.

5. Bayesian Inference Revisited

With (6) and (9), we express Bayes's theorem (5) as $\Pr[A|\bar{G}] \propto e^Q$ where

$$Q = \alpha S - \frac{1}{2}\chi^2 . \tag{11}$$

Finding the $A(\omega)$ that maximizes $\Pr[A|\bar{G}]$ is thus equivalent to maximizing Q with respect to $A(\omega)$

$$\left.\frac{\delta Q}{\delta A}\right|_{A=\hat{A}} = 0 \tag{12}$$

The result, however, will depend on α, and in what is called ***historic maximum entropy*** [1, 6, 7], α is adjusted to make $\chi^2 = N$. This choice is the expected value of χ^2 when the errors are Gaussian noise but is otherwise *ad hoc*.

We now describe a recent development, called ***classic maximum entropy***[6, 7]. We start by rewriting (5) as

$$\begin{aligned} \Pr[A, \alpha|\bar{G}] &= \Pr[\bar{G}|A, \alpha] \Pr[A, \alpha] / \Pr[\bar{G}] \\ &= \Pr[\bar{G}|A, \alpha] \Pr[A|\alpha] \Pr[\alpha] / \Pr[\bar{G}] \end{aligned} \tag{13}$$

to make explicit the dependence of our results on α. The normalization factor $\Pr[\bar{G}]$ equals $Z_L Z_S$ where the factors of Z normalize the likelihood function and the entropic prior. Z_S depends on α [7]. Introduced is a new function $\Pr[\alpha]$, which we will usually take to be what is called the Jeffrey prior [1]. In (13), the first two functions on the right-hand side are just (9) and (6).

Using (4) and then (3), we can also write the posterior probability $\Pr[A|\bar{G}]$ as

$$\begin{aligned} \Pr[A|\bar{G}] &= \int d\alpha \Pr[A, \alpha|\bar{G}] \\ &= \int d\alpha \Pr[A|\bar{G}, \alpha] \Pr[\alpha|\bar{G}] . \end{aligned} \tag{14}$$

The distribution function $\Pr[\alpha|\bar{G}]$ follows from functionally integrating (13) with respect to $A(\omega)$

$$\begin{aligned}\Pr[\alpha|\bar{G}] &= \int \mathcal{D}A \Pr[\bar{G}|A,\alpha]\Pr[A|\alpha]\Pr[\alpha]/\Pr[\bar{G}] \\ &= \Pr[\alpha]\int \mathcal{D}A\, A(\omega)\frac{e^{Q}}{Z_L Z_S}\,.\end{aligned} \tag{15}$$

If the number of data is large, then $\Pr[\alpha|\bar{G}]$ is often sharply peaked at some value of $\alpha = \hat{\alpha}$ [7], and (14) becomes $\Pr[A|\bar{G}] \propto \Pr[A|\bar{G},\hat{\alpha}]$. With the $\hat{\alpha}$, the $A = \hat{A}$ satisfying (12) becomes the classic maximum entropy solution. The ill-posed nature of the our problem, however, restricts the space in which we can search for a solution so that our effective number of data is usually small and we cannot legitimately use the above.

To proceed, for a fixed value of α, we find the solution $\hat{A}_\alpha$ that satisfies (12). Then, we compute numerically the solution $\bar{A}(\omega)$ defined by

$$\bar{A}(\omega) \equiv \int d\alpha\, A(\omega)\Pr[A|\alpha] \tag{16}$$

where $\Pr[\alpha|\bar{G}]$ is given by (15). In the absence of data, this procedure produces the default model as the answer.

These are the major details of classic maximum entropy method and its application to quantum Monte Carlo data; the seminal works [6, 7, 10] should be discussed for a more detailed description. The main point is that α is placed on a similar footing as $A(\omega)$. This allows $A(\omega)$ to be determined without needing to adjust α or using properties of α not supported by the data.

6. Quantum Monte Carlo

The procedures just discussed are not tied the problem of extracting spectral densities from imaginary-time quantum Monte Carlo data. They are quite general and can be applied to the solution of almost any ill-posed problem where the function sought is non-negative. To tie them to our problem, we must make certain that our measurements $\bar{\mathbf{G}}^{(j)}$ are statistically independent and distributed by a multivariant Gaussian probability function. These conditions have to be promoted because they are not even roughly mimicked by the quantum Monte Carlo data [1].

Normally, the simulation produces a sequence of measurements that are partitioned into M groups (bins), each of length k. The distribution from which these measurements are drawn is not known. The basic strategy is to make k large enough so that the means of the measurements in each bin, i.e., the $\bar{\mathbf{G}}^{(j)}$, are statistically independent and to make M large enough so that the mean of these measurements is approximately distributed according the the Gaussian distribution predicted by the central limit theorem.

Auto-correlation measures exists that are helpful in estimating the distance between correlated measurements [11], but more than one correlation length is most likely present, so such estimates are rough and the actual bin size k effective for decorrelation needs to be 10 to 20 times larger than the estimated correlation length. Additionally, the behavior at $\tau = 0$ generally differs from the behavior at $\tau = \frac{1}{2}\beta$, so this measure, as the ones to be discussed below, needs to be done at several different τ values and the worst case determines the bin-size to be used.

With M independent measurements, their mean, their variance, and an histogram of the measurements, a standard χ^2 test can be applied to test the consistency of the histogram with a normal distribution with the same mean and variance [1]. Such a test is helpful but is more useful for establishing inconsistency with a normal distribution than consistency. The lowest order corrections to the normal distribution predicted by central limit theorem are controlled by the skewness and kurtosis of the distribution of the measurements [12]. If this distribution is Gaussian, then these deviations would be zero, but in a statistical sampling process, these measures, as the standard deviation, are expected to scale as $1/\sqrt{M}$ [13]. It is important to increase M until this scaling is approximately obeyed [14].

There are two matrices in the problem: the $N \times N$ covariance matrix C and the $N \times L$ discretized form of the kernel K. Both matrices can have ratios of singular values that exceed what can handled accurately on a finite precision computer. When the singular values become too small, the rank of these matrices has to be reduced to maintain numerical stability. These reductions represent a loss of information. For C, the reduction says that not all the data represent independent *information*. For K, it is a reflection of the ill-posed nature of the problem. For C, the reduction is greatly minimized if $M > L$ [14]. For K, it is generally severe, and K becomes a matrix of rank s where $s \ll L, N$.

The covariance matrix is poorly approximated by a diagonal matrix. In the illustrative example of Monte Carlo integration, the estimates of $G(\tau)$ at different values of τ were produced independently which is not the same as saying they are independent. The independent production is a consequence of the algorithm. In quantum Monte Carlo algorithms, the Green's function at one value of τ is generally produced from all the others.

7. Concluding Remarks

We have tried to present a discussion bridging the seminal works in Bayesian methods and maximum entropy [6, 7, 10] with their application to the extraction of dynamical information from imaginary-time quantum Monte Carlo data [1, 2]. These methods provide a consistent logical framework to approach the problem with the assumptions in the approach clearly defined. The Bayesian methods lead to a procedure (16) with no adjustable parameters.

Ill-posedness causes the need for such sophisticated procedures. By the nature of ill-posedness, the solution of an ill-posed problem is sensitive to the size of the errors on the input data. The procedures described require not only small errors but also Gaussian-distributed ones. The need for Gaussian behavior is a result of a particular choice of the likelihood function. Other choices could have been made, but they would be inconsistent with the way quantum Monte Carlo data, or data from other simulation methods, are usually analyzed. Since producing data with the proper statistical qualities is the hardest part of the problem, investigating other likelihood functions is certainly warranted. Because a specific function is unlikely to follow from analytic considerations, such a function will most likely need to be generated empirically from the quantum Monte Carlo data itself by use of non-parametric, robust statistical methods.

References

1. J. E. Gubernatis, M. Jarrell, R. N. Silver, and D. S. Sivia, "Quantum Monte Carlo simulations: dynamics from imaginary-time data," Phys. Rev. B, to appear.
2. R. N. Silver, D. S. Sivia, and J. E. Gubernatis, Phys. Rev. B **41**, 2380 (1990); R. N. Silver, J. E. Gubernatis, D. S. Sivia, and M. Jarrell, Phys. Rev. Lett. **65**, 496 (1990); M. Jarrell, J. E. Gubernatis, R. N. Silver, and D. S. Sivia, Phys. Rev. B **43**, 1206 (1991); M. Jarrell, J. E. Gubernatis, and R. N. Silver, Phys. Rev. B **44**, 5347 (1991); J. Deisz, M. Jarrell, D. L. Cox, Phys. Rev. B **42**, 4869 (1990); M. Jarrell, D. S. Sivia, B. Patton, Phys. Rev. B **42**, 4804 (1990); J. Deisz, K.-H. Luc, M. Jarrell, D. L. Cox, "Spin and charge dynamics for the one-dimensional t-J model," Phys. Rev. B, to appear.
3. J. D. Doll, D. L. Freeman, and T. L. Beck, Adv. Chem. Phys. **77**, 61 (1990), and references therein.
4. A. Papoulis, *Probability and Statistics* (Prentice-Hall, Englewood Cliffs, 1990).
5. This identification follows from functionally integrating Bayes theorem with respect to A, using $\int \mathcal{D}A\, \Pr[A|\bar{G}] = 1$, and then observing that $\Pr[\bar{G}] = \int \mathcal{D}A\, \Pr[\bar{G}|A] \Pr[A]$.
6. J. Skilling, in *Maximum Entropy and Bayesian Methods*, edited by J. Skilling (Kluwer, Dordrecht, 1989), pg. 45.
7. S. F. Gull, in *Maximum Entropy and Bayesian Methods*, edited by J. Skilling (Kluwer, Dordrecht, 1989), pg. 53.
8. M. H. Kalos and P. A. Whitlock, *Monte Carlo Methods, Volume 1: Basics* (Wiley, New York, 1986), chap. 4.
9. S. Brandt, *Statistical Computational Methods in Data Analysis* (North-Holland, New York, 1983.
10. R. K. Bryan, Eur. Biophys. J. **18**, 165 (1990).
11. G. Jacucci and A. Rahman, Nuovo Cimento D **4**, 341 (1983).

12. D. T. Gillespie, Am. J. Phys. **51**, 520 (1983).
13. W. H. Press, B. P. Flannery, S. A. Teukolsky, W. T. Vetterling, *Numerical Recipes* (Cambridge University Press, Cambridge, 1986), chap. 13.
14. M. Jarrell, unpublished.

Anyons on a Lattice

Y. Hatsugai[1], *M. Kohmoto*[1], *and Y.-S. Wu*[1,2]

[1]Institute for Solid State Physics, University of Tokyo,
Roppongi 7-22-1, Minato-ku, Tokyo 106, Japan

[2]Department of Physics, University of Utah, Salt Lake City, UT 84112, USA

Abstract. Anyons on the square lattice are studied numerically. We treat several different boundary conditions. These boundary conditions define different topological structures such as plane(plate), annulus (cylinder) and torus. To realize the system consistently, we start from a braid group analysis. The system is characterized not only by the anyon statistics θ but also by the magnetic fluxes Φ_x (and Φ_y) threading through the holes of the system. We generalize the construction to get a system where several species of anyons with $\theta=0, \pm 2\pi/q, \pm 4\pi/q, \cdots, (\pi)$, *exist simultaneously*. Explicit numerical calculations are demonstrated.

1. Introduction

A system of anyons [1, 2] is a typical example where phases in quantum mechanics play an essential role. The fractional statistics is first realized in the fractional quantum Hall effect [3]. Superconductivity in the anyon system is proposed to explain the high-T_C[4]. Although it is not clear yet whether anyons are relevant to the real high-T_C materials, their behavior is an interesting theoretical problem in its own right There are several numerical studies of anyons[5-7]. Here we show a careful constructions of anyon systems based on the braid group analysis and explicit results are given.

2. Braid group and anyons

A representation of the fundamental group of the N-body configuration space $\pi_1(C_N)$ determines the allowed statistics. In two dimensions, $\pi_1(C_N)$ is the braid group B_N. If the system is a plate (fixed boundary condition), it is generated by the local exchange operators σ_i, $(i=1,\cdots, N-1)$ where σ_i represents an interchange of i th and i+1 th particle counter clockwise. For an annulus (fixed boundary condition along y-direction and periodic boundary condition along x-direction), there is another type of generators ρ_j, which represents moving j th particle around the hole once in the positive direction. In this case, a relation $\rho_{j+1} = \sigma_j \rho_j \sigma_j$ holds. When we consider one-dimensional representation which corresponds to the usual one-component wavefunction, we get the representation $\sigma_i = e^{i\theta}$ for σ_i and $\rho_j(\Phi) = \exp[i \{2\theta (j-1)+2\pi\Phi\}]$ where θ is the statistics and Φ is the flux through the hole. When we consider a torus (periodic boundary condition along x- and y- directions), there is another generator τ_j and the system is characterized by θ, Φ_x and Φ_y. As discussed by several authors[6-8], we need multicomponent wave functions to put anyons on the torus. The most important consequence is that the usual period for Φ_x and Φ_y, reduces. If $\theta = \pi$ (p/q) with p and q mutually prime, then the smallest period is actually 1/q rather than 1. Further if one considers a q-component wave function on a plate, we can treat anyon systems with $\theta=0, \pm 2\pi/q, \pm 4\pi/q, \cdots$ simultaneously [9]. To realize it, we replace the usual effect of strings to give phases by shifting an index of a component

3. Numerical method

We perform exact numerical diagonalizations on a finite lattice. The dimension of the matrix is ${}_NC_M$ where N is the number of sites and M is the number of anyons. For a q-component calculation, the dimension is q times larger. We confirm that the mean field(MF) results are good unless a level crossing occurs. After the level crossing, however, the MF approximation is not so good. In the MF calculation on the annulus, the region where the MF is good in the MF from bosons is wider than the region where the MF is good in the MF from fermions, because there are fewer level crossings near the hard-core boson point [7].

We also confirm that eigen energies E_n satisfy $E_n(\Phi_x,\Phi_y) = E_n(\Phi_x + 1/q,\Phi_y) = E_n(\Phi_x ,\Phi_y + 1/q)$ on a torus when $\theta=\pi/q$. Since in general a given energy level does *not* necessarily have this period (it may have a larger period), we naturally expect the emergence of spectral flow and level crossing. In particular, energy levels may be rearranged as Φ_x or Φ_y changes by 1/q[7].

We also perform a q-sheet calculation on a plate with small magnetic field and get that $\theta=0, \pm 2\pi/q, \pm 4\pi/q, \cdots$ appears and each state is characterized by a different statistics [9]. (Table 1)

Table 1: 6 lowest energies of a 3 anyon system on a 4×3 lattice by q(=5)sheet calculation with a uniform external magnetic field $\phi_p=2\pi/50$. The statistics of each state is given by $\exp(i2\pi p/q)$ where p is a q-ality in the table.

	Energy	q-ality
E_1	-7.4407	0
E_2	-7.1974	1
E_3	-6.9946	4
E_4	-6.6477	3
E_5	-6.5514	2
E_6	-6.3857	2

References

[1] F. Wilczek, Phys. Rev. Lett. **49**, 957(1982).

[2] Y. S. Wu,. Rev. Lett. **52**, 2103 (1984).

[3] B.I.Halperin, Phys. Rev. Lett. **52**, 1583 (1984).

[4] R. B.Laughlin, Phy. Rev. Lett. **60**, 2677 (1988).

[5] G.S.Canright, S.M.Girvin and A.Brass, Phy.Rev.Lett. **63**, 2291 (1989); ibid. 2295 (1989).

[6] X. G. Wen, E. Daggoto and E. Fradkin,, Phys. Rev. **B42**, 6110 (1990).

[7] Y. Hatsugai, M. Kohmoto and Y.S. Wu, Phys. Rev. **B43**, 2661 (1991); ibid., 10761, Y. S. Wu, Y. Hatsugai and M. Kohmoto, Phys. Rev. Lett **66**, 659 (1991).

[8] T. Einarsson, Phys. Rev. Lett. **64**, 1995 (1990).

[9] Y. Hatsugai, M. Kohmoto and Y. S. Wu, preprint, ISSP Ser. A 2456 August.

Ground State Wave Function of an Extended Hubbard Model

T. Nishino

Department of Physics, Faculty of Science,
Osaka University, Toyonaka, Osaka 560, Japan

Energy level structure of an extended Hubbard model, which contains an additional spin flip term, is investigated. At half filling the system is insulating, where the ground state is easily obtained. When a hole is introduced, a bound state appears in the middle of the Hubbard gap.

The Hubbard model[1-2] is a simple model that describes the magnetism and the metal-insulator transition in narrow band. Since the model has not been solved yet in two or three dimensions, many approximations had been applied on it. Recently Montorsi and Rasetti proposed a generalized Hubbard model that is solvable in any dimensions within a certain range of parameter set.[3] I investigate a special case of their model numerically, outside the solvable region of parameter set.

Let us consider a Fermion model described by a Hamiltonian

$$H = -\sum_{<ij\sigma>} (c^+_{i\sigma} c_{j\sigma} + c^+_{j\sigma} c_{i\sigma}) + U \sum_i n_{i\uparrow} n_{i\downarrow} - \nu \sum_{<ij\sigma>} (c^+_{i\sigma} c_{j-\sigma} + c^+_{j\sigma} c_{i-\sigma}),$$

where the first two terms give the Hubbard Hamiltonian. The last term represents the electron hopping with spin flipping. The term destructs the antiferromagnetic spin correlation between nearest neighboring sites for positive ν. One naturally introduces two Fermion operators $a_i = (c_{i\uparrow} + c_{i\downarrow})/\sqrt{2}$ and $b_i = (c_{i\uparrow} - c_{i\downarrow})/\sqrt{2}$, and rewrites the Hamiltonian as

$$H = -(1+\nu)\sum_{<ij>} (a^+_i a_j + a^+_j a_i) - (1-\nu)\sum_{<ij>} (b^+_i b_j + b^+_j b_i) + U \sum_i a^+_i a_i b^+_i b_i .$$

First I calculated lower lying eigenstates of 8-site ring at half filling, with the parameters $U = 8$ and $0 < \nu < 1$, by the Lanzcos method.(Fig.1) There is a group of eigenstates within the range $-2 < E < 2$. There are some level crossings in the group, but all eigenstates stay within the

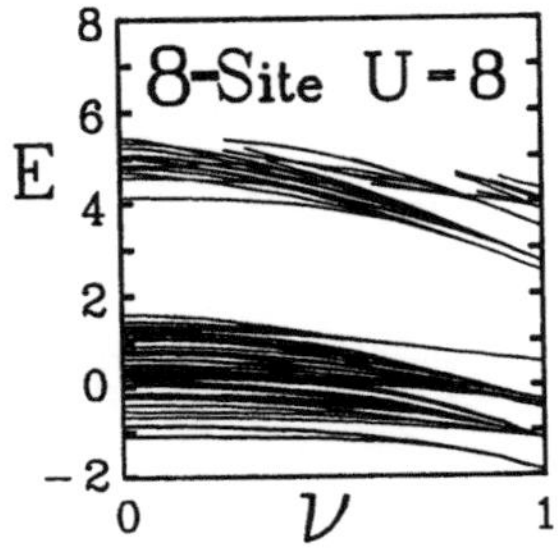

Fig.1: Several lower lying energy levels of the 8-site extended Hubbard ring when U=8.

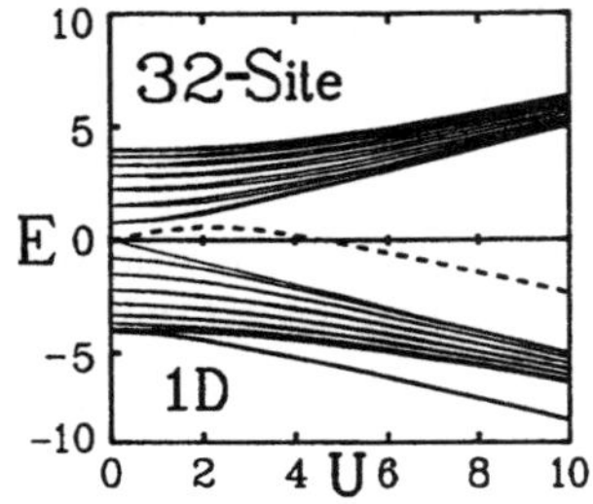

Fig.2: One a-particle Energy Level of the 32-site extended Hubbard ring at $\nu = 1$, where fifteen b-particles distribute alternatingly.

group. Above the group there is a finite excitation gap which represents the energy cost for a charge excitation.

The mobility of the b-particle is lost when ν is one. At this limit, the Hamiltonian is the same as that of the two component alloy[2]; the occupation of b-particles decides the energy level structure of a-particles. When both a and b-particles are half filled, one can get the ground state wave function exactly; each b-particle alternatingly occupies the lattice sites. If a b-particle is removed, an a-particle is bounded around this point defect. Figure 2 shows the one a-particle energy levels at $\nu = 1$, where a b-particle is removed. Bonding and anti-bonding state levels are shown by thick and thick dotted lines, respectively. The anti-bonding state is created above the lower band by absorbing a degree of freedom from upper band.

In conclusion I investigated an extended Hubbard model, which gives an intuitive picture for the insulating phase and the charge excitation of the Hubbard model. Note that one can investigate several lattice fermion models such as the d-p model, by use of the extension. More detailed discussion will appear elsewhere. The present research was supported partially by Grants-in-Aid for Scientific Research of the Ministry of Education, Science and Culture.

References

1. J.Kanamori: Prog. Theor. Phys **30** 275(1963).
2. J.Hubbard: Proc. Roy. Soc **A281** 401(1964).
3. A.Montorsi and M.Rasetti: Phys. Rev. Lett **66** 1383(1991).

Numerical Study on the t-t'–J-J' Model

K. Takano[1], *K. Sano*[2], *and I. Doi*[3]

[1]Toyota Technological Institute, Tenpaku-ku, Nagoya 468, Japan
[2]Faculty of Education, Mie University, Tsu, Mie 514, Japan
[3]Central Research Laboratory, Hitachi Ltd., Kokubunji, Tokyo 185, Japan

Abstract. We examined the one-dimensional t-t'-J-J' model in the one-hole case by the numerical diagonalization. We found a ***soliton*** (bound state) consisting of the hole and a singlet pair in a ferromagnetic background for $t' > 0$ and $J < 0$ and another soliton of the hole and a triplet pair in a singlet background for $t' < 0$ and $J > 0$. We also found that a *dimer* state persists for $J' \approx J/2(> 0)$ and $t' = 0$.

Various simplified models are useful to understand high-T_c superconductors and other highly correlated electron systems. In particular it is important to examine electronic structures in the nearly half-filled case[1]. The t-t'-J-J' model is one of the simplest models with strong correlation. To clarify a relation between a hole and electronic spins around the hole, we confine ourselves in the simplest case: i.e. the one-dimensional and the one-hole case. The Hamiltonian is written as

$$H = \sum_{ij}^{\mathrm{NN}}(-t\sum_{\sigma} c_{i\sigma}^{\dagger}c_{j\sigma} + \frac{1}{2}J\mathbf{s}_i\cdot\mathbf{s}_j) + \sum_{ij}^{\mathrm{NNN}}(-t'\sum_{\sigma} c_{i\sigma}^{\dagger}c_{j\sigma} + \frac{1}{2}J'\mathbf{s}_i\cdot\mathbf{s}_j) \qquad (1)$$

on condition that doubly occupied states are excluded. Here the nearest-neighbor (NN) and the next-nearest-neighbor (NNN) terms are considered.

We numerically diagonalized Eq.(1) for finite systems in the case of $J' = (t'/t)^2 J$[2]. The phase diagram of the ground state for a system of 13 sites (12 electons) is shown in Fig.1. For $J < 0$ ($J > 0$), the ground state is a ferromagnetic (singlet) state over almost all regions. On the other hand, there exist two characteristic phases of $S = S_{MAX} - 1$ and of $S = 1$. The shapes and the sizes of these regions are almost independent of the system size. In particular we found that the phase diagram with 5 sites (4 electrons) is essentially the same as that in Fig.1. This result suggests that the phase of $S = S_{MAX} - 1$ ($S = 1$) originates from a localized object or ***soliton*** which is smaller than 5 lattice spacings.

We propose a soliton picture as shown in Fig.2. In the phase of $S = S_{MAX} - 1$ ($S = 1$), the soliton consists of the hole and a singlet (triplet) pair and is bound within 3 sites. To confirm this soliton picture we calculated the projection of the ground state to the subspace spanned only by states of the same type as that in Fig.2. In the case of $t' = 0.4|t|$ and $J = -0.4|t|$ ($t' = -0.4|t|$ and $J = 0.4|t|$), the extrapolated value is 0.984 (0.887) for $S =$

Springer Proceedings in Physics, Vol. 70 **Computational Approaches in Condensed-Matter Physics**
Editors: S. Miyashita · M. Imada · H. Takayama

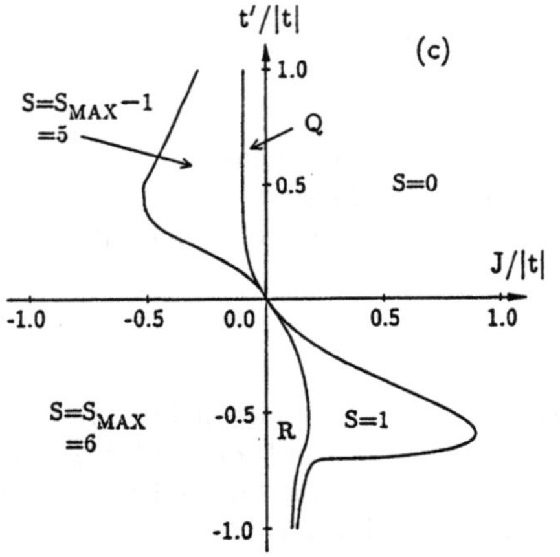

Fig.1. The phase diagram of the ground state in the case of 13 sites and 12 electrons. There are two soliton phases of $S{=}S_{MAX} - 1$ and of $S = 1$. The regions of Q and R consists of many phases with intermediate values of S.

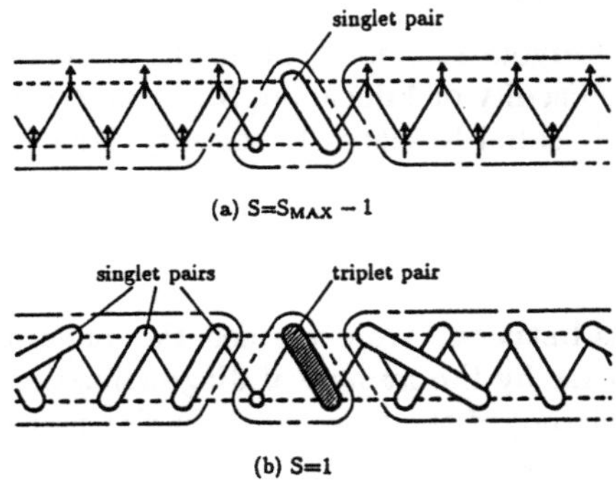

Fig.2. Picture of solitons for (a) $S = S_{MAX} - 1$ and (b) $S = 1$. A straight (dashed) line represents the NN (NNN) transfer $t(t')$ and the NN (NNN) exchange $J(J')$.

$S_{MAX} - 1$ ($S = 1$) and is very close to unity, confirming the soliton picture. These soliton phases are also found in a small two-dimensional system[3].

We also examined the case of $t' = 0$ and arbitrary J'. If there is no hole (just half-filled), the system reduces to a frustrated Heisenberg model and a dimer state appears for $0.3 \lesssim J'/J \leq 0.5$[4]. We found numerically that, for $J'/J \approx 0.5$, the dimer state are stable with the minimum modification even when one hole is introduced[5]. Recently Ogata et. al. examined the case of finite hole density[6].

References

1. P.W. Anderson, Science **235**, 1196 (1987).
2. I. Doi, K. Sano and K. Takano, to be published in Phys. Rev. **B**.
3. K. Takano and K. Sano, Phys. Rev. **B39**, 7367 (1989).
4. C.K. Majumdar, J. Phys. **C3**, 911 (1970); C.K. Majumdar and D.K. Gosh, J. Math. Phys. **10**, 1399 (1969); F.D.M. Haldane Phys. Rev. **B25**, 4925 (1982); T. Tonegawa and I. Harada, J. Phys. Soc. Jpn. **56**, 2153 (2153).
5. I. Doi, K. Sano and K. Takano, Physica **B165&166**, 1025 (1990).
6. M. Ogata, M.U. Luchini and T.M. Rice, preprint.

Random Magnetic Field and Anderson Localization

T. Sugiyama and N. Nagaosa

Department of Applied Physics, The University of Tokyo,
Bunkyo-ku, Hongo, Tokyo 113, Japan

Abstract. The Anderson localization is investigated for the tight binding model on the 2D square lattice where the phase of each transfer integral is an independent random variable between 0 and 2π. This corresponds to the model of the electronic states in the random magnetic field. We calculate the Thouless number by diagonalizing the Hamiltonian matrix of the square sample and the localization length for the stripe sample by the MacKinnon's method. We find that the localization length ξ behaves approximately as $\xi \propto \exp(14.5E)$ when the energy E is near the band edge E_0. In the finite energy region around $E = 0$ any symptom of the localization is not seen up to the width of 64 sites. This suggests the possibility of the existence of the mobility edge in this 2D model.

Recently the concept of the gauge invariance and the gauge field have been introduced to the theory of the high-T_c superconductors and other highly correlated electronic systems. In this approach the strong repulsion between the electrons on the same atomic orbital is expressed in terms of the local constraint which is closely related to the local gauge invariance. Hence the gauge field is naturally introduced, and the problem is reduced within a certain approximation to the motion of a particle in the gauge field configurations. One of the possibility is the so called flux phase where the nonzero uniform gauge flux is spontaneously created. Another possibility is the random fluctuating gauge flux. In this paper we focus on the latter possibility and investigate the following tight binding Hamiltonian

$$H = -\sum_{ij} t_{ij} C_i^\dagger C_j. \tag{1}$$

The transfer integral t_{ij} is nonzero only for the nearest neighbor pair ij and is given by

$$t_{ij} = \exp[\mathrm{i}\theta_{ij}], \tag{2}$$

where the phase θ_{ij} is an independent random variable between 0 and

2π and $\theta_{ij} = -\theta_{ji}$. A special attention is paid to the localization effect. Hereafter we call the gauge flux the magnetic field because its effect on the particle is the same as that of the magnetic field. The localization effect arises from the quantum interference between the two paths which are related to each other by the time reversal operation. This enhances the probability of the particle staying at the starting point. It is now well known that all the states are localized in 2D system when the symmetry of the Hamiltonian matrix is orthogonal which corresponds to the usual random potential problem with time reversal symmetry [1]. The magnetic field, which breaks the time reversal symmetry, suppresses the interference and hence the localization effect. In the limit of strong magnetic field the state at the center of the Landau level is believed to be extended which is relevant to the quantum Hall effect [2]. In the present case there is no random potential and the magnetic field introduces the randomness and the suppression of the interference effect simultaneously. Then it is a nontrivial problem wheather the states are localized or extended even in 2D.

We investigate this problem in two ways. One is the diagonalization of the Hamiltonian matrix to calculate the Thouless number and the other is the MacKinnon's method to calculate the localization length [3]. First we describe the results of the former one. The details of the calculation will be published elsewhere. Because the magnetic field introduces the frustration and increases the kinetic energy, the lower edge of the density of states is lifted to about -3.5 from -4.0 for the unperturbed case. Hence the total band width is reduced by $\sim$ 0.9. The density of states has a broad and slight minimum at $E = 0$. Fig. 1 shows the averaged Thouless number as a function of the energy. The localization length ξ is estimated from the dependence of the Thouless number g on the sample size L ($g \sim \exp(-L/\xi)$). The localization length is rapidly increasing as a function of E, and the inset in Fig. 1 shows the semi log plot which shows that $\xi \sim \exp(14.5E)$ holds near the band edge. For higher energy the present method is not enough to estimate the localization length, and we turn to the MacKinnon's method next.

By the MacKinnon's method the localization length ξ of the 1D system with finite width M can be calculated from the Green's function. The length along the 1D direction is very large (300000 for $M = 16$, 80000 for $M = 32$ and 50000 for $M = 64$), and the calculations have been done for three different random confugurations to calculate the average as well as the deviation. For $E = -3.2$, ξ/M decreased as M increases which indicates the localization. For $E = 0.0, -1.0, -2.0$, however, ξ/M seems to approach a finite value or slightly increase up to $M = 64$. This suggests that ξ is infinite for $M = \infty$.

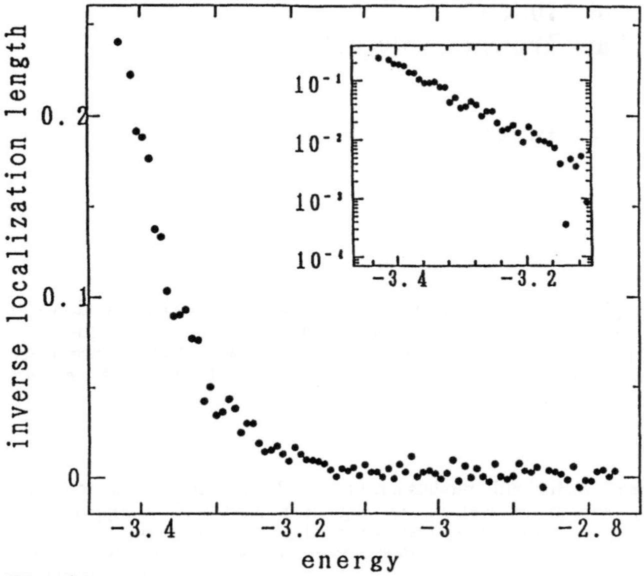

Fig. 1
The inverse of the localization length ξ^{-1} as a function of the energy E evaluated from the sample size dependence of the Thouless number. The inset is the semi-log plot which shows the approximate relation $\xi \sim \exp(14.5E)$.

In conclusion we have investigated the localization effect of the random magnetic field in the 2D tight binding model. Up to the width of the sample $M = 64$ any symptom of localization was not found in a finite energy range around the middle of the band, while the states near the band edge are localized. This suggests the possibility of the mobility edge in 2D.

References

[1] E.Abrahams, P.W.Anderson, D.C.Licciardello, and T.V.Ramarkrishnan, Phys. Rev. Lett. **42**, 673(1979).
[2] T.Ando, J. Phys. Soc. Jpn. **52**, 1740(1983); T.Ando and H.Aoki, J. Phys. Soc. Jpn. **54**, 2238(1985).
[3] A.MacKinnon and B.Kramer, Z. Phys. B-Condensed Matter 53, 1(1983).

Numerical Studies of Models with Suhl-Kondo Pair Transfer Interaction

Y. Asai

Electrotechnical Laboratory, Tsukuba, Ibaraki 305, Japan

Abstract. We have been applying a quantum monte carlo method to the two band models with Suhl-Kondo pair transfer interaction. These two band models may be applicable to high-T_c superconductivity in the copper oxides and the doped C_{60}. We have found clear enhancements of the s-wave pair-field susceptibilities at low temperature in these models. We will review our results and give some justifications of these models in the two superconductors.

1. Introduction

The Hubbard and dp (three band Hubbard) models have received much attention as models of the high-T_c copper oxides. These models well explain magnetic properties of the copper oxides. Published monte carlo results of theses models show that there may be no superconducting parameter region in these two models. Something additional may be required to explain the superconductivity. On the other hand, it has been pointed out that there may be the second band of oxygen character which is close to the Fermi level.[1] Recently, Yoshinari et al have measured anisotropies of the Knight shift on oxygen site and they have obtained the result which supports this point of view.[2] It is necessary to include this two quasiparticle bands nature in the model of the copper oxides. The importance of multiband nature is clearer in the doped C_{60}.

2. Suhl-Kondo pair transfer interaction

Suhl et al [3] and Kondo [4] had used this interaction to explain the T_c enhancements in transition metals. This interaction can be expressed as follows:

$$K\sum_j (a^\dagger_{j\uparrow}a^\dagger_{j\downarrow}b_{j\downarrow}b_{j\uparrow}+\text{h.c.}),$$

where a and b are destruction operators in the two different Wannier orbitals on the same site. J.Kondo had shown that there is an electronic interaction of this type. In this case, the coupling constant K takes the same value with the direct exchange integral between the two atomic orbitals. Suhl et al assumed that phonon exchange mediates this type of interaction. In his case, K is a BCS type coupling constant with the energy cut-off. I have shown that nonadiabatic electron-phonon coupling includes this type of interaction in the higher order and that the coupling constant can be very large if some states are close to the Fermi level. [5]

3. dp+π model as a model of the copper oxides

The dp+π model represents the two bands nature on oxygen sites:

$$H = -t_{pd}\sum_{<i,j>\sigma} (d^\dagger_{i\sigma}p_{j\sigma}+\text{h.c.})+\varepsilon_p\sum_j n^p_j+\varepsilon_d\sum_i n^d_i+U_d\sum_i n^d_{i\uparrow}n^d_{i\downarrow} -t_{\pi\pi}\sum_{<j,l>\sigma} (\pi^\dagger_{j\sigma}\pi_{l\sigma}+\text{h.c.})+\varepsilon_\pi\sum_j n^\pi_j$$
$$+K\sum_j (\pi^\dagger_{j\uparrow}\pi^\dagger_{j\downarrow}p_{j\downarrow}p_{j\uparrow}+\text{h.c.}).$$

d,p,π are destruction operators for copper, oxygen $2p_\sigma$ and the second oxygen hole state: oxygen $2p_\pi$ orbitals, respectively. t_{pd},$t_{\pi\pi}$ are the transfer integral between Cu $d_{x^2-y^2}$ and $O2p_\sigma$ orbitals and between the two $O2p_\pi$ orbitals. ε_d,ε_p,ε_π are site energies on Cu $d_{x^2-y^2}$, $O2p_\sigma$ and $O2p_\pi$ orbitals. U_d is the on-site Coulomb repulsion at each Cu site. K is the coupling constant of the pair transfer interaction between the two orbitals at each oxygen site.

Springer Proceedings in Physics, Vol. 70 Computational Approaches in Condensed-Matter Physics
Editors: S. Miyashita · M. Imada · H. Takayama

The temperature dependencies of the s-wave pair-field susceptibilities in a one dimensional lattice have been calculated in various parameter regions.[6-8] Fig.1 shows the susceptibility on O2p_σ site obtained with the parameter set $t_{pd}=t_{\pi\pi}=1$, $U_d=4$, $K=1$, $\varepsilon_p-\varepsilon_d=0$, $\varepsilon_\pi-\varepsilon_d=2$. Occupation numbers on each site at $T=0.1t_{pd}$ are $\langle n\rangle_{Cu}=0.5$, $\langle n\rangle_{p\sigma}=0.7$ and $\langle n\rangle_{p\pi}=0.0$, for copper, O2$p_\sigma$, and O2$p_\pi$ sites, respectively. The result with the same parameters but with different carrier concentration has been published in ref.(8). No holes are doped into the O2p_π sites, in these two cases. We have found an enhancement of the s-wave pair-field susceptibility on O2p_σ sites at low temperature.

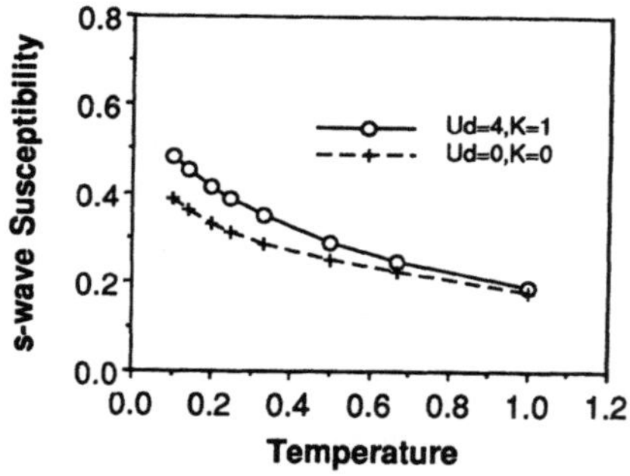

Fig.1 The superconducting susceptibility. The broken line is the result with no Fermion interaction.

This suggests that there may be a superconducting parameter region in this model in the higher dimensions. We have found a similar result in two dimension. When we dope more holes (0.6 holes per unit cell) [8], we find a clearer enhancement. It is sufficient but not necessary that holes are actually doped into the second band for the superconductivity in this model. A similar conclusion has been obtained in a simpler model by a diagonalization study.[9]

4. The two bands model as a model of the doped C_{60}

In the doped C_{60}, intramolecular electronic interactions are expected to be very small . The doped C_{60} may be described by the two band model given as follows:

$$H=\sum_{i=1}^{m} \varepsilon^{i}_{k\sigma}c^{\dagger}_{ik\sigma}c_{ik\sigma}+\sum_{\langle i,j\rangle}\sum_{kk'}K_{ij}(k,k')(c^{\dagger}_{ik\uparrow}c^{\dagger}_{i-k\downarrow}c_{j-k'\downarrow}c_{jk'\uparrow}+h.c.),$$

where $K_{ij}(k,k')$ is a coupling constant for the pair transfer interaction and $c_{ik\sigma}$ is the destruction operator for the i-th band. In this case, the nonadiabatic electron-phonon interaction is important. If we set m=2 and $K_{ij}(k,k')=\pm K$, we find clear enhancements of the s-wave pair-field susceptibilities again in a two dimensional square lattice. Competitions with other instabilities should be studied in a more realistic lattice structure.

5. Conclusions

There are superconducting parameter regions in the dp+π model and the two bands model. These models may be applicable to the copper oxides and the doped C_{60}, respectively.

References

[1] Y.Asai, J.Phys.Soc.Jpn.**58**,3264(1989)
[2] Y.Yoshinari, H.Yasuoka and Y.Ueda, Proc. M²S-HTSCIII: a special issue of Physica C
[3] H.Suhl, B.Matthias and L.R.Walker, Phys.Rev.Lett.**3**,552(1959)
[4] J.Kondo, Progr. Theor. Phys.**29**,1(1963)
[5] Y.Asai, submitted to Phys.Rev.B Rapid Communications
[6] Y.Asai, Proc. M²S-HTSCIII: a special issue of Physica C
[7] Y.Asai, Proc. Int. Workshop on Electronic Properties and Mechanisms in high-T_c Superconductors (Elsevier)
[8] Y.Asai, Proc. 4-th Int. Symp. on Superconductivity (Springer, Tokyo)
[9] K.Yamaji, T.Yanagisawa and K.Harigaya, Proc. M²S-HTSCIII: a special issue of Physica C

Part III

Quantum Spin Systems and Others

Numerical Methods for Quantum Spin Systems

M. Makivić

Department of Physics, The Ohio State University,
Columbus, OH 43210, USA

Abstract. We discuss numerical methods used in studies of quantum spin systems. The Monte Carlo method for quantum spin systems based on Suzuki-Trotter mapping of a D-dimensional quantum system onto a (D+1)-dimensional equivalent classical system is reviewed. We present the development and applications of an algorithm, based on this transformation, which is designed for a hypercube parallel computer. The issues related to the constraints imposed by conservation laws and methods of measurement of physical quantities are dicussed. The application of the Maximum Entropy Method to the world-line spin simulations is outlined. We also give a short review of the results of static and dynamic simulations of non-frustrated Heisenberg antiferromagnet and XY model on a square lattice. A self-consistent t-matrix approximation for magnons is outlined, and a method for numerical determination of single-magnon Green's function is proposed.

1. Introduction

The discovery of high temperature superconductors [1] has brought about a resurgence of interest in two-dimensional quantum antiferromagnets, following P. W. Anderson's suggestion [2] that novel kinds of excitations which exist in strongly fluctuating spin systems may ultimately lead to superconductivity in doped materials such as $La_{2-x}Sr_xCuO_4$. There is a consensus now that a valid theoretical description of spin correlations in the undoped compound La_2CuO_4 is provided by the spin-1/2 antiferromagnetic Heisenberg model (AFHM) [3,4], which is the strong coupling limit of the Hubbard model at half-filling (*i.e.*, with charge fluctuations integrated out).

Besides Heisenberg antiferromagnet, XY model has been one of the standard paradigms of statistical mechanics. The classical 2D XY model exhibits a finite temperature, topological defect driven phase transition to a state with no broken symmetry, but with a diverging correlation length. This is the celebrated Kosterlitz-Thouless transition [5]. It is important to understand the effects of quantum fluctuations on a marginal transition like this one [6-9].

Heisenberg model is the isotropic limit ($\lambda = 1$) of the Hamiltonian:

$$H = J \sum_{\langle ij \rangle} (S_i^x S_j^x + S_i^y S_j^y + \lambda S_i^z S_j^z) \tag{1}$$

while the XY model corresponds to $\lambda = 0$. The summation goes over all the

Springer Proceedings in Physics, Vol. 70 Computational Approaches in Condensed-Matter Physics
Editors: S. Miyashita · M. Imada · H. Takayama

nearest neighbor pairs on a square lattice and $\mathbf{S}_i$ is the spin operator at the i-th site. The energy scale is set by the effective exchange coupling J.

Two-dimensional systems in the extreme quantum limit are an excellent target for simulations. The exact results, which are available for one-dimensional systems, are virtually non-existent. On the other hand, unlike three-dimensional systems, it is difficult to design reasonably controlled perturbative approaches, since strong fluctuations due to low dimensionality are coupled with pronounced quantum fluctuations. Monte Carlo methods have been successfully applied to study both ground state and finite temperature properties of two-dimensional spin models [10-18].

Recent gains in computational power, particularly the advance of parallel supercomputers, made it feasible to perform simulations on systems which are almost two orders of magnitude larger than those achieved before [18]. This paper will review the development and applications of a Monte Carlo algorithm, based on a generalized path integral approach to quantum statistics, for a hypercube parallel computer. The algorithm is designed with speed and efficiency on this particular architecture in mind [19]. Recently developed maximum entropy method for extracting real time dynamical properties from imaginary time data will be outlined. Finally, the results of simulations of Heisenberg and XY models will be discussed.

We will also present a method for complementing Monte Carlo studies of dynamics with "numerical many body theory" using non-trivial self-consistent approximations [20]. This approach was initiated by Bickers and Scalapino in the context of the Hubbard model [21].

2. Computational Method

2.1. The Suzuki-Trotter Transformation

The Suzuki-Trotter method [22] is based on M. Suzuki's generalization of the Trotter formula for exponential operators. The first step towards obtaining a controlled approximation of the density operator $\rho = exp(-\beta H)$ $(\beta = 1/k_B T)$ is to write the exponential as a product of exponentals with "small" arguments, by dividing the whole imaginary time interval β into small time slices.

In the next step, one approximates the short-term propagators in such a manner that the matrix elements are readily evaluated. There is a considerable freedom in choosing the approximation, but they can be classified according to their error dependence on the number of time slices M (Trotter number), for large values of M.

This can be formally written as [22]:

$$e^{-\beta H} = (e^{-\beta H/M})^M = lim_{M\to\infty}(f_n[-\beta H/M])^M \quad , \tag{2}$$

where the n-th order approximant f_n satisfies: $\exp[-\beta H/M] = f_n[-\beta H/M] + O(1/M^n)$. This relation then implies for the partition function:

$$Z = Tre^{-\beta H} = Tr(f_n[-\beta H/M])^M + O(\frac{1}{M^{n-1}}) = Z_n + O(\frac{1}{M^{n-1}}) \quad . \tag{3}$$

It can be shown that for the exponent of a sum of operators [22]:

$$f_2[\sum_{j=1}^{p} A_j] = \prod_{j=1}^{p} e^{A_j} \quad . \tag{4}$$

It is not difficult to generate much higher order approximants.

It was proved by Suzuki that, under rather general conditions, the approximant for the partition function is an even function of M [23], so that $Z = Z_{2n}(M) + O(1/M^{2n})$. A similar result is valid if one calculates average values of operators which satisfy $Q^T = Q$, namely

$$< Q >= Tr(Q[f_{2n}(\{A_j\})]^M)/Z_{2n}(M) + O(1/M^{2n}) \quad . \tag{5}$$

Therefore, the systematic error of using a finite value of M, which is what one has to do in practice, is of the order of $1/M^2$ or higher [24]. The exact result then may be obtained by calculating the quantities of interest for a few large values of M, and then extrapolating to $M \rightarrow \infty$, using the formula: $Q(M) = Q(\infty) + a/(MT)^2 + b/(MT)^4 + ...$, where $Q(\infty)$ is the correct value.

To use equation (4) we break the Hamiltonian into 4 pieces [18,22], $H = \sum_{i=1}^{4} H_i$, each containing a commuting subset of nearest neighbor bonds on a square lattice. H_1 contains odd bonds in x-direction, H_2 contains odd bonds in y-direction, H_3 contains even bonds in x-direction, and H_4 contains even bonds in y-direction. The subHamiltonians H_i themselves do not commute, but it is important that their constituent bonds do. Then,

$$Z = \mathrm{Tr} \lim_{M\rightarrow\infty} (\prod_{i=1}^{4} U_i)^M = \lim_{\Delta\tau\rightarrow 0} \sum_{\{C\}} \langle C_1|U_1|C_2\rangle \cdots \langle C_{4M}|U_4|C_1\rangle \quad , \tag{6}$$

where M is the Trotter number, $\Delta\tau = \beta/M$, $U_i = e^{-\Delta\tau H_i}$ and C_i denote $4M$ resolutions of unity inserted between adjacent exponentials.

The intermediate states, C_j, which are chosen to diagonalize S_z, may be regarded as belonging to different time slices. Therefore, the original 2D quantum system of size LxL is mapped onto equivalent 3D classical system of size $LxLx4M$, where the labels of intermediate quantum states, being ordinary numbers, now serve as classical degrees of freedom. The propagators U_j decompose into products of exponentials involving only 2 spins on a bond [25]. Hence, with this approximation of the short-term propagators one has to solve only a two-body problem. We have to calculate 16 matrix elements of the operator $\rho_{ij} = \exp[-J\Delta\tau(S_i^x S_j^x + S_i^y S_j^y + \lambda S_i^z S_j^z)]$, between two adjacent time slices, τ and $\tau+1$: $\rho_{ij}(\tau, \tau+1) = \langle S_{i,\tau}^z S_{j,\tau}^z|\rho_{ij}|S_{i,\tau+1}^z S_{j,\tau+1}^z\rangle$, where spatial indices i, j correspond to the particular bond. The four spin labels associated with this transfer matrix element may be identified with four classical spins sitting on the corners of an interacting 4-spin plaquette, which is then the basic building block of the interaction in the equivalent classical system.

The "local density matrix" $W = \rho_{ij}$, which, in this context is the "local Boltzman factor," is constrained by conservation laws, unlike real classical system. In the models with x-y and full rotational symmetry, S_z is conserved. Because of that, it is easy to see that only 6 out of 16 spin configurations of a 4-spin plaquette have nonzero Boltzmann weights. In the case of XY model, it is sometimes advantageous to work in the basis which diagonalizes S_y. Since S_y is conserved only modulo 2, there will be 8 nonzero matrix elements. The off-diagonal matrix elements of an antiferromagnet are negative. To interpret them as Boltzman factors, one has to render them positive by performing a canonical transformation on one of the sublattices, say B, [26]: $H \Rightarrow UHU^\dagger$ where the unitary operator U=exp($i\pi \sum_{i\in B} S_i^z$) maps $S^x \Rightarrow -S^x$ and $S^y \Rightarrow -S^y$.

2.2. The Updating Procedure

Special care has to be exercised to ensure ergodicity for quantum problems [27] due to complications arising from conservation laws. For example, a single spin flip *always* generates a configuration with infinite energy. It is not difficult to see that one has to flip a *closed loop of spins*. This would be sufficient in a case where there are 8 nonzero matrix elements (*e.g.*, XY model in S_y representation). In the Heisenberg case, the conservation of S_z imposes the following constraints. Two spins on a horizontal plaquette edge (same time slice) are updated only if they are opposite. Two spins on a vertical edge of an interacting plaquette are updated only if they are the same.

The ergodicity requirement then reduces to the problem of generating all possible closed loops. This could be done stochastically, in principle, by generating loops of arbitrary shape and length on a computer, but it is not efficient. A better way to generate all possible closed loops under these constraints is to exponentiate a set of 4 types of elementary updates which mimic the generators of the fundamental homotopy group on a torus, with the conservation law built in. The toroidal geometry is the consequence of spatial and temporal periodic boundary conditions.

We have two types of local updates and two types of global updates [18]. The local updates are homotopic to the unit loop. There are two types of smallest possible loops which are bounded by edges of interacting plaquettes. One such loop belongs on a sigle time slice, while the other one spans four time slices. They are shown in Figures 1 and 2. Due to the constraint, one may attempt to flip the spins on these loops only if they are in a Néel configuration. The global update in time direction is shown in Fig. 3. One searches for a straight line of all up or all down spins, and flips them all. This move is responsible for generating fluctuations of the total magnetization. In Fig. 4a, the other type of global update, extending in spatial directions, is shown.

By employing the connection between spin and boson algebra this updating procedure can be given a useful alternative interpretation. Using a projector $P_i(0)$ onto a state with 0 particles on site i, we defined the restricted Bose operators $a_i = P_i(0)b_i$, $a_i^\dagger = b_i^\dagger P_i(0)$ which vanish outside the subspace with 0 or 1 particle. After the identification $|1> \Rightarrow |+>$ and $|0> \Rightarrow |->$, one has the equivalence $a^\dagger \Rightarrow \sigma_-$ and $a \Rightarrow \sigma_+$, where σ_- and σ_+ are spin lowering and raising operators. Since $N = a^+a = (1-\sigma^z)/2$, the conservation of S_z may be interpreted as number conservation in this language.

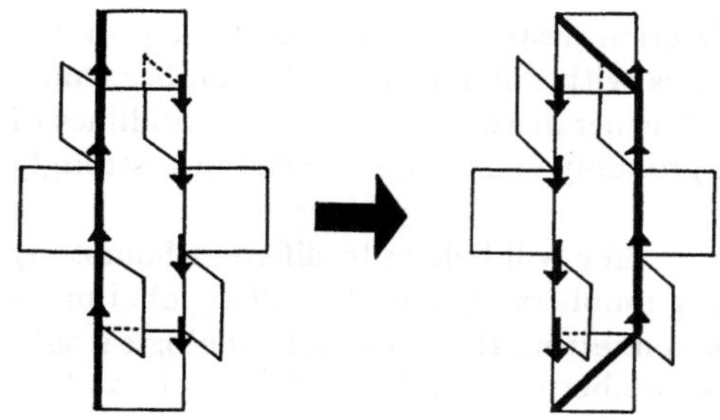

*Fig.*1. A local flip distorts a worldline.

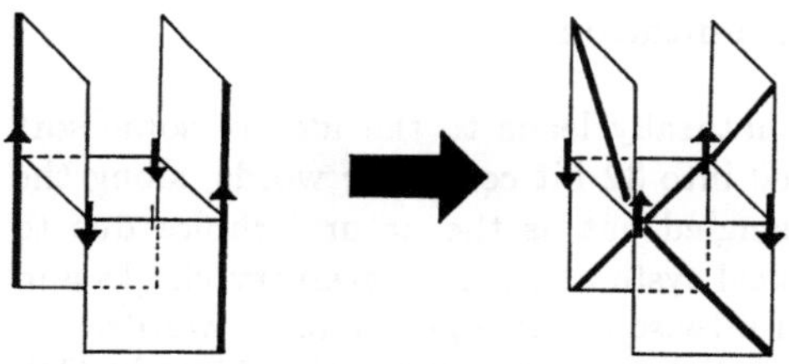

*Fig.*2. The other local flip twists two worldlines around one another.

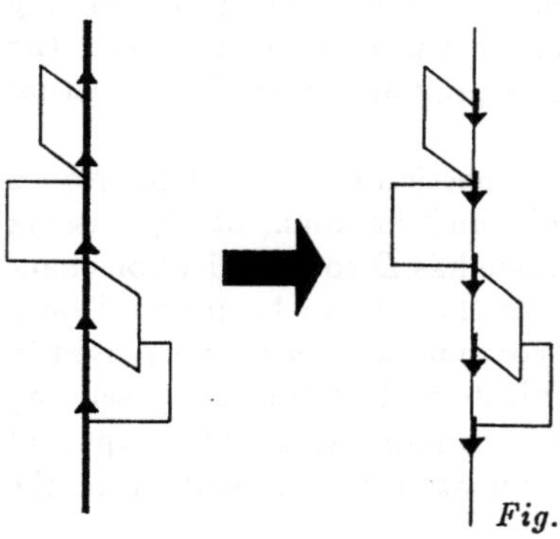

*Fig.*3

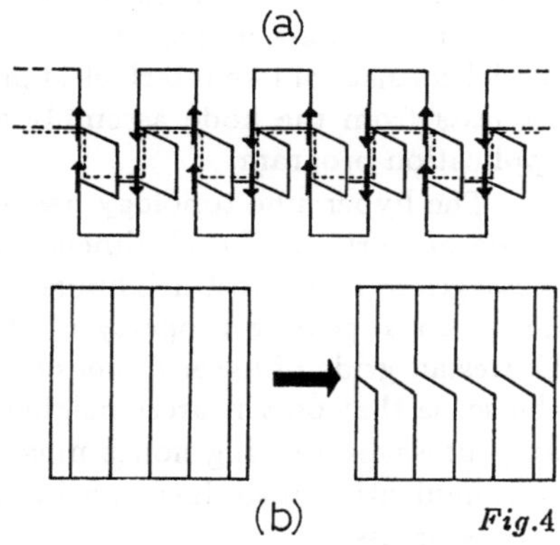

*Fig.*4

*Fig.*3. A global flip in the imaginary-time direction.

*Fig.*4. (a) A global flip in spatial directions. The dashed line indicates the string of spins being flipped. (b) A configuration with different winding number is reached.

By connecting the sites with spins down (or up, due to particle-hole symmetry), one obtains the worldlines of bosons with infinite on-site repulsion propagating in imaginary time. The updating of the spin system is equivalent to generating all possible configurations of worldlines.

The elementary moves have a nice interpretation in the boson worldline picture. One of the local flips just distorts a worldline (Fig. 1). The effect of the other local flip is to exchange and wrap two nearby worldlines around one another (Fig. 2).

The global move in the time direction destroys or creates static particles (holes) (Fig. 3). It connects subspaces with different particle number (magnetization). This global move is easily generalized to tracing the worldlines of any shape (mobile particles), which probably cannot be avoided in a strongly fluctuating system.

Since the worldlines are on a torus they will belong to different homotopy classes, characterized by their winding numbers N_x and N_y. The global move in spatial directions is responsible for connecting these distinct topological subspaces. It breaks and reconstructs a number of worldlines of order L, and its effect is shown in Fig. 4b.

3. Algorithm and Parallel Implementation

Ising spins are 1-bit objects, which naturally leads to the idea of some sort of spin packing. The spins are packed into 32 bit computer words, along the imaginary time direction, which is singled out as the natural choice due to the anisotropy of the equivalent classical system. All the necessary checks and updates can be implemented through bitwise logical operations on words.

The code was adapted for a parallel supercomputer, the 64-node Caltech/JPL MarkIIIfp Hypercube [19] at Caltech. MarkIIIfp is an assembly of rather sophisticated processors (nodes) which are connected as a 6-dimensional hypercube and are managed by a separate computer (host). The programming model consists of two interlocked programs: the host program which serves the requests from the node assembly and the node program which is the actual application program.

The hypercube topology has a rich interconnection structure [19]. Nodes reside on vertices of a D dimensional hypercube and communicate by passing messages to its nearest neighbors. Thus each node has D communication channels. An appealing property of the topology is that it can be mapped onto Cartesian grids of lower dimensionality. The communactions are very rigid in the sense that only nearest neighbors communicate and both message-sending and message-receiving nodes must expect the communication. This expected communication introduces an implicit synchronization in the working of the processor assembly.

The nearest neighbor interaction in the spin system allowed for an efficient parallelization [18]. The node system is first mapped onto a 2D grid and then configured as disconnected rings, by selecting a subset of channel masks. Each processor uses only 2 channel masks to communicate with processor to the left and the one to the right in its ring. Each ring corresponds to an independent system, so trivial parallelism is achieved just by configuring the node assembly in this way.

Each processor is assigned a piece of the 3D lattice, consisting of a number of (x,τ)-planes, as shown in Fig. 5. In other words, the system is divided into slabs stacked along the y-direction. The communication between processors is required only if the boundary spins are updated and involves only nearest neighbor processors. The ring decomposition is optimal because it minimizes

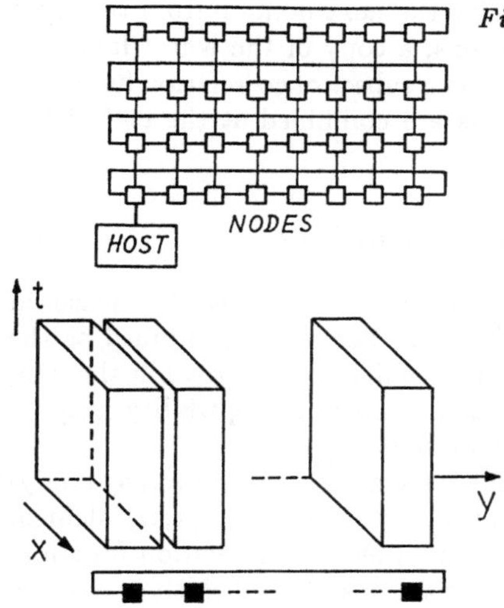

Fig.5. The decomposition of the physical space of a lattice among the nodes in a ring.

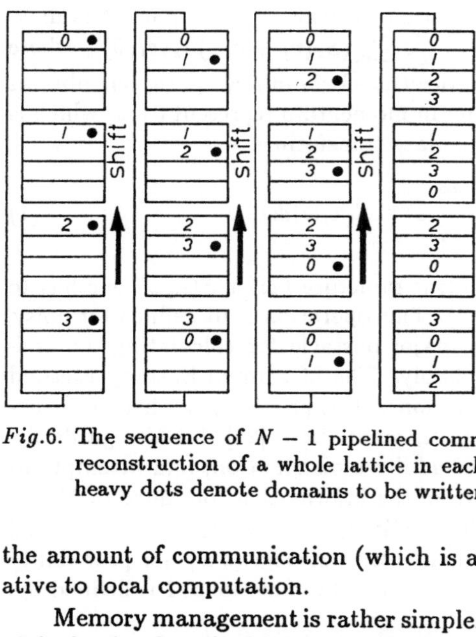

Fig.6. The sequence of $N-1$ pipelined communication requests executed during the reconstruction of a whole lattice in each node of an (N=4) processor ring. The heavy dots denote domains to be written to a node's neighbor.

the amount of communication (which is a bottleneck in this architecture) relative to local computation.

Memory management is rather simple. Each node operates on its own piece of the lattice, but the lattice coordinates are shifted according to the position of the node in a ring, so that a node's own domain begins with coordinate $y = 0$. The measurements of correlation function involve global data exchanges, which are handled in a pipelined manner for the ring decomposition (Fig. 6). In a single communication call, all nodes send their domains to their left neighbor, and receive the domain of the right neighbor. This domain is stored adjacent to processor's own domain, and, in the next communication call they forward that domain to their left neighbor and receive the next domain from the right neighbor. Hence, after $N-1$ communications, a copy of the whole lattice is stored in every node. Each node then computes a subset of correlation functions on the whole spin system and partial results are combined at the end of the simulation.

4. Measurement of Static and Dynamic Quantities

There are, in general, two types of quantities that can be easily measured with a path integral Monte Carlo method: those that can be obtained as the derivatives of the partition function, and those which are diagonal in the chosen representation. Energy, specific heat and uniform susceptibility may be calculated as the derivatives of the partition function [25].

The most detailed information about the system may be acquired by studying dynamic spin correlations. The dynamic structure factor $S(\mathbf{k},\omega)$, defined as the Fourier transform of the time-dependent spin-pair correlation function, contains important information about the excitations. It is directly measured in inelastic neutron scattering experiments.

Unfortunately, in a path-integral-like simulation, the quantity that is easily measured is the *imaginary* time correlation function $S(\mathbf{r},\tau)$, since it is diagonal in S_z. It is defined as: $S(\mathbf{r},\tau) = \frac{1}{L^2}\sum_{\mathbf{r}_j}\langle S_z(\mathbf{r}_j+\mathbf{r},\tau)S_z(\mathbf{r}_j,0)\rangle$, where $S_z(\mathbf{r},\tau) = e^{H\tau}S_z(\mathbf{r})e^{-H\tau}$. After a Fourier transform in real space, one obtains the imaginary time correlation function in momentum representation, which is related to the structure factor via a Laplace transform:

$$S(\mathbf{k},\tau) = \frac{1}{2\pi}\int_{-\infty}^{\infty} e^{-\omega\tau} S(\mathbf{k},\omega)d\omega \quad . \tag{7}$$

To calculate the frequency-dependent structure factor $S(\mathbf{k},\omega)$ one has to invert the Laplace transform of Eq. (7). This operation is ill-defined for noisy, finite set of data, and used to be a major obstacle for calculating dynamical properties of quantum systems reliably. The Maximum Entropy Method (MEM) was recently suggested as a solution to this problem, and has been succesfuly applied to a variety of Monte Carlo simulations [28]. MEM regards the dynamic structure factor as a positive additive distribution, and attempts to infer the most adequate distribution from the information contained in Monte Carlo data and the prior knowledge about $S(\mathbf{k},\omega)$. Following general statistical inference and information-theoretic arguments, it associates a prior probability $e^{\alpha I[S]}$ with a particular choice for $S(\mathbf{k},\omega)$, where the functional I is defined by:

$$I = \int_0^{\infty} [-S(\mathbf{k},\omega)\log(S(\mathbf{k},\omega)/m(\mathbf{k},\omega)) + S(\mathbf{k},\omega) - m(\mathbf{k},\omega)]d\omega \quad . \tag{8}$$

This entropic functional is defined with respect to the measure $m(\mathbf{k},\omega)$, called the default model. The default model may be used to incorporate prior knowledge about the structure factor, beyond its positivity. The χ^2 measure, with the assumption of independent Gaussian errors, is used to calculate the likelihood

$e^{-\chi^2/2}$ of the outcome of Monte Carlo simulation, $S(\mathbf{k},\tau)$, given a particular $S(\mathbf{k},\omega)$. Special care in the error analysis must be exercised to ensure that the results are indeed uncorrelated and Gaussian distributed. The posterior probability of $S(\mathbf{k},\omega)$, given the Monte Carlo data, follows from Bayes theorem: $P \propto e^{\alpha I-\chi^2/2}$. The MEM image $S(\mathbf{k},\omega)$ maximizes this posterior probability. The parameter α, which weighs the relative importance of the misfit statistic χ^2 and entropy I, may be determined by maximizing its own posterior probability [28]. Alternatively, one can also integrate α out of the problem by calculating the average $S(\mathbf{k},\omega)$, weighted by the posterior probability of α. The advantage of the latter procedure is that we then obtain the posterior probability $P[S]$ of the image $S(\mathbf{k},\omega)$ given just the data and the model. The knowledge of $P[S]$ may be used to pick the "best" model for a given data set.

We experimented with a default model where the relaxation function $F(\mathbf{k},\omega)$, defined by $S(\mathbf{k},\omega) = \frac{1}{2}\chi(\mathbf{k},T)[\beta\omega/(1-e^{-\beta\omega})]F(\mathbf{k},\omega)$, is assumed to be a sum of Lorentzians centered at excitation frequencies [29] and, also, we used the result of Schwinger boson mean field theory [30]. However, the best model turns out to be the sum rule default described below. The images generated by the sum rule default have a few orders of magnitude higher posterior probabilities than the other two. The sum rule default $m(\mathbf{k},\omega)$ is obtained by maximizing the information-theoretic entropy, $\int_0^\infty m(\mathbf{k},\omega)\log[m(\mathbf{k},\omega)]d\omega$, subject to the detailed balance condition $m(\mathbf{k},-\omega) = e^{-\beta\omega}m(\mathbf{k},\omega)$ and the three sum rules [31], $\int \omega^n S(\mathbf{k},\omega)d\omega = 2\pi A_n(\mathbf{k})$. The sum rule quantities are static and can be directly and accurately measured in the simulation: $A_{-1} = \chi(\mathbf{k})$ (static momentum-dependent susceptibility), $A_0(\mathbf{k}) = S(\mathbf{k},\tau=0)$, and $A_1(\mathbf{k}) = -(<E>/3N)[2-\cos(k_x)-\cos(k_y)]$ (energy sum rule). This leads to:

$$m(\mathbf{k},\omega) = e^{-1}\exp\{-\lambda_0[1+e^{-\beta\omega}] - \lambda_1[1-e^{-\beta\omega}]\omega - \lambda_{-1}[1-e^{-\beta\omega}]/\omega\} \quad ,(9)$$

where the Lagrange multipliers $\lambda_i(\mathbf{k})$ are determined numerically from the sum rules [32].

If the dynamic structure factor has unimodal character, spin wave dispersion and damping may be computed from the frequency moments of $F(\mathbf{k},\omega)$ [29]. The spin wave frequency is calculated as $\omega_\mathbf{k} = \int_0^\infty \omega F d\omega / \int_0^\infty F d\omega$, while the damping is obtained as: $\gamma_\mathbf{k}^2 =<\omega^2> -\omega_\mathbf{k}^2$, where $<\omega^2>= \int_{-\infty}^\infty \omega^2 F d\omega / \int_{-\infty}^\infty F d\omega$.

For the XY model we calculated additional quantities. The topological order in the model may be quantified by vortex density. One definition of total density of vortices and antivortices is due to Betts [33]:

$$<V>=\frac{1}{4}(1-2<\sigma_1^x\sigma_3^x>+<\sigma_1^x\sigma_2^y\sigma_3^x\sigma_4^y>) \quad , \qquad (10)$$

where the spins are labeled in a clockwise direction along an elementary plaquette on the square lattice. This operator cannot be simultaneously diagonalized by a single basis for the whole lattice, but it is fortunate that it has the same selection rules as the propagator product, U_1U_2 or U_3U_4. It can be shown after some algebra that one has to measure the expectation value of the quantity $\langle C_\tau | V U_1 U_2 | C_{\tau+2\Delta\tau}\rangle / \langle C_\tau | U_1 U_2 | C_{\tau+2\Delta\tau}\rangle$ between a pair of time slices which are $2\Delta\tau$ apart [34].

The anticipated Kosterlitz-Thouless transition is characterized by a universal jump in the spin stiffness at T_c [35]. One can measure spin stiffness (ρ_s) using the relation between ρ_s and winding number fluctuations [36], $\rho_s = \frac{1}{2} k_B T \langle W^2 \rangle$. The winding number, for a system of linear size L, is defined as: $\sum_{i=1}^{N} (\mathbf{r}(\beta) - \mathbf{r}(0)) = L\mathbf{W}$, where $\mathbf{r}(0)$ is the position of a worldline at $\tau = 0$, $\mathbf{r}(\beta)$ is its position at $\tau = \beta$, and the summation is performed over all worldlines.

5. Overview of results

The most important conclusions from the simulations of Heisenberg antiferromagnet concern the nature of spin correlations. The low energy sector of the model is very well described by the quantum generalized non-linear σ model, as proposed by Chakravarty *et al* [37]. Their description leads to an essentially classical picture, with the parameters of the theory renormalized by quantum fluctuations. This is a consequence of Néel ordering in the ground state. The exponential divergence of the correlation length is found to be quantitatively well described by the renormalized classical picture at 2-loop level of approximation [18]. From the divergence of correlation length, $\xi(T) = A\exp(2\pi\rho_s/T)$, we deduced the value of quantum renormalized spin stiffness, $\rho_s/J = 0.199(2)$. By comparing the computed correlation lengths with neutron scattering experiments on La_2CuO_4 [38], we obtained the value of the effective exchange coupling $J = 1450 \pm 30\ K$.

By combining Monte Carlo with the maximum entropy method we calculated the dynamic structure factor $S(\mathbf{k}, \omega)$ on lattices of up to 64 X 64 spins [39]. Dynamics is governed by renormalized damped spin waves and we calculated their frequencies (Fig. 7) and linewidths. Short wavelength ($\lambda \ll \xi$) propagating modes are supported by local ordering on scales comparable to the correlation length ξ, and at low temperatures their dispersion is rather accurately given by the spin wave theory, while the behavior in the immediate vicinity of the zone boundary is diffusive. At long wavelengths and low frequencies, the system exhibits scaling behavior. The scaling plot of spin wave frequencies is shown in Fig. 8. Results qualitatively agree with the scaling theory based on the non-linear sigma model, except for the characteristic frequency scale, related to the width of the quasielastic mode, which is found to be inversely proportional to the correlation length.

Simulations of the XY model show rather convincingly that the extreme quantum limit is in the Kosterlitz-Thouless universality class. The correlation length and susceptibility can be very well fitted with the essential singularity $\propto e^{1/(T-T_c)^{1/2}}$ typical of a KT transition, while the specific heat remains finite [9]. Spin stiffness rises sharply below the transition temperature, and agrees, within statistical errors, with the universal jump prediction [34]. The exponent η which governs correlations below T_c, also agrees with the universal value $\eta(T_c) = 1/4$. We have very good agreement between the values of η deduced from correlation

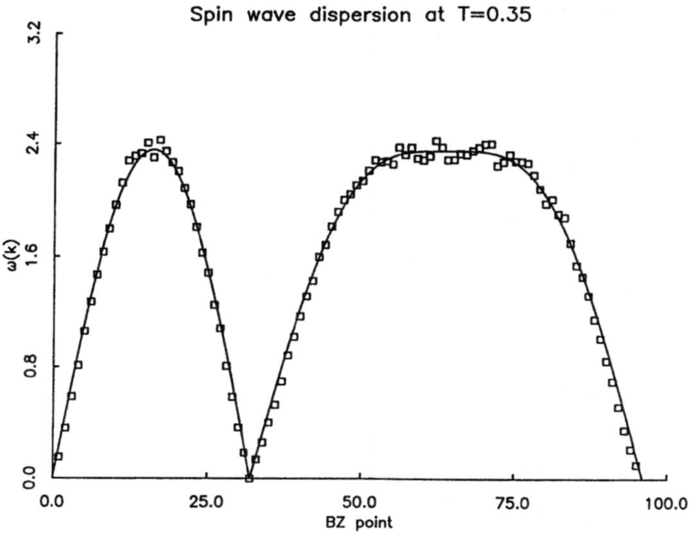

*Fig.*7. Spin wave frequencies at $T = 0.35J$. The solid line is the $T = 0$ spin wave dispersion, uniformly renormalized by the quantum factor of $Z_c = 1.18$. The integers on the abscissa enumerate points in the Brillouin zone of a 64X64 system, along the triangle spanned by $\mathbf{k}_1 = (0,0)$, $\mathbf{k}_2 = (\pi,\pi)$ and $\mathbf{k}_3 = (\pi,0)$.

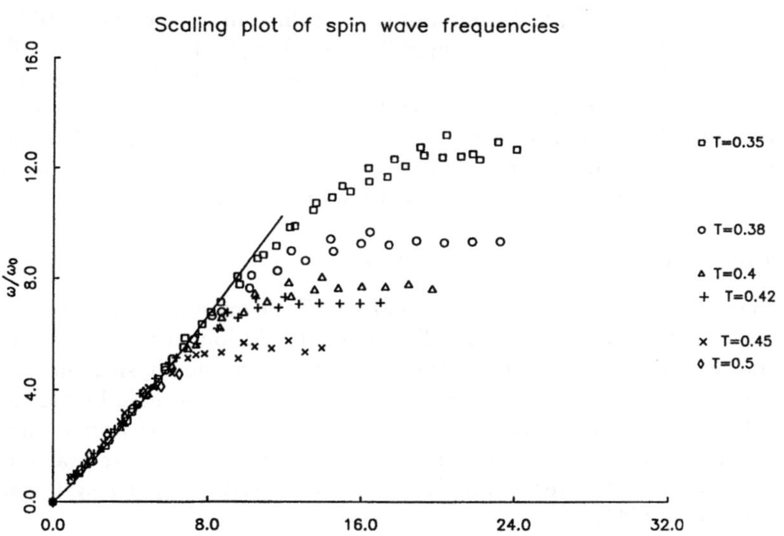

*Fig.*8. The scaling plot of spin wave frequencies $\omega_{\mathbf{q}} = \omega_0(T)\Omega(\mathbf{q}\xi)$.

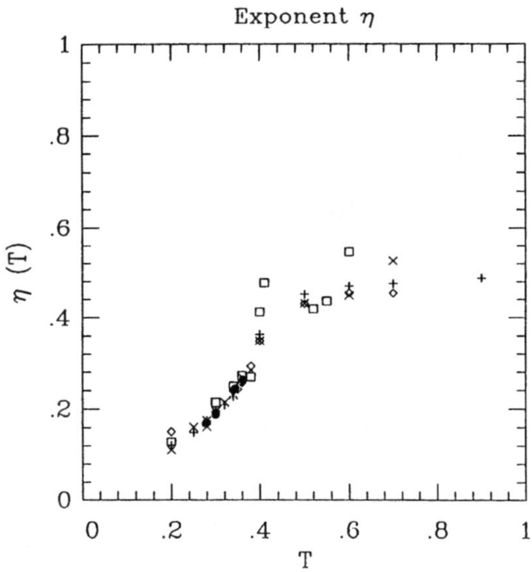

Fig.9. The exponent η obtained from $C(r)$ fits and from spin stiffness via $\eta(T) = T/(2\pi\rho_s(T))$ (solid symbol). Above T_c, we measured Ornstein-Zernike exponent, $\lambda \simeq 1/2$.

function fits, $C(r) \propto r^{-\eta(T)}$, and from finite size scaling analysis of susceptibility below T_c, $2 - \eta(T) = \lim_{L_1,L_2\to\infty} \ln(\chi(T, L_1)/\chi(T, L_2))/\ln(L_1/L_2)$ [34,40]. In addition, assuming that the low temperature phase is a spin wave fixed line, an exact relation between ρ_s and η follows: $\eta(T) = T/(2\pi\rho_s(T))$ [35], which is verified within the statistical acuracy of the simulation (Fig. 9) [34]. The transition temperature also marks a steep rise in vortex density, which is in agreement with the idea of a vortex dissociation driven transition.

6. Self-consistent treatment of spin waves

As a complementary method to dynamical Monte Carlo simulations, we would like to obtain numerically non-trivial self-consistent solutions for spin correlations. These are non-perturbative solutions built on top of the physical picture of spin wave (magnon) excitations, the validity of which is confirmed by Monte Carlo calculations. We ultimately hope to be able to calculate the Raman line shape. This is particularly interesting in view of the fact that two-magnon scattering treated in RPA approximation works very well for spin-1 systems, but fails for spin-1/2 [41].

We will describe the method for the ferromagnetic Heisenberg hamiltonian, because it is technically simpler, but it can be easily adapted to the antiferromagnet. The first step is to map the spin hamiltonian onto a boson hamiltonian using the Dyson-Maleev transformation [42]: $S_j^+ = \sqrt{2S}(1 - b_j^\dagger b_j/(2S))b_j$, $S_j^- = \sqrt{2S}b_j^\dagger$, $S_j^z = S - b_j^\dagger b_j$, and ignoring the kinematic con-

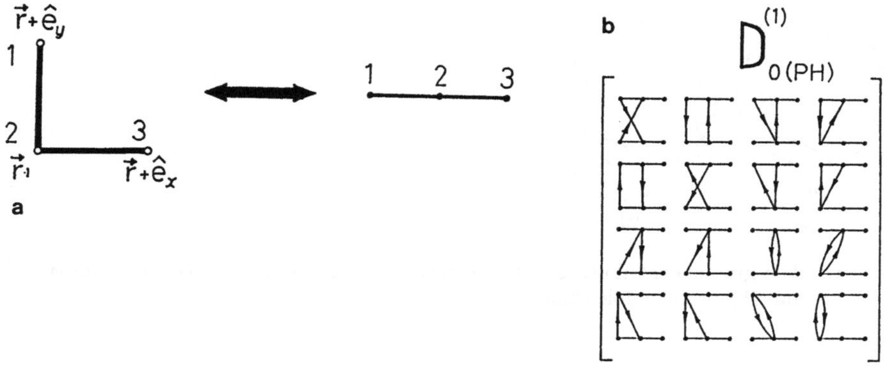

*Fig.*10. (a) A bond cluster involves two bonds attached to a lattice site. The schematic representation of a bond cluster used in (b) is also shown. b) The first quadrant of the propagator D^0_{PH} matrix between two bond clusters. The other quadrants are obtained by exchanging bonds in a cluster.

straint $n_i \leq 2S$ [42]. One then carries out finite temperature perturbation theory for bosons, treating the quartic term as a perturbation ($K = JS$):

$$H = K\sum_{\langle ij\rangle}(b_i^\dagger b_j + b_i b_j^\dagger - n_i - n_j) + \frac{K}{S}\sum_{\langle ij\rangle}(\frac{1}{2}(b_i^\dagger b_i b_i b_j^\dagger + b_i^\dagger b_j^\dagger b_j b_i) - n_i n_j). \quad (11)$$

It is customary at this stage to switch to momentum space where the quadratic (spin wave) part is diagonal. However, the bare interaction vertex acquires non-local and non-trivial dependence on momentum variables. This should be contrasted to the bare vertex in the Hubbard model, which is constant in momentum space due to contact interaction. This feature forces us to do perturbation theory in real space.

The basic constituents of the perturbation theory are matrices of interaction vertices in particle-particle $V^{\alpha\beta}_{PP}$ and particle-hole $V^{\alpha\beta}_{PH}$ channels and two-particle propagators in particle-particle $D^{\alpha\beta}_{PP}(\mathbf{r}\tau, \mathbf{r}'\tau')$ and particle-hole $D^{\alpha\beta}_{PH}(\mathbf{r}\tau, \mathbf{r}'\tau')$ channels. The "bare" two-particle propagators are denoted by D^0. These objects are defined on bond clusters (Fig. 10a). The spatial arguments in two-particle propagators denote the location of the central site of a bond cluster in a discretized space-time mesh of the system, while the matrix indices correspond to different processes that are allowed on a given bond cluster. We show explicitly the "bare" particle-hole propagator matrix in Fig. 10b. We put quotation marks around bare and dressed to remind that all single particle Green's functions which appear in the graphs are actually self-consistently determined dressed Green's functions.

The short range of the interaction in real space makes it possible to reduce the number of degrees of freedom and to consider only processes where an excitation pair originates on one bond cluster and terminates on another. Using translational invariance, the memory required to store a two-particle propaga-

*Fig.*11. The graphs contributing to the generating potential $\Phi[G]$ in (a) particle-particle and (b) particle-hole channel.

tor scales as $\propto N$, where N is the number of points in the spacetime mesh. In momentum space, memory requirement scales as $\propto N^3$. Thus, switching to real space makes the difference between managable and non-managable.

It was shown by Baym and Kadanoff [43] that a conserving self-consistent approximation for the single particle Green's function may be obtained if the self-energy takes the form of a perfect functional derivative $\Sigma(1,2) = \delta\Phi/\delta G(2,1)$ where $\Phi[G]$ is a generating functional and $1 = (\mathbf{r}_1, \tau_1)$, $2 = (\mathbf{r}_2, \tau_2)$. In the t-matrix approximation, the generating functional $\Phi[G]$ contains contributions shown in Fig. 11, which may be in the particle-particle (a) or particle-hole (b) channels.

The corresponding contributions to Σ are given below. Note that "second" order contribution cannot distinguish between the two channels. We use condensed notation, $\mathbf{x} = (\mathbf{r}, \tau)$, Q stands for PP or PH, and ϵ is $+1(-1)$ in $PH(PP)$ channel:

$$\begin{aligned} \Sigma^{(2)}(\mathbf{x}) &= \sum_{\alpha\delta} V_{PP}^{\alpha\beta} T^0(\mathbf{x};0) V_{PP}^{\gamma\delta} G(-\mathbf{x}) \ , \\ \Sigma_Q(\mathbf{x}) &= \sum_{\alpha\delta} V_Q^{\alpha\beta} [D_Q(\mathbf{x};0) - D_Q^0(\mathbf{x};0)] V_Q^{\gamma\delta} G(\epsilon\mathbf{x}) \ . \end{aligned} \tag{12}$$

"Dressed" two-particle propagators are solutions of the equations:

$$D_Q(\mathbf{x},0) = D_Q^0(\mathbf{x};0) + \sum_{\mathbf{y}} D_Q^0(\mathbf{x};\mathbf{y}) V_Q^{\gamma\delta} D_Q(\mathbf{y},0) \ . \tag{13}$$

To obtain a closed set of equations, we need Dyson's equation for the single particle propagator, $G = G^0 + G^0 \Sigma G$. The equations (12) and (13) and Dyson's equation can now be solved iteratively on a computer until self-consistency. Fast Fourier transforms at every iteration are used to switch to momentum space where the last two are easily solved, and then to move back to real space where D^0 are updated. We have also included chemical potential which is used to keep average boson density at 1/2 per site.This idea was previously used by Takahashi [44] in his variational spin wave theory and it is an ad hoc device to enforce Mermin-Wagner theorem on the average.

7. Goals

Spin wave spectrum will be obtained and compared with the results of Monte Carlo calculations. The computation will be extended to include multi-spin correlation functions which determine the Raman spectrum. The results from self-consistent spin wave calculations may be used to formulate better default models, particularly at high frequencies, where little information comes from Monte Carlo data. We will also apply the method to other systems with short range interactions, e.g. weakly interacting bose systems and extended Hubbard model.

8. Acknowledgments

The author is thankful for valuable interactions with F. Buda, D. L. Cox, M. C. Cross, H.-Q. Ding, G. C. Fox, S. Homma, M. Jarrell, S. Miyashita, Y. Okabe, and P. B. Weichman. I would also like to express my gratitude to S. Homma, S. Miyazawa and other members of The Faculty of Electrical Engineering at Gunma University, where a part of this work was completed, for their hospitality. This work is supported in part by the NSF grant No. ASC-9015000.

References

[1] J.G. Bednorz and K.A. Müller, Z.Phys. **B64**, 189 (1986).
[2] P.W.Anderson, Science **235**, 1196 (1987).
[3] S.Chakravarty, in *The Los Alamos Symposium Proceedings-1989, High Temperature Superconductivity*, p. 136, (Addison-Wesley, 1990).
[4] E. Manousakis, Rev. Mod. Phys. **63**, 1 (1991).
[5] J.M. Kosterlitz and D.J. Thouless, J. Phys. **C6**, 1181 (1973); J.M. Kosterlitz, J. Phys. **C7**, 1046 (1974).
[6] M. Suzuki, S. Miyashita, A. Kuroda and C. Kawabata, Phys. Lett. **A60**, 478 (1977).
[7] H. De Raedt, B. De Raedt, J. Fivez and A. Lagendijk, Phys. Lett **104A**, 430 (1984); H. De Raedt, B. De Raedt and A. Lagendijk, Z. Phys. **B57**, 209 (1984).
[8] E. Loh, Jr., D. J. Scalapino and P. M. Grant, Phys. Rev. **B31**, 4712 (1985).
[9] H.-Q. Ding and M. Makivic, Phys. Rev. **B42**, 6827 (1990).
[10] J. D. Reger and A. P. Young, Phys. Rev. B **37**, 5978 (1988).
[11] M. Gross, E. Sanchez-Velasco, and E. Siggia, Phys. Rev. B **39**, 2484 (1989).
[12] S. Miyashita, J. Phys. Soc. Jpn. **57**, 1934 (1988).
[13] Y. Okabe and M. Kikuchi, J. Phys. Soc. Jpn **57**, 4351 (1988).
[14] E. Manousakis and R. Salvador, Phys. Rev. Lett. **60**, 840 (1988); Phys. Rev. B **39**, 575 (1989); G. Gomez-Santos, J. D. Joannopoulos and J. W. Negele, Phys. Rev. B **39**, 4435 (1989).
[15] S. Liang, Phys. Rev. **B42**, 6555 (1990).
[16] N. Trivedi and D. Ceperley, Phys.Rev. B **40**, 2737 (1989).

[17] T. Barnes and A. S. Swanson, Phys. Rev. B **37**, 9405 (1988).
[18] H.-Q. Ding and M. S. Makivic, Phys. Rev. Lett. **64**, 1449 (1990).
[19] G. Fox, M. Johnson, G. Lyzenga, S. Otto, J. Salmon, and D. Walker, *Solving Problems on Concurrent Processors*, Prentice Hall, Englewood Cliffs, New Jersey (1988).
[20] M. Makivic and F. Buda, unpublished.
[21] N. E. Bickers and D. J. Scalapino, Ann. of Phys. **193**, 206 (1989).
[22] M. Suzuki, Prog. Theor. Phys. **56**, 1457 (1976); M. Suzuki, J. Stat. Phys. **43**, 883 (1986).
[23] M. Suzuki, J. Math. Phys. **26**, 601 (1985).
[24] M. Suzuki, Phys. Lett. **113A**, 299 (1985).
[25] J. J. Cullen and D. P. Landau, Phys. Rev. B **27**, 297 (1983).
[26] J. W. Negele, H. Orland, in *Quantum Many-Particle Systems* (Addison Wesley, New York 1988).
[27] M. Marcu, in *Quantum Monte Carlo Methods in Equilibrium and Nonequilibrium Systems*, M. Suzuki ed. (Springer, Berlin, 1987), p. 64.
[28] J.E. Gubernatis, M.Jarrell, R.N.Silver and D.S.Sivia, to appear in Phys. Rev. B (September 1991).
[29] W. Jones and N. H. March, in *Theoretical Solid State Physics*, Chapter 10, (Dover Publications, Inc., New York, 1973), vol. 1.
[30] A. Auerbach and D. P. Arovas, Phys. Rev. Lett. **61**, 617 (1988).
[31] P.C.Hohenberg and W.F.Brinkman, Phys.Rev. B **10**, 128 (1974).
[32] J. Deisz, M. Jarrell and D. Cox , Phys.Rev. B **42**, 4869 (1990).
[33] D. D. Betts, F. C. Salevsky, and J. Rogiers, J. Phys. A **14**, 531 (1981).
[34] M. Makivic, unpublished.
[35] D. R. Nelson and J. M. Kosterlitz, Phys. Rev. Lett. **39**, 1201 (1977).
[36] E. L Pollock and D. M. Ceperley, Phys.Rev. B **36**, **36**, 8343 (1987).
[37] S. Chakravarty, B. I. Halperin, and D. Nelson, Phys. Rev. Lett. **60**, 1057 (1988).
[38] Y. Endoh, K. Yamada, R. J. Birgenau, D. R. Gabbe, H. P. Jenssen, M. A. Kastner, C. J. Peters, P. J. Picone, T. R. Thurston, J. M. Tranquada, G. Shirane, Y. Hidaka, M. Oda, Y. Enomoto, M. Suzuki, and T. Murakami, Phys. Rev. B **37**, 7443 (1988).
[39] M. Makivic and M. Jarrell, unpublished.
[40] T. Onogi, S. Miyashita and M. Suzuki, in *Quantum Monte Carlo Methods in Equilibrium and Nonequilibrium Systems*, M. Suzuki ed. (Springer, Berlin, 1987), p. 75.
[41] W. Hayes and R. Loudon,*Scattering of Light by Crystals*, 1978 (Wiley, New York); J. B. Parkinson, J. Phys. C **2**, 2012 (1969).
[42] F. J. Dyson, Phys. Rev. **102**, 1217 (1956); **102**, 1230 (1956); S. V. Maleev, Sov. Phys., JETP **64**, 654 (1958).
[43] G. Baym and L. P. Kadanoff, Phys. Rev. **124**, 287 (1961); G. Baym, Phys. Rev. **127**, 1391 (1962).
[44] M.Takahashi, Phys.Rev.Lett. **58**, 168 (1987).

Quantum Effects in Random Spin Systems

S. Miyashita[1] *and J. Behre*[2]

[1]Department of Physics, CLAS, Kyoto University, Kyoto 606, Japan
[2]Institut für Theoretische Physik, Universität Hannover,
W-3000 Hannover, Fed. Rep. of Germany

Abstract. The effect of inhomogeneity is studied in diluted Heisenberg antiferromagnets ($S=\frac{1}{2}$) on the square lattice by the use of a quantum Monte Carlo simulation and a diagonalization method. Spatial distributions of the spin correlation function in various lattice configurations are investigated. Spin correlations near non-magnetic sites have a tendency to be enhanced. On the other hand, correlations between largely separated spins are found to be reduced. Degree of the enhancement strongly depends on the lattice configuration. Shape dependence of the enhancement is also studied in one-dimensional lattices.

1. Introduction

The ground state of the Heisenberg antiferromagnet with $S=\frac{1}{2}$ on the square lattice, $\mathcal{H}=2J\sum_{<ij>}\mathbf{S}_i\cdot\mathbf{S}_j$, has been studied extensively [1-3]. There seems to exist finite long range staggered order in the ground state, but its amount is much reduced from the corresponding classical value by the quantum fluctuation due to the non-commutativity between the Hamiltonian and the order parameter $N_z \equiv \sum_{i\in\mathrm{A}}\sigma_i^z-\sum_{i\in\mathrm{B}}\sigma_i^z$. We are interested in some peculiar phenomena due to the combination of effects of inhomogeneity and quantum effect. Such effects have been studied for various types of inhomogeneity [4-8]. In the present work, we study the effect in systems diluted by non-magnetic impurities [7]. We have investigated microscopic distribution of the enhancement of the correlations by the use of a quantum Monte Carlo simulation (QMC) based on Suzuki-Trotter decomposition and a diagonalization method. It is found that the correlations near the impurities are likely enhanced. Actually, the nearest neighbor correlation is equivalent to the energy and the enhancement can be understood from the variational principle. The range of enhancement is, however, rather short and the correlations between largely separated spins are reduced. We also found that the enhancement near the impurities is not uniform and some of the nearest neighbor correlations are reduced even at neighbors of impurities. The distribution of enhancement strongly depends on the configuration of sites, which suggests that this non-uniformness is due to the strongly correlated quantum fluctuations. We also study the dependence of the spin correlations on lattice shapes in one dimensional lattices, where the spin correlations at large distances change systematically with the shapes. Thus, it reminds us a kind of "quantum interference" in a spin system. This effect may cause a change the percolation threshold of impurity concentration for the long range order in the ground state.

We study systems with up to 24 spins, by a diagonalization method using a simple power method and also TIT2-package [9] where program codes with the Lanchos algorithm are generally prepared for spin systems with $S = \frac{1}{2}$. For larger lattices, we performed quantum Monte Carlo simulations. It has been found that the existence of non-magnetic sites causes difficulty for the ergodicity problem in QMC. In particular at low temperatures, we found a tendency that the antiferromagnetic order is enhanced too much by a kind of freezing effect due to lack of ergodicity of QMC. In order to avoid this difficulty we introduced a new type of global flip. QMC with the new algorithm gives good agreement with results by an exact diagonalization method even for systems with highly concentrated impurity (non-magnetic sites) [7].

In §2, spatial distributions of enhancement of the spin correlation function in diluted systems are studied. In §3, dependence of the spin correlation function on lattice shapes is investigated in one dimensional lattices.

2. The Spin Correlation Functions of Diluted Systems

The Hamiltonian of the model studied here is

$$\mathcal{H} = 2J \sum_{<ij>} \epsilon_i \epsilon_j \mathbf{S}_i \cdot \mathbf{S}_j,$$

where ϵ_i =1 or 0. We choose the positions where $\epsilon = 0$ randomly. We first investigated the ground state of 4×4 systems by a diagonalization method. The nearest neighbor (nn) correlations around the impurities are enhanced and we find that some further neighbor correlations are also enhanced.

In order to investigate how far separated correlation functions are enhanced, we study larger lattices. For large systems such as 8×8 or 12×12, we used quantum Monte Carlo method based

```
           I          I          I          I          I          I *        I
         1.00       1.09        .96        .96       1.00        .94   *   1.05
           I          I          I          I          I          I        * I
  o        x- 1.16  -x-   .99  -x-   .99  -x- 1.00  -x-   .95  -x- 1.09  -x
           I                     I          I          I          I          I
         1.03                  1.04        .97        .99       1.08       1.02
           I                     I          I          I          I          I
----x- 1.19  -x        o         x- 1.12  -x-   .94  -x- 1.06  -x- 1.03  -x-   .95
    I          I                 I          I          I                     I
  1.07        .97               .98        .97        .97                   .95
    I          I                 I          I          I                     I
----x-   .93  -x- 1.12  -x- 1.10  -x-   .96  -x- 1.08  -x         o          x- 1.07
    I          I                 I          I          I                     I
   .94        .99               .96        .91       1.19                  1.21
    I          I                 I          I          I                     I
----x- 1.10  -x        o         x- 1.09  -x-   .92  -x-   .96  -x-   .95  -x-   .99
    I        * I                 I *        I                     I
  1.01   *  1.01              1.13    *  1.13                  1.32
    I *        I                 I        * I                     I
    x- 1.09  -x- 1.01  -x-   .88  -x- 1.34  -x        o           x          o
    I          I        * I                                       I
  1.04        .95   *  1.22                                      .97
    I          I *        I                                       I
----x-   .96  -x- 1.07  -x        o          o          x- 1.14  -x-   .97  -x- 1.03
    I          I          I                           I *        I        * I
   .95        .99        .97                        1.18    *   .94    *  1.13
    I          I          I                           I        * I *        I
----x-   .98  -x- 1.00  -x-   .97  -x- 1.20  -x-   .93  -x-   .97  -x-   .99  -x-   .95
    I          I          I          I *        * I          I          I          I
  1.12        .98        .98       1.09    *  1.09          .94       1.02        .95
    I          I          I          I *        * I          I          I          I
----x- 1.02  -x-   .97  -x-   .96  -x-   .98  -x-   .97  -x- 1.01  -x-   .98  -x- 1.03
               I          I          I          I          I          I *        I
               I          I          I          I          I          I     *    I
```

Fig. 1 Nearest neighbor spin correlation functions at $T = 0.2$ normalized by the value of pure case (-0.462).

on Suzuki-Trotter decomposition [10] using the checker-board decomposition [11]. Here we have three states on a site k in the (2+1)-dimensional lattice. Namely, $S_k = 0$, or ± 1, corresponding to impurity, spin-up and spin-down, respectively. The most important new point is how to recover the ergodicity of Monte Carlo time-evolution efficiently. A 'Néel chains', which is a sequence of up-down spins in the real space in the same Trotter layer, is hardly flip near impurities. In order to avoid this difficulty we introduced a kind of global flip around the impurities (loop flip) and also a non-straight Marcu-flip [12].

The nn correlations are given in Fig. 1, where we find qualitatively similar enhancements to those in the 4×4 systems. Number of enhanced correlations at larger distances are also listed in Table 1. Here we find the range of enhancement rather short.

In order to see the distance dependence of correlation, we study the square of the staggered magnetization which is a sum-

Table 1 Number of enhanced bonds more than 2% in the lattice shown in Fig. 1

distance	(1,1)	(2,0)	(2,1)	(2,2)	(3,0)	(3,1)	(3,2)	(4,0)
numbers	29	22	41	15	19	11	18	4

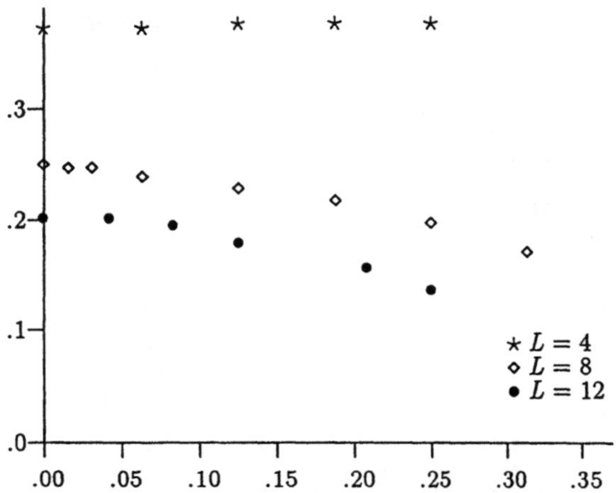

Fig. 2 Size- and δ-dependence of $< N_z^2 > /N_s^2$.

mation of all staggered correlations $< N_z^2 >= \sum_{i,j} \epsilon_i \epsilon_j < S_i S_j >$, where $< \cdots >$ denotes the expectation value in the ground state. Thus, if we study the size-dependence of this quantity per spin, $\mathcal{A} \equiv < N_z^2/N_s^2 >$, where N_s is the number of spins, the distance dependence of correlation function can be known. If the correlation is uniform over the system, $\mathcal{A}$ does not depend on N_s. The size-dependence of $\mathcal{A}$ is plotted in Fig.2, where δ is the concentration of impurity. We find that $\mathcal{A}$ decreases as the size increases and the decrease becomes large as δ increases. Thus we conclude that the correlation functions separated largely are reduced by the existence of impurities.

3. Spin Correlation Function and Lattice Geometry

In Fig. 1, we find reduced correlation function even at neighbors of impurities. Enhancement and reduction of nn correlations seem to happen somehow alternately. This observation reminds us the interference phenomena. Its precise mechanism will be studied in the future but here we demonstrate strong dependence of

```
[O]
1    2    3    4    5    6    7    8    9   10   11   12   13   14   15   16
o----o----o----o----o----o----o----o----o----o----o----o----o----o----o----o
       0.066     0.037     0.025     0.018     0.014     0.011     0.007
-0.218    -0.073    -0.042    -0.029    -0.022    -0.018    -0.014    -0.011

[A]
1    2    3    4    5    6    7    8    9   10   11   12   13   14   15   16
o----o----o----o----o----o----o----o----o----o----o----o----o----o----o----o
     |    |              |    |                                  |    |
     o----o              o----o                                  o----o

      0.071     0.049     0.036     0.035     0.034     0.033     0.031
-0.169    -0.070    -0.043    -0.051    -0.046    -0.042    -0.041    -0.061
[B]
1    2    3    4    5    6    7    8    9   10   11   12   13   14   15   16
o----o----o----o----o----o----o----o----o----o----o----o----o----o----o----o
     |    |              |    |                             |    |
     o----o              o----o                             o----o

      0.071     0.050     0.039     0.038     0.032     0.022     0.012
-0.172    -0.075    -0.049    -0.061    -0.054    -0.047    -0.024    -0.017
[C]
1    2    3    4    5    6    7    8    9   10   11   12   13   14   15   16
o----o----o----o----o----o----o----o----o----o----o----o----o----o----o----o
          |    |         |    |         |    |
          o----o         o----o         o----o

      0.058     0.028     0.017     0.012     0.008     0.005     0.003
-0.228    -0.049    -0.024    -0.025    -0.012    -0.009    -0.007    -0.005
[D]
1    2    3    4    5    6    7    8    9   10   11   12   13   14   15   16
o----o----o----o----o----o----o----o----o----o----o----o----o----o----o----o
          |    |              |    |    |    |
          o----o              o----o    o----o

      0.057     0.023     0.011     0.005     0.003     0.002     0.001
-0.230    -0.050    -0.026    -0.011    -0.004    -0.003    -0.002    -0.0015
```

Fig. 3 The spin correlation functions $< S_1 S_i >$, $i = 2 \cdots 16$

the spin correlations on the lattice geometry. In Fig. 3, we show three typical samples of 16 spin chain with decorations. The correlations, $< S_1 S_i >$, $i = 2 \cdots 16$ are listed in the Figures. Here we find that the decorations sometimes enhance the correlations very largely (e.g., (A)), but also reduce it to 1/8 (e.g., in (D)). In a very long chain the above effects come randomly and we may imagine that the correlation may decay exponentially in the analogy to the Anderson localization. In the previous section, we found that the correlations are reduced at large distances. It might give a change of the percolation threshold of δ for the long range spin order in the ground state. which will be studied in the future.

We would also like to point out the following: So far we studied $< \mathbf{S}_i \mathbf{S}_j >$, which is the direct correlation and is often enhanced near the impurity. We find, however, that the loop flips around impurities occur very often in QMC. This suggests that the configurations around the impurities change very often keeping the relative antiferromagnetic order. The direct correlation is not necessarily an intuitive (classical or static) correlation but it may be an instantaneous (dynamical) correlation. The difference between these is just as the difference between $< N_z^2 >$ and the staggered susceptibility multiplied by the temperature at finite temperatures. In this context the behavior of the canonical correlation is interesting, which will be reported elsewhere.

Acknowledgments

The present work is partially supported by the Grant-in Aid from Ministry of Education, Sciences and Culture. One of the authors (J.B.) likes to thank the BMFT for financial support in Germany and the DAAD for financial support in Japan.

References

[1] J. D. Reger and A. P. Young, Phys. Rev. **B37**, 5978 (1988)

[2] S. Miyashita, J. Phys. Soc. Jpn. **57**, 1934 (1988)

[3] E. Manousakis, Rev. Mod. Phys. **63**, 1 (1991)

[4] E. F. Talbot, Phys. Rev. **B33**, 4906 (1986)

[5] R. E. Camley, W. von der Linden and V. Zevin, Phys. Rev. **B40**, 119 (1989)

[6] N. Nagaosa N, Y. Hatsugai and M. Imada J. Phys. Soc. Jpn. **58**, 978 (1989)

N. Bulut, D. Hone, D. J. Scalapino and E. Y. Loh, Phys. Rev. Lett. **62**, 2192 (1989)

K. J. B. Lee, P. Schlottmann, Phys. Rev. **B42**, 4426 (1990)

[7] J. Behre , S. Miyashita and H.-J. Mikeska, J. Phys. A: Math. Gen. **23**, L1175 (1990)

J. Behre and S. Miyashita, in preparation

[8] D. D. Betts and J. Oitmaa, private communication

[9] H. Nishimori *TIT2 package*, (1990) private communication

[10] M. Suzuki, *Quantum Monte Carlo Methods in Equilibrium and Nonequilibrium Systems*, ed M. Suzuki (Springer 1987) p 2

[11] E. Y. Loh, D. J. Scalapino and P. M. Grant, Phys. Rev. **B31**, 4712 (1895)

[12] M. Marcu, *Quantum Monte Carlo Methods in Equilibrium and Nonequilibrium Systems*, ed M. Suzuki (Springer 1987) p 64

[13] S. Miyashita and J. Behre, in preparation.

Numerical Calculations of S=1 Heisenberg Antiferromagnetic Chain

M. Takahashi

Institute for Solid State Physics, University of Tokyo,
Roppongi 7-22-1, Minato-ku, Tokyo 106, Japan

The properties of $S = 1$ Heisenberg antiferromagnetic chains are completely different with those of $S = 1/2$ Heisenberg antiferromagnetic chains. Numerical methods such as projector Monte Carlo method, world line Monte Carlo method and exact diagonalization method are very useful for the investigation of these systems.

For the Heisenberg antiferromagnet in one dimension:

$$\mathcal{H} = J\sum_i \mathbf{S}_i \cdot \mathbf{S}_{i+1}, \tag{1}$$

only S=1/2 case is soluble using the Bethe ansatz. The linear spin-wave theory gives the elementary excitation spectrum:

$$\epsilon(K) = 2JS|\sin K|. \tag{2}$$

On the contrary it is known that the ground state of this Hamiltonian for $S = 1/2$ has the gapless excitation[1)]:

$$\epsilon(K) = \frac{\pi}{2}J|\sin K|. \tag{3}$$

This means that the gap between the ground state energy and the first excited energy becomes zero in the limit of infinite systems. This fact is rigorously proved by Lieb, Schultz and Mattis[2)]. Their proof can be extended to the S =half integer cases. Unfortunately this proof cannot be extended to the case of integer S. Haldane[3)] predicted that in the integer S cases the system has a finite energy gap using the mapping of Hamiltonian (1) to the classical 2-D $O(3)$ non-linear σ model. It is expected that the energy gap decrease as $e^{-\pi S}$. As far as we know, no one has proved rigorously the existence of a gap for integer S cases. But we have many numerical evidence for the existence of the gap especially for the $S = 1$ case. The most direct numerical method for this problem is the exact diagonalization method[4)]. At the first time the length of the

Springer Proceedings in Physics, Vol. 70 Computational Approaches in Condensed-Matter Physics
Editors: S. Miyashita · M. Imada · H. Takayama

chain is $N = 12$. But recently the diagonalization of $N = 16$ chain is very popular. $N = 18$ chain is also sometimes used. About the quantum Monte Carlo calculation, the Hamiltonian (1) has no negative sign problem unlike to the triangular antiferromagnet or 2-3D fermion problems. Nightingale and Blöte[5)] calculated the energy gap at $S = 1$ case with $N = 32$ spins. They concluded that the energy gap is $0.41J$. A rough estimation of the correlation length of this system was done by the author using the world line Monte Carlo method[6)]. Nomura[7)] did a more elaborate calculation of the correlation length and obtained $\xi = 6.1$ in units of atomic spacing. The author obtained the lowest energy eigen value for the given momentum of $N = 32$ chain using the projector Monte Carlo method[8)]. It is expected that the difference from the ground state energy gives the elementary excitation spectrum.

Recently several substances which are described by the Hamiltonian (1) are found such as $CsNiCl_3$[9)] and NENP[10)]. The experimental results of these substances coincide with theoretical predictions and numerical results for the Hamiltonian (1) and strongly suggest the existence of the energy gap. In this paper we review Monte Carlo results for the Hamiltonian (1) at $S = 1$ and discuss the comparison with the experiment.

In the quantum Monte Carlo method there are two categories. The first one is for the finite temperature case. This is sometimes called as world line MC or path-integral MC[11)]. The second is for the zero temperature cases[12)]. It is called projector MC method or Green function MC method.

In the first method the low temperature properties can be derived by using the long imaginary time axis. The checker-board decomposition is based on the exact solution of two spins. One can use the exact solution of larger clusters. This is called the large cluster decomposition. The program becomes complicated. But one can put the time slice longer in the imaginary time axis and the acceptance ratio becomes higher. In these world line Monte Carlo, the calculation of the correlation function $\rho(l) \equiv < S_i^z S_{i+l}^z >$ is very easy. So one can calculate the correlation function and correlation length. The Fourier transform of $\rho(l)$ is the structure factor $S(K)$:

$$S(K) \equiv \sum_{l=1}^{N} \exp(ilK)\rho(l). \tag{4}$$

Using the variational wave function we can calculate the upper limit of the elementary excitation:

$$\epsilon(K) \le g(K) \equiv 2J(1 - \cos K)[-\rho(1)]/S(K). \tag{5}$$

The structure factor $S(K)$ is shown in Fig.1. $S(\pi)$ has the finite value

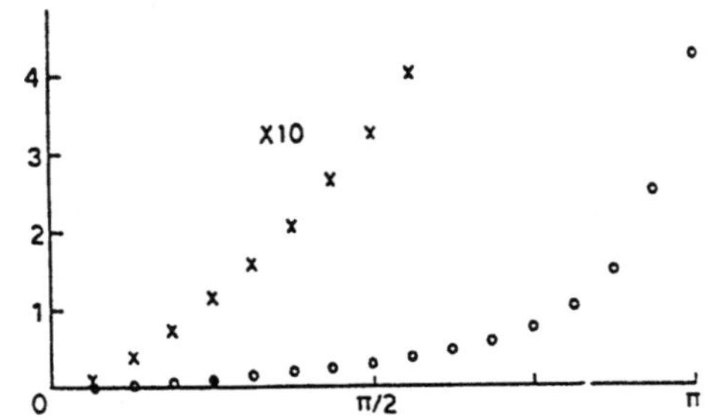

Fig.1
Structure factor $S(K)$ for $S = 1$, $N = 32$ chain at the ground state. Circles give its net values. Crosses are ten times of net values. For small K, $S(K)$ behaves as K^2.

about 4.2. This is different from $S = 1/2$ case where $S(K)$ diverges logarithmically at $K = \pi$. Near $K = 0$ $S(K)$ behaves as K^2. On the contrary it behaves as $|K|$ near $K = 0$ for $S = 1/2$ chain.

On the other hand the projector Monte Carlo method is useful to determine the ground state energy. The ground state energy is determined by using the following formula:

$$E_0 = \lim_{\tau\to\infty} \frac{< j|\mathcal{H}e^{-\tau\mathcal{H}}|i >}{< j|e^{-\tau\mathcal{H}}|i >}, \tag{6}$$

if the ground state is not orthogonal to the initial state $|i >$. If the state $|i >$ has the momentum K, the r.h.s. of (6) should converge to the lowest energy in momentum space K if it is not orthogonal to $|i >$. Thus we consider the following function $B(K,\tau)$:

$$B(K,\tau) \equiv \frac{< \psi|R(-K)\mathcal{H}e^{-\tau\mathcal{H}}R(K)|\varphi >}{< \psi|R(-K)e^{-\tau\mathcal{H}}R(K)|\varphi >},$$

$$R(K) \equiv \sum_{l=1}^{N}(S - S_l^z)\exp(iKl). \tag{7}$$

Here $|\varphi >$ is the ground state wave function. $|\psi >$ is a state with zero momentum. Actually the estimation of $B(K,\tau)$ becomes harder as τ becomes big because the statistical error becomes larger. This is a kind of negative sign problem. In Fig.2 it is shown that the limiting value of $B(K,\tau)$ gives the correct excitation energy at momentum K for $N = 14$, $S = 1$ chain. In Fig.3 the elementary excitation obtained in the above method coincides with Bethe ansatz solutions for $S = 1/2$, $N = 32$

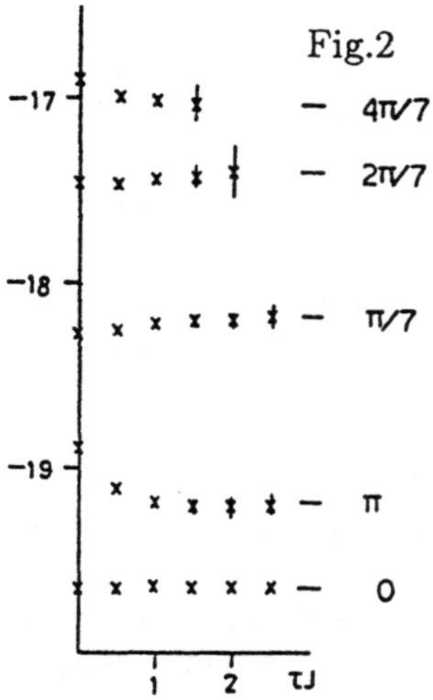

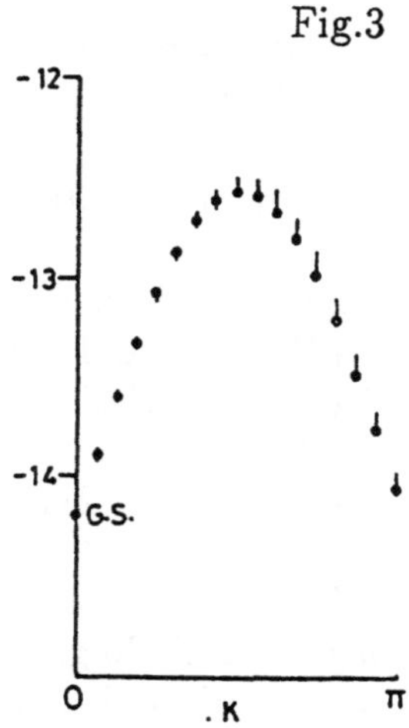

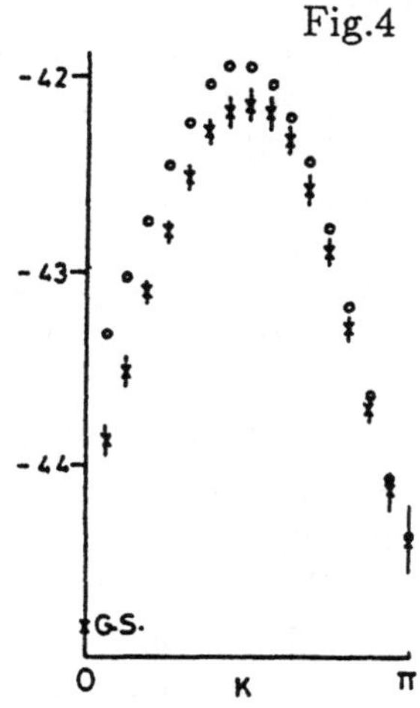

Fig.2
Function $B(K,\tau)/J$ for $N = 14, S = 1$ chain at $K = 0,\ \pi/7,\ 2\pi/7,\ 4\pi/7,\ \pi$. $E(K)/J$'s are plotted as horizontal bars. MC results and exact diagonalization results are compared. The coincidence of $E(K)$ and $B(K,\tau)$ at $\tau J = 1 - 2$ is very good. But statistical error grows exponentially as τ becomes large.

Fig.3
$E(K)/J$ for $N = 32, S = 1/2$ chain. Circles are results of Bethe Ansatz equation . Bars are results of MC calculation.

Fig.4
$E(K)/J$ for $N = 32, S = 1$ chain. The spectrum has a gap at $K = \pi$. The value of gap is about $0.4J$ and coincides with NB's calculation[5]. Small circles are upper bound of $E(K)/J$ given by the variational method. This upper bound was calculated from structure factor and variational relation.

chain. The elementary excitation for $S = 1$, $N = 32$ chain satisfies the variational relations. It has apparently the gap. The elementary excitation is not symmetric about the inversion at $K = \pi/2$. At $K = \pi$ the gap is $\Delta = 0.4J$. On the other hand the gap at $K = 0$ is about 2Δ (Fig.4).

References

1) J. des Cloiseaux and J.J. Pearson, Phys. Rev. **128** 2131 (1962).
2) E.H. Lieb, T. Schultz and D.J. Mattis, Ann. Phys. NY **16** 407 (1961).
3) F.D.M. Haldane, Phys. Rev. Lett. **50** 1153 (1983); Phys. Lett. A **93** 464 (1983).
4) J.B. Parkinson and J. Bonner, Phys. Rev. B **32** 4703 (1985).
5) M.P. Nightingale and H.W.J. Blöte, Phys. Rev. B **33** 659 (1986).
6) M. Takahashi, Phys. Rev. B **38** 5188 (1988)
7) K. Nomura, Phys. Rev. B **40** 2421 (1989).
8) M. Takahashi, Phys. Rev. Lett. **62** 2313 (1989).
9) W.J.L. Buyers, R.M. Morra, R.L. Armstrong, P. Gerlach and K. Hirakawa, Phys. Rev. Lett. **56** 371 (1986).
10) J.P. Renard M. Verdaguer, L.P. Regnault, W.A.C. Erkelens, J. Rossat-Mignod and W.G. Stirling, Europhys. Lett. **3** 945 (1987).
11) M. Barma and B.S. Shastry, Phys. Lett. **61A** 15 (1977); Phys. Rev. B **18** 3351 (1978). J.E. Hirsch, R.L. Sugar, D.J. Scalapino and R. Blankenbecler, Phys. Rev. B **26** 5033 (1982). R. Blankenbecler and R.L. Sugar, Phys. Rev. D **27** 1304 (1983). K.Sogo and M. Uchinami, J. Phys. A **19** 493 (1986). M.Uchinami, Phys. Lett. A **127** 151 (1988).
12 J.H. Hetherington, Phys. Rev. A **30** 2713 (1983). K. Nomura and M. Takahashi, J. Phys. Soc. Jpn. **57** 1424 (1988). D.M. Ceperley and M.H. Kalos, Monte Carlo Methods in Statistical Physics, Edited by K. Binder (Springer, Berlin, 1979).

Frustrated Quantum Antiferromagnets in Two Dimensions

H. Nishimori

**Department of Physics, Tokyo Institute of Technology,
Oh-okayama, Meguro-ku, Tokyo 152, Japan**

Abstract. I discuss how quantum effects modify the classical picture of the ground states of frustrated spin systems, such as the triangular lattice antiferromagnet and the square lattice model with next nearest neighbor interactions. I show that much more powerful numerical methods than available today are required to settle outstanding controversies about the type of ordering in these systems.

1. Introduction

The ground state of antiferromagnetic spin systems is generally easy to describe if the spins are allowed to be treated classically. Quantum effects introduce fluctuations caused by the uncertainty principle. These fluctuations appear as the reduction of the expectation value of a spin operator from the classical value S,

$$< S_i^z >= S - \Delta S. \tag{1}$$

The ratio of ΔS to S is a measure of quantum effects.

Further complication is caused frustration which modifies the ground state into a non-conventional one even in the classical limit. Typical examples are found in the antiferromagnets on the triangular lattice and on the square lattice with next nearest neighbor interactions. The latter will be called the J_1-J_2 model hereafter.

A main purpose of the present contribution is to discuss how classical ground states are affected by the interplay of frustration and quantum fluctuations. Both analytical and numerical approaches are explained.

2. Spin Wave Theory

The basic idea of the spin wave theory is to evaluate quantum corrections by the asymptotic expansion in powers of $1/S$; the limit $S \to \infty$ corresponds to the classical state. This method was first applied to the frustrated triangular antiferromagnet by Oguchi [1]. He found that the spin reduction ΔS is equal to 0.26, a finite value less than 1/2. This result implies that the expectation value (1) remains positive, and the classical state, the three-sublattice Néel state, is qualitatively stable against quantum fluctuations.

The situations are more complicated for the J_1-J_2 model. Chandra and Doucot [2] calculated the first order quantum correction and found that the spin reduction exceeds the classical value S if the ratio α of the second neighbor interaction to the first neighbor counterpart, J_1/J_2, is close to 1/2, irrespective of S as long as it is finite. They suggested that a new, purely quantum mechanical phase may exist in this region of intermediate values of α.

Their suggestion lead to a number of new proposals of candidates of the non-classical phase, such as the dimer state, the chiral state, and the twisted state [3]. However, results from the naive, first-order spin wave theory should be taken with sufficient care, in particular when higher order corrections give similar divergences with opposite sign [4]. We thus analyzed this problem by the modified spin wave theory [5]. In this theory the condition that the ground state has a vanishing value of the z-component of the total magnetization is imposed by the Lagrange multiplier method [6]. Our result indicates that the classical states are stable for large S even at $\alpha = 1/2$. The two-sublattice Néel state and the classical columnar state exchange their dominance by a first order transition around $\alpha = 1/2$.

3. Numerical Diagonalization Studies

The negative-sign problem is quite serious in these frustrated quantum spin systems, and quantum Monte Carlo simulations have not yet yielded conclusive answers. Direct diagonalization of the Hamiltonians is the only possible way to investigate these systems numerically.

We performed numerical diagonalization of the triangular-lattice spin systems with $S = 1/2$ up to the systems size $N = 27$ [7]. In order to determine the thermodynamic properties, one has to extrapolate the finite-size data to the infinite-size limit. Concerning the values of short-range correlations, this extrapolation works in a quite stable fashion. We were able to predict the ground-state nearest neighbor correlation $\langle \sigma_0^x \sigma_1^x \rangle$, which is proportional to the ground-state energy, as -0.273 for the XY antiferromagnet and -0.243 for the Heisenberg antiferromagnet on the triangular lattice.

The situations are much worse for the extrapolation of long-range correlations. Let us explain the difficulties by the example of the ferromagnetic, thus unfrustrated, XY model on the triangular lattice. To measure the degree of ordering, we calculate the squared total magnetization

$$M^2 = \langle (\sum_j \mathbf{S}_j)^2 \rangle. \tag{2}$$

Finite-size data of $N = 7, 9, 13, 19, 21$ and 27 for $4M^2/N^2 - 2/N$ were fitted to several extrapolation formulas to the thermodynamic limit. The subtraction of $2/N$ is required to exclude the term corresponding to the self-multiplication in the expansion of the square in (2), since it has nothing to do with long-range spin correlations. Although this type of corrections may be neglected for large N, it is essential to extract only non-trivial terms in the analysis of small-size data. Among various extrapolation formulas, the following two showed especially excellent fit to the finite-size data.

$$f_a = N^{-0.11906/2}(1.15896 - 0.25321/N)$$
$$f_b = 0.81719 + 0.85108/\sqrt{N} - 0.97219/N.$$

The first formula f_a corresponds to a power-decaying correlation function in the thermodynamic limit, while the second f_b indicates that there exists finite long-range order with spin-wave excitations. We found that both formulas account for the finite-size data equally well in that the residuals are almost comparable between f_a and f_b. It is then virtually impossible to settle the question of the existence of long-range order in the thermodynamic limit from these small-size data; short-range effects are still too large. We can tell at most that the value of the order parameter would not be very much different from 0.81719 (which is equivalent to $\langle S_i^z \rangle = 0.452$) *if* there exist finite long-range order.

Similar arguments apply to the frustrated models. We may be able to say that *if* the antiferromagnetic XY model has finite long-range order on the triangular lattice, its value would be close to $\langle S_i^z \rangle = 0.256$, but it is impossible to predict the existence or non-existence of long-range order. In the case of the antiferromagnetic Heisenberg model, the predicted squared long-range order turns out to be negative, a quite inconsistent result with the spin-wave prediction. Therefore, the problem of long-range order on the triangular lattice antiferromagnet still remains unsolved.

Extrapolations are difficult also in the J_1-J_2 model. Most of numerical calculations were carried out for small systems with $N = 16$ or $N = 20$ [3]. In these small systems, short-range effects dominate the numerical values of the sums of correlation functions from which the existence of long-range order is discussed. Typically, a sum of this sort has a form

$$\sum_i \langle O_0 O_i \rangle, \tag{3}$$

where O_i represents an operator to express the appropriate order, such as the two-sublattice Néel state or the chirality order. The summation (3) on the square lattice is dominated by short-range effects if N is as small as 16; There exist one self-term, $i = 0$, and four nearest neighbor terms out of sixteen terms. As a matter of fact, these five short-range terms account for most part of the numerical value of the summation (3). Therefore, it is dangerous to try to extract information on the thermodynamic properties from finite-size data for $N = 16$ or $N = 20$. Recently Schulz and Ziman [8] diagonalized the case of $N = 36$, which is a significant value both in view of computational efforts and from the purpose of extracting long-range effects. I should stress, however, that another larger system size must be investigated to extrapolate data to the thermodynamic limit; Schulz and Ziman used the 16-site and 36-site data to infer the infinite-system properties, but the former are dominated by short-range effects as mentioned before.

4. Rigorous Results

A few rigorous results are available on the present problem. Kishi and Kubo proved the existence of two-sublattice Néel order in the J_1-J_2 model for $\alpha \leq 1/2$ and S small [9]. Their estimation of long-range fluctuations implies at the same time that the twisted order, which is one of the proposed non-classical states, does not exist if α is smaller than or equal to 1/2. We proved, along the same line of arguments, that there does not exist spontaneous symmetry breaking corresponding to the twisted order [10]. Thus recent numerical search for twisted order has significance only in the region $\alpha > 1/2$. It is highly unlikely, however, that a non-classical twisted order exists in this large-α region when it does not for $\alpha \leq 1/2$.

Almost no rigorous results have been obtained for the triangular lattice antiferromagnet with the exception of the proof of absence of vector chiral order at finite temperatures [11].

5. Conclusion

Numerical investigations of the ground-state long-range order of frustrated quantum systems are still at a primitive level. Short-range effects are too large in data obtained by numerical diagonalization techniques, and the negative-sign problem sets limits to the practical applicability of quantum Monte Carlo simulations. New ideas are required to settle the outstanding problems by numerical methods as well as by reliable analytical approaches. In particular reconciliation of the perturbative spin-wave picture and numerical results should urgently be pursued.

This work was supported by the Grant-in-Aid for priority area by the Ministry of Education, Science and Culture.

References

[1] T. Oguchi, J. Phys. Soc. Jpn. **52**, Suppl. 183 (1983).
[2] P. Chandra and B. Doucot, Phys. Rev. B**38**, 9335 (1988).
[3] D. Poiblanc, E. Gagliano, S. Bacci and E. Dagotto, Phys. Rev. B**43**, 10970 (1991) and references therein.
[4] S. Chakravarty, B. Halperin and D.R. Nelson, Phys. Rev. B**39**, 2344 (1989).
[5] H. Nishimori and Y. Saika, J. Phys. Soc. Jpn. **59**, 4454 (1990).
[6] M. Takahashi, Phys. Rev. B**40**, 2494 (1989).
[7] H. Nishimori and H. Nakanishi, J. Phys. Soc. Jpn. **57**, 626 (1988); *ibid* **58**, 3433 (1989). This work extends the calculations by S. Fujiki and D.D. Betts, Can. J. Phys. **64**, 876 (1986), **65**, 76 (1987), Prog. Theor. Phys. Suppl. **87**, 268 (1986).
[8] H.J. Schulz and T.A.L. Ziman, preprint.
[9] T. Kishi and K. Kubo, J. Phys. Soc. Jpn. **58**, 2547 (1989), Phys. Rev. B**43**, 10844 (1991).
[10] H. Nishimori and Y. Saika, preprint.
[11] T. Kawarabayashi and M. Suzuki, Phys. Lett. A**143**, 17 (1990).

Symmetry Breaking in a Quantum Double-Well Chain

J.E. Gubernatis[1], *D.K. Campbell*[2], *and Xidi Wang*[2]

[1]Theoretical Division, Los Alamos National Laboratory,
Los Alamos, NM 87545, USA
[2]Center for Nonlinear Studies, Los Alamos National Laboratory,
Los Alamos, NM 87545, USA

Abstract. We present numerical evidence that quantum fluctuations can produce a symmetric ground-state in the double-well chain, restoring the symmetry that is broken classically. In particular, we present the phase diagram for this model that shows the symmetry restoration occurs more easily than predicted by a perturbation theory calculation of the continuum limit of the model.

1. Introduction

The double-well chain is a system of particles that move on a lattice of symmetric, double-well potentials which are coupled harmonically to their nearest neighbors. The physical properties of this system are invariant to changing the sign of the displacement variables, and this symmetry leads to a doubly-degenerate ground-state. Classically, this symmetry is always broken, and the particles all sit in the left- or the right-hand side of their individual wells. Quantum mechanically, however, we find a range of model parameters for which symmetry is restored by quantum fluctuations.

Our interest in this model was motivated by a series of quantum Monte Carlo simulations we are performing on one-dimensional, electron-phonon models of conducting polymers. Several years ago, for similar models, Fradkin and Hirsch [1] investigated how the electron motion can generate an effective double-well potential for the phonons and thereby cause the lattice to dimerize. They also argued that for certain models quantum fluctuations at low temperatures restore symmetry (i.e., destroy the dimerization). The quantum double-well chain is a simpler problem on which to study similar issues and to test new numerical methods.

The model, however, is also interesting on its own. It is a discretized version of a $1+1$ dimensional quantum ϕ^4 field theory. In the continuum version of the model, kink/anti-kink pairs are elementary excitations above the ground state [2]. One of our objectives was to learn more about the nature and consequence of these excitations.

The model is also relevant to the study of the structural properties of such chain-like, hydrogen-bonded materials as the hydrogen halides [3]. In many cases, the energy as a function of hydrogen position has been obtained by electronic structure calculations and fitted to the physical parameters of the double-well chain. Thus, with the phase diagram of the model, we can suggest whether these materials exist in the broken-symmetry (hydrogen-bonded) or restored-symmetry (covalently bonded) phase.

Springer Proceedings in Physics, Vol. 70 Computational Approaches in Condensed-Matter Physics
Editors: S. Miyashita · M. Imada · H. Takayama

2. Model Hamiltonian

The Hamiltonian we are considering is

$$\hat{H} = \sum_{n=1}^{N} \left[\frac{\hat{\pi}_n^2}{2m} + \frac{w}{2}(\hat{\phi}_n - \hat{\phi}_{n+1})^2 - \frac{k}{2}\hat{\phi}_n^2 + \frac{g}{4}\hat{\phi}_n^4 \right] \tag{1}$$

where $\hat{\phi}_n$ and $\hat{\pi}_n$ are the coordinate and momentum operators of a particle with mass m on n^{th} site of a chain of length N. The last two terms on the right-hand side of (1) are those for the symmetric, on-site double-well potentials. These potentials have two absolute minima displaced $\pm\sqrt{k/g}$ relative to each lattice site and have an energy barrier $E_b = k^2/4g$ between these minima. Using the frequency $\omega = \sqrt{w/m}$, the energy $\hbar\omega$, and the displaced distance of minima to scale time, energies, and lengths, we rewrite (1) in the following dimensionless form

$$\hat{H} = \epsilon \sum_{n=1}^{N} \left[\frac{\gamma}{2}\hat{\pi}_n^2 + \frac{\gamma}{2}(\hat{\phi}_n - \hat{\phi}_{n+1})^2 + \frac{1}{4}(\hat{\phi}_n^4 - 1)^2 \right] \tag{2}$$

where $\epsilon = 4E_b/\hbar\omega$ and $\gamma = w/k$, and a constant term has been added.

We will calculate the ground-state properties of (2) by doing quantum Monte Carlo simulations at successively lower and lower temperatures. The formalism on which the simulations are based requires the partition function Z of the model in terms of the Feynman path integral [4, 5]. In general, $Z = \int \mathcal{D}\phi \, e^{-S\{\phi\}}$ where $S\{\phi\}$ is the action associated with the scalar field ϕ. In terms of (2), we find that

$$S\{\phi\} = \epsilon \sum_n \int_0^\beta d\tau \left[\frac{\gamma}{2} \left(\frac{\partial \phi_n}{\partial \tau} \right)^2 + \frac{\gamma}{2}(\phi_n - \phi_{n+1})^2 + \frac{1}{4}(\phi_n^2 - 1)^2 \right] \tag{3}$$

where τ is the imaginary-time variable and β is the inverse of the temperature T.

An implicit parameter in (1) is the lattice constant a. Letting a vanish, we can obtain the continuum version of the model. If treated classically, this version of the model has in the ground state $\phi = \pm 1$, an energy $E_g = 0$, and soliton solutions representing excitations from the ground state. If treated quantum mechanically, its ground state properties are renormalized, and the lowest order correction (at the one-loop level) [2] leads to

$$E_g = \frac{2\sqrt{2}}{3}\epsilon\sqrt{\gamma} + \frac{1}{\sqrt{\gamma}} \left(\frac{-3}{\pi\sqrt{2}} + \frac{1}{2\sqrt{2}} \right). \tag{4}$$

When

$$\epsilon < \frac{18 - \sqrt{3}\pi}{8\pi\gamma} \simeq \frac{0.5}{\gamma} \tag{5}$$

E_g becomes negative, then the broken symmetry ground state is unstable to the formation of kink/anti-kink pairs. This instability is an indication of the restoration of symmetry. We will find that the above estimate, while not very useful quantitatively, does point to the correct physical picture.

3. Numerical Methods

For the simulations, we discretize the integration in imaginary time into L steps of size Δ defined by $L\Delta = \beta$ and express the action $S\{\phi\}$ in terms of variable $\phi_{i,j}$,

$$S\{\phi\} = \Delta\epsilon \sum_{i,j}^{L,N} \left[\frac{\gamma}{2} \left(\frac{\phi_{i+1,j} - \phi_{i,j}}{\Delta} \right)^2 + \frac{\gamma}{2}(\phi_{i,j} - \phi_{i,j+1})^2 + \frac{1}{4}(\phi_{i,j}^4 - 1)^2 \right] \quad (6)$$

Here, the subscripts i and j are the labels for the imaginary time and space dimensions. Formally, this action is similar to the discrete, classical, two-dimensional ϕ^4 system [6, 7]. In the present case, however, to insure we are describing quantum behavior, we have to require Δ to be much smaller than the reciprocal of γ.

The conventional, path integral Monte Carlo procedure evaluates the expectation value of a physical quantity $A = A\{\phi\}$, $\langle A \rangle = \int \mathcal{D}\phi\, A\{\phi\}\, e^{-\beta S\{\phi\}}/Z$ by generating a sequence ϕ_i of independent configurations of the ϕ-fields with weight $e^{-\beta S}/Z$. The procedure reduces the computation of the expectation values to the simple summation, $\langle A \rangle = \sum_{i=1}^{M} A\{\phi_i\}/M$ where M is the number of independent configurations generated.

Instead of using the conventional Path Integral Monte Carlo method to generate the ϕ fields, we used the Hybrid Monte Carlo Method [8]. This method combines molecular dynamics and Monte Carlo methods. One consequence is more rapid equilibration of the system and less correlation between successive measurements. Another feature is the high inherent degree of vectorization and hence parallelization. The combination of these features allowed us to simulate efficiently long chains at low temperatures.

To study whether the symmetry of the ground state is broken, we adopted the following strategy: for successively lower values of the temperature T, we study the behavior of the order parameter, the energy, and their mean-squared fluctuations as functions of ϵ and γ. In terms of the discretized action (6), we define a "quasi" inverse temperature $\beta_q \equiv \epsilon$ and a "quasi" (classical) two-dimensional Hamiltonian

$$H_q \equiv \sum_{i,j} \left[\frac{\gamma}{2\Delta}(\phi_{i+1,j} - \phi_{i,j})^2 + \frac{\Delta\gamma}{2}(\phi_{i,j+1} - \phi_{i,j})^2 + \frac{\Delta}{4}(\phi_{i,j}^2 - 1)^2 \right] \quad (7)$$

such that $\beta_q H_q\{\phi\} = S\{\phi\}$. At each value of the physical temperature, we fix γ and then use standard methods to determine whether the system defined by H_q undergoes a transition from the broken to the restored symmetry state at some critical value of β_q, $\beta_{qc} \equiv \epsilon_c$.

Within this strategy, finding the condition for the broken symmetry at a fixed physical temperature for our one-dimensional quantum model is equivalent to finding the critical inverse quasi-temperature ϵ_c for the two-dimensional Hamiltonian H_q. In the absence of infinite-ranged interactions, however, a true phase transition in one-dimension can only occur for an infinite-sized system at zero temperature (i.e., for N and $L \to \infty$). Within our strategy, we search for a phase transition in a two-dimensional system whose inverse temperature is ϵ. Again, a true transition will only occur in an infinite system, but it can occur at a finite value of inverse quasi-temperature ϵ. We seek to determine if such a transition is indicated and if these indications remain as we increase L and N. What distinguishes our quantum simulations from those for the classical system H_q is the need to require that the physical quantities we compute to be independent of Δ to within the accuracy of our calculation.

4. Results

As one indicator of the symmetry state of the system, we took for the order parameter

$$X(\epsilon) \equiv \frac{1}{LN}\langle|\sum_{i,j}^{L,N} \phi_{i,j}|\rangle . \tag{8}$$

As the size of the system N becomes very large and the true temperature $T = 1/\Delta L$ approaches zero, the value of the order parameter will change from a finite positive value to zero if quantum fluctuations are restoring the symmetry. We also studied the susceptibility associated with the order parameter, the energy, and the specific heat.

We simulated a variety of system sizes, ranging from $N = 7$ ($\beta = 4.95$) to $N = 128$ ($\beta = 45.25$). The $N = 128$ calculations were done on a Thinking Machines CM-2 computer. The remainder were done on a Sun Sparcstation 1, a Convex 230, and a Cray X-MP. The simulations on the CM-2 took about 65 minutes of computation time for 40,000 measurements and about 120 minutes for 80,000. This was at least 10 times the speed of the X-MP.

To estimate ϵ_c, we used the Cumulant Intersection Method [6, 9]. To reduce the computation time near the critical point, we used the Histogram Method of Ferrenberg and Swendsen [10] to help estimate the moments of ϕ needed in the cumulant method.

Our principal result is shown in Fig. 1 where we plot the phase diagram as a function of the model parameters ϵ and $1/\gamma$. The straight line with a slope of approximately 2 is the phase boundary estimated from the continuum theory results of (4) and (5). It predicts the existence of a restored symmetry phase above this line. The markers in this figure are the results from the quantum Monte Carlo simulation. Above the curve represented by these points lies the restored symmetry phase. We see that the continuum theory qualitatively predicts the correct physics, but quantitatively the the transition actually occurs more easily.

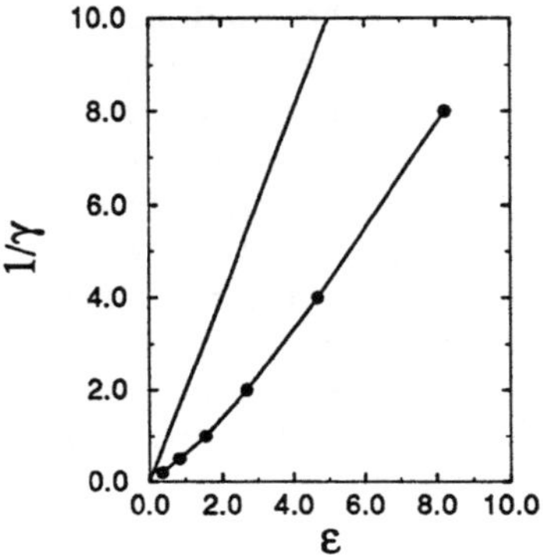

Fig. 1. The phase diagram. The straight line is the phase boundary predicted from the perturbation analysis of the continuum limit; the markers are the results obtained from the simulations. The restored symmetry phase lies above the curve represented by these markers.

5. Concluding Remarks

We have presented numerical evidence that quantum fluctuations can produce a symmetric ground-state in the double-well chain, restoring the symmetry that is broken classically. In particular, we presented the phase diagram for this model that shows the symmetry restoration occurs more easily than predicted by a perturbation theory calculation of the continuum limit of the model. In another paper will report the full details of our analysis and results [11].

We are currently trying to determine the universality class for the model by conformal-charge methods [12, 13]. The implications of our results the hydrogen-halide materials will be discussed elsewhere [13].

Acknowledgments

We thank W. R. Somsky and H. Q. Lin for helpful conversations. The work was supported by the U. S. Department of Energy. We also thank the Advanced Computing Laboratory at the Los Alamos National Laboratory for the use of its facilities.

References

1. E. Fradkin and J. E. Hirsch, Phys. Rev B **27**, 1680 (1982); J. E. Hirsch and E. Fradkin, Phys. Rev. B **27**, 4032 (1983).
2. R. Dashen, B. Hasslacher, and A. Neveu, Phy. Rev. D **10**, 4114, 4139 (1974).

3. For example, R. W. Jansen, R. Bertoncini, D. A. Pinnick, A. I. Katz, R. C. Hanson, O. F. Sankey and M. O'Keeffe, Phys. Rev. B **35**, 9830 (1987).
4. R. P. Feynman and A. R. Hibbs, *Quantum Mechanics and Path Integrals* (McGraw-Hill, New York, 1965).
5. M. Creutz and B. Freedman, Ann. Phys. **132**, 427 (1981).
6. A. Milchev, D. W. Heermann and K. Binder, J. Stat. Phys. 44 , 749 (1986).
7. R. Toral and A. Chakrabari, Phys. Rev. B **42**, 2445 (1990).
8. S. Duane, A. D. Kennedy, B. J. Pendelton, and D. Roweth, Phys. Lett. B **195**, 216 (1987).
9. K. Binder, in *Applications of the Monte Carlo Method to Statistical Physics*, edited by K. Binder (Springer-Verlag, Berlin, 1984), Chap. 1.
10. A. M. Ferrenberg and R. Swendsen, Phys. Rev. Lett. **61**, 2635 (1988); Phys. Rev. Lett. **63**, 1195 (1989).
11. Xidi Wang, D. K. Campbell, J. E. Gubernatis, "Symmmetry breaking in a quantum double-well chain," unpublished.
12. Rajiv R. P. Singh and G. A. Baker, Jr., Phys. Rev. Lett. **61**, 1 (1991).
13. Xidi Wang, D. K. Campbell, G. A. Baker, Jr., and J. E. Gubernatis, "Conformal charge of the two-dimensional ϕ^4 field theory," unpublished.

Numerical Experiments on Frustrated Quantum Spin Systems

S. Miyazawa[1], *M. Makivić*[2], *T. Horiki*[3], and *S. Homma*[1]

[1]Department of Computer Science, Department of Physics, Gunma University, Kiryu 376, Japan
[2]Department of Physics, The Ohio State University, Columbus, OH 43210, USA
[3]The Institute of Physical and Chemical Research, Wako 351-01, Japan

Abstract. Thermodynamic properties of the antiferromagnetic Heisenberg model with frustration on the square lattice are investigated using quantum simulation based on the decoupled cell Monte Carlo method. The effect of frustration is introduced through an additional next-nearest neighbor antiferromagnetic interaction J_2. With increasing J_2, certain spin ordering was observed at low temperatures.

The discovery of high T_c superconductivity renewed the interest in two- dimensional quantum spin systems [1]. Among these systems the antiferromagnetic Heisenberg model with frustration on the square lattice (FAFH) is an important one due to its relation to the doped oxide superconductors. The frustration is introduced through an additional next-nearest neighbor exchange interaction along the diagonal of the plaquettes of the lattice. In this short note, we study the thermodynamic properties of the spin-1/2 FAFH on the square lattice, defined by the Hamiltonian

$$H = 2J_1 \sum_{\mathbf{i},\epsilon} \mathbf{S_i S_{i+\epsilon}} + 2J_2 \sum_{\mathbf{i},\delta} \mathbf{S_i S_{i+\delta}} \tag{1}$$

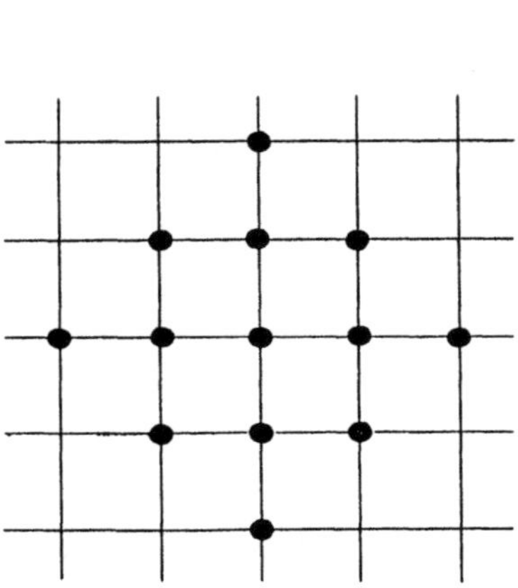

Fig.1

Springer Proceedings in Physics, Vol. 70 Computational Approaches in Condensed-Matter Physics
Editors: S. Miyashita · M. Imada · H. Takayama

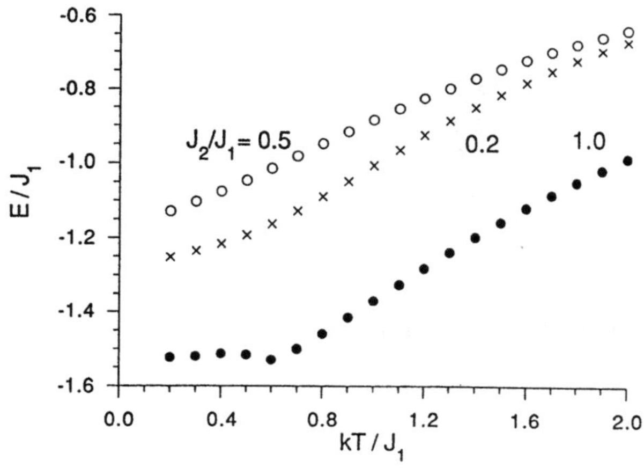

Fig.2

Here, $\mathbf{i}$ denotes a site of the square lattice, while ϵ are vectors along x or y directions, and δ are along the diagonals of a plaquette. The energy scale is fixed by setting $J_1 = 1$. We performed a quantum Monte Carlo simulation based on the decoupled cell Monte Carlo method (DCM) [2]. The basic ingredient of DCM is to calculate transition probabilities from a spin state to another one on a cell of finite size, where the central spin of the cell makes a transition from up(down) to down(up). Transition probabilities thus derived generate a sample of the Markov chain. The time average of this Markov chain gives relevant thermodynamic quantities. We calculated energy E, uniform magnetic susceptibilty χ and spin pair correlation function $C(\mathbf{r})$, defined by:

$$E = \frac{1}{N} < H >, \quad \chi = \frac{1}{NT} < (\sum_{\mathbf{i}} S_{\mathbf{i}}^z)^2 >, \quad C(\mathbf{r}) = \frac{1}{N} \sum_{\mathbf{i}} < S_{\mathbf{i}}^z S_{\mathbf{i+r}}^z > \quad (2)$$

where $< \Omega >$ denotes the canonical average of Ω. The decoupled cell (DC) here includes 13 spins, with free boundaries, as depicted in Fig. 1. In DC, the central spin is flipped with all the remaining spins being fixed. The total numbers N of lattice sites are 32 X 32 and 64 X 64, with periodic boundary conditions. The simulation begins at $T = 2.0J_1$ from a random initial spin configuration. The system is then cooled in steps of $\Delta T = 0.1J_1$.

We now present results for the 64 X 64 system. In Fig. 2 we show the temperature dependence of E for various values of J_2. For $J_2 = 0.5$, the results of frustration are so strong that they prevent formation of some spin ordering. For $J_2 = 1.0$ we obtain nonuniform behavior of E near $kT/J_1 \simeq 0.6$, resulting in negative specific heat. At the present time, we do not have conclusions whether this is due to finitness of DC or an indication of the formation of spin ordering depicted in Fig. 4, which is observed below $KT/J_1 = 0.5$ for $J_2 = 1.0$ in our calculation. In Fig. 3 we show the dependence of χ on J_2. With increasing J_2, χ decreases. We could not find a nonuniform behavior of the susceptibility in the temperature region of our simulations. Present work is a preliminary one, and further, more detailed calculations will be published in the near future.

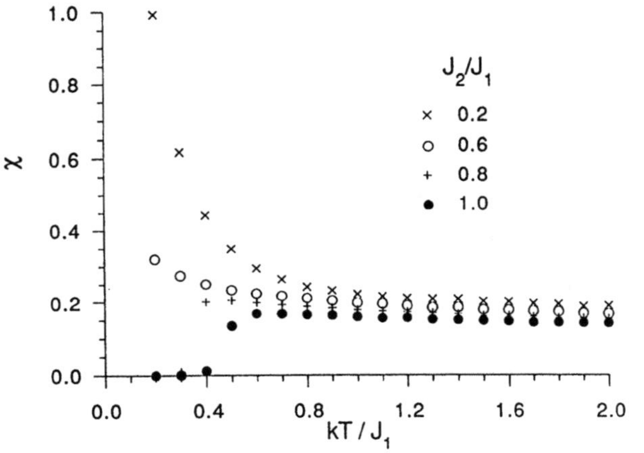

Fig.3

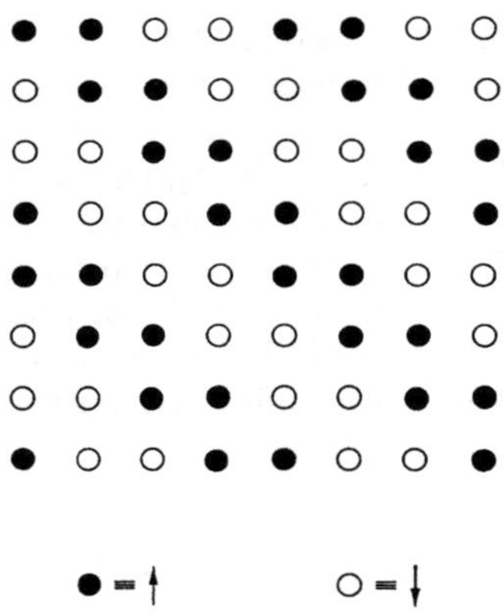

Fig.4

References

[1] E. Manousakis, Rev. Mod. Phys. **63**, 1 (1991).
[2] S. Homma, H. Matsuda, and N. Ogita, Prog. Theor. Phys. **75**, 1058 (1986).

Haldane Antiferromagnet in a Magnetic Field

T. Sakai[1] *and M. Takahashi*[2]

[1]Faculty of Science, Himeji Institute of Technology, Kamigori-cho, Ako-gun, Hyogo 678-12, Japan
[2]Institute for Solid State Physics, University of Tokyo, Roppongi 7-22-1, Minato-ku, Tokyo 106, Japan

Abstract. $S = 1$ antiferromagnetic Heisenberg chain in a magnetic field at $T = 0$ is studied by numerical diagonalizations up to $N = 16$.

Antiferromagnetic quantum spin chains have attracted great interest, since Haldane predicted [1] that an energy gap exists for integral S. We study the $S = 1$ antiferromagnetic Heisenberg chain in a magnetic field H at $T = 0$ by numerical diagonalizations up to $N = 16$. The Hamiltonian is

$$\mathcal{H} = \mathcal{H}_0 + \mathcal{H}_1, \qquad \mathcal{H}_0 = \sum_j \boldsymbol{S}_j \cdot \boldsymbol{S}_{j+1}, \qquad \mathcal{H}_1 = -H \sum_j S_j^z. \tag{1}$$

Using the Lanczos algorithm for the N-site system (P.B.C.), we calculate $E_k(M)$, which is the lowest energy of $\mathcal{H}_0$ for $\sum_j S_j^z = M$ and wave vector k, and define the lowest one among all $E_k(M)$'s as $E(M)$. We define two critical fields H_{c1} and H_{c2}; the ground state has $M = 0$ for $H < H_{c1}(= \Delta)$ and $M = N$ for $H > H_{c2}(= 4)$, where Δ is the Haldane gap, estimated as $\Delta = 0.411 \pm 0.001$ [2].

We use the asymptotic form of the spin-excitation energy

$$E(M+1) - E(M) \sim \epsilon'(m) + \epsilon''(m)/2N^{-1} + O(N^{-2}), \tag{2}$$

where $\epsilon(m) \equiv \lim_{N\to\infty} E(M)/N$ ($m = M/N$ is fixed), and the relations $\epsilon'(m) = H$ and $dm/dH = 1/\epsilon''(m)$, to get the magnetization curve and dm/dH curve (Fig. 1) [2].

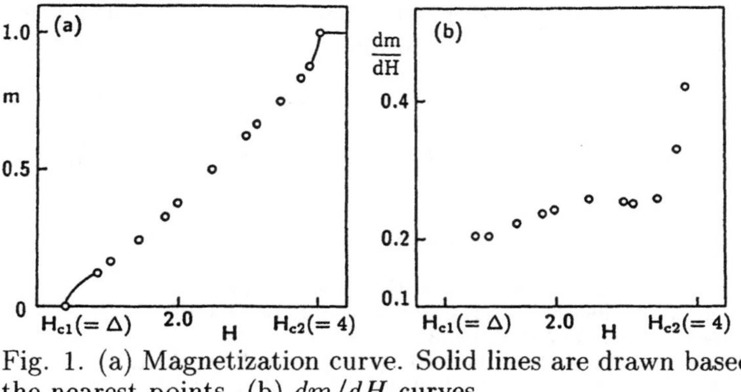

Fig. 1. (a) Magnetization curve. Solid lines are drawn based on (5) from the nearest points. (b) dm/dH curves.

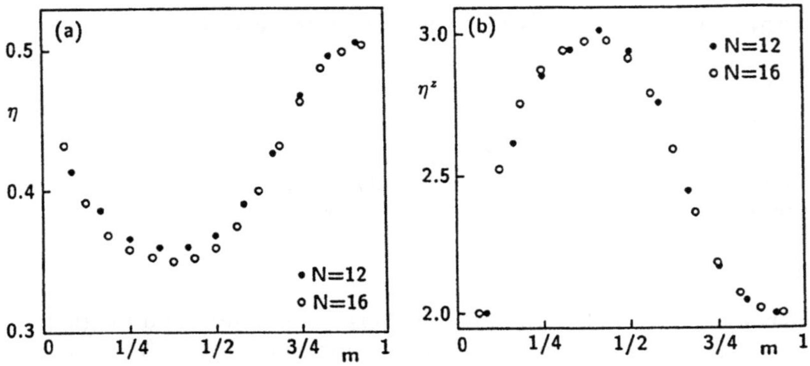

Fig. 2. (a) η vs m (b) η^z vs m plotted for N =12 and 16.

Next we study spin correlations[3]. As the ground state of $\mathcal{H}$ is massless for $0 < m < 1$[2], we assume that spin correlations have the asymptotic forms

$$\langle S_0^x S_r^x \rangle \simeq (-1)^r r^{-\eta}, \qquad \langle S_0^z S_r^z \rangle - m^2 \simeq \cos(2k_F r) r^{-\eta^z}, \tag{3}$$

where $2k_F$ is a soft mode, which is given by $2k_F = 2\pi M/N$ for even M and $2k_F = \pi - 2\pi M/N$ for odd M. Conformal invariance predicts [4] that the energy gaps for the spin excitation and the soft mode depend on N as

$$E(M+1) - E(M) - H \simeq \pi v_S \eta \frac{1}{N}, \qquad E_{2k_F}(M) - E(M) \simeq \pi v_S \eta^z \frac{1}{N}, \tag{4}$$

where v_S is the sound velocity. We use (4) to estimate exponents η and η^z for N =12 and 16 (Fig. 2). Those converge well with respect to N. In addition we checked that $\eta\eta^z = 1$ is satisfied within a few percent for $N = 16$. It suggests that the Luttinger liquid concept [5] is satisfied for $0 < m < 1$. By extrapolations we get $\eta = 1/2$ and $\eta^z = 2$ at m =0 and 1. It means that the elementary excitation is described by a free-Fermion system, which has a dispersion linear to k^2. Thus the forms of the magnetization curve near the two limits are

$$m \simeq (H - H_{c1})^{1/2} \quad (m \to 0), \quad 1 - m^2 \simeq (H_{c2} - H)^{1/2} \quad (m \to 1). \tag{5}$$

We draw the curves based on (5) from the nearest points in Fig. 1 (a).

References

[1] F. D. M. Haldane: Phys. Lett. **93A** (1983) 464 ; Phys. Rev. Lett. **50** (1983) 1153.
[2] T. Sakai and M. Takahashi: Phys. Rev. **B43** (1991) 13383.
[3] T. Sakai and M. Takahashi: J. Phys. Soc. Jpn. **60** (1991) 3615.
[4] J. L. Cardy: J. Phys. **A17** (1984) L385.
[5] F. D. M. Haldane: Phys. Rev. Lett. **45** (1980) 1358; Phys. Lett. **81A** (1981) 153; J. Phys. **C14** (1981) 2585.

Spin Correlations in the S=1 XXZ Chain

K. Kubo

Institute of Physics, University of Tsukuba, Ibaraki 305, Japan

Abstract. The spin correlations in the S=1 XXZ chain are studied numerically by using quantum transfer matrix. The temperature variation of the correlation lengths and the susceptibilities have been examined.

Since Haldane predicted a novel ground state in the S=*integer* antiferromagnetic(AF) chains, much effort has been done to clarify the ground state of the S=1 XXZ chain described by the Hamiltonian

$$H = \sum_i (S_i^x S_{i+1}^x + S_i^y S_{i+1}^y + \Delta S_i^z S_{i+1}^z),$$

where S_i^α is the α-component of the spin operator with size one at the site i and Δ denotes the anisotropy parameter. As the result it has become a general consensus that the system has four ground state phases, i.e. the ferromagnetic Ising phase for $\Delta < -1$, the XY phase for $-1 < \Delta < \Delta_1$ without long range order(LRO) and with gapless excitations, the Haldane phase for $\Delta_1 < \Delta < \Delta_2$ without usual LRO but with string LRO and an excitation gap and the AF Ising phase with an AF LRO. The value of Δ_2 is known to be about 1.2 while the estimates of Δ_1 still scatter from 0 to 0.3.

We have calculated the usual and the canonical correlation functions as well as the string ones[1] $-\langle S_0^\alpha \exp(i\pi \sum_{l=1}^{i-1} S_l^\alpha) S_i^\alpha \rangle$ at finite temperatures for $\Delta = 0$, 1 and 1.4. We have used the quantum transfer matrix[2] with Trotter slicings n up to seven. The quantum mechanical limit has been obtained by extrapolating the sets of approximants with n=(5,6,7) and (4,5,6,7). The results of two extrapolations agree within two percent at temperatures studied.

We have examined the correlation length ξ_x of S^x-correlations and found that ξ_x^{-1} linearly vanishes at $T = 0$ for $\Delta = 0$ while it approaches a constant for $\Delta = 1$. For $\Delta = 1$ $\xi_x(T = 0)$ is estimated to be about 6.5 and it decreases with Δ.

In Fig.1(a) the correlation length ξ_{sz} of the S_z-string correlations is shown. At $\Delta = 0$ ξ_{sz} diverges as T^{-1} while it diverges apparently as

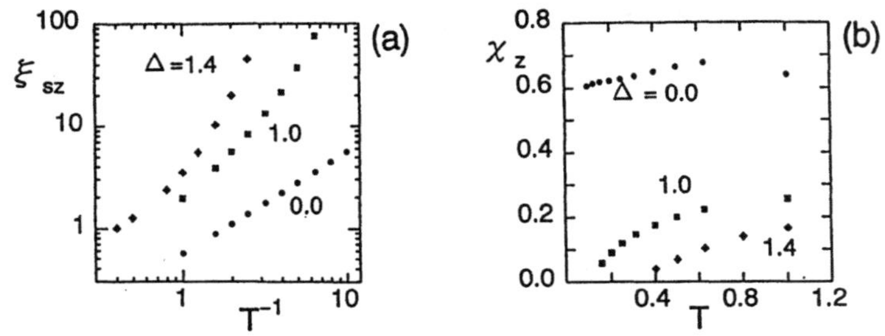

Fig. 1 The correlation length ξ_{sz} of the S_z-string correlations(a) and χ_z(b) for Δ=0(circles), 1(squares) and 1.4(diamonds).

$T^a \exp(E_0/T)$ at $\Delta = 1$ and 1.4. We estimate $a = -1$ and $E_0 = .33$ for $\Delta = 1$ and $a = -1/2$ and $E_0 = 1.41$ for $\Delta = 1.4$.

The susceptibility in z-direction χ_z approaches a finite value at $\Delta = 0$ as $T \to 0$ while it decreases rapidly and vanishes at $T = 0$ as $T^b \exp(-E_1/T)$ at $\Delta = 1$ and 1.4 as is shown in Fig. 1(b). We estimate $b = -1/2$ and $E_1 = 0.40$ for $\Delta = 1$ and $b = -1$ and $E_1 = 1.56$ for $\Delta = 1.4$. We have also calculated the staggerd susceptibility in x-direction which diverges as $T^{-1.80}$ at $\Delta = 0$ and approaches a finite value at $\Delta = 1$ as $T \to 0$.

The present results at $\Delta = 1$ and 1.4 are consistent with the existence of an exctation gap as well as the ground state string LRO. It is puzzling, however, that the energy gap estimated from ξ_{sz} and χ_z disagree. Also the temperature dependence of the prefactor at $\Delta = 1$ differs from that at $\Delta = 1.4$. We need more study at lower temperatures to see whether these disagreements are real behavior in the low temperature limit.

References

[1] M. den Nijs and K. Rommelse, Phys. Rev. **B 40**, 4709 (1989).
[2] S. Takada and K. Kubo, J. Phys.Soc. Jpn. **55**, 1671 (1986).

Crossover Between the Haldane Gap Phase and the Dimer Phase in the Spin 1/2 Alternating Heisenberg Chain

K. Hida

Department of Physics, College of Liberal Arts,
Saitama University, Urawa, Saitama 338, Japan

Abstract. The ground state of the alternating spin 1/2 Heisenberg chain with couplings $J(>0)$ and J' is studied. This model interpolates the S=1 and S=1/2 antiferromagnetic Heisenberg chains continuously. The behavior of the string order and energy gap indicates that the ground state over the whole range $-\infty \leq J' < J$ can be regarded as a single phase. The physical picture of the ground state and the first excited state is also discussed.

The antiferromagnetic Heisenberg chain (AFHC) with spin S has two essentially different ground states depending on whether $S =$ integer (Haldane state) or $S =$ half-integer(spin liquid state)[1]. As a model interpolating the $S = 1$ and $S = 1/2$ AFHC, we study the alternating Heisenberg chain with spin 1/2;

$$H = 2J' \sum_{i=1}^{N} \boldsymbol{S}_{2i-1}\boldsymbol{S}_{2i} + 2J \sum_{i=1}^{N} \boldsymbol{S}_{2i}\boldsymbol{S}_{2i+1}, \ (J > 0) \tag{1}$$

As J' increases from $-\infty$ to J, the ground state changes continuously from the Haldane phase to the gapless spin liquid phase ($J = J'$). Extending the definition of the string order O_{str}[2,3] to the present case, we study the behavior of O_{str} and the energy gap to clarify the nature of these phases.

Hamiltonian (1) is diagonalized for $N \leq 24$ using the program package TITPACK ver. 2. The values of O_{str} extrapolated to $N \to \infty$ are finite for $-\infty \leq J' < J$ (Fig. 1). For $J \simeq J'$ (dimer phase), the bosonization method shows $O_{\mathrm{str}} \sim (1 - J'/J)^{1/6}$ within the self consistent harmonic approximation. For $J' = 0$, O_{str} takes the maximum(=1). Therefore O_{str} can be regarded as the measure of the local singlet correlation on J-bond. It survives the perturbation by J'-bonds both in the Haldane and dimer phases, while it is killed in the spin liquid phase.

The energy gap is finite except $J' = J$ and has a cusp at $J' = 0$. This can be understood by the first order perturbation in J'. This calculation reveals that the first excited state is the extended plane wave of a local triplet pair for small J'. Because no discontinuity is found for $0 > J' >$

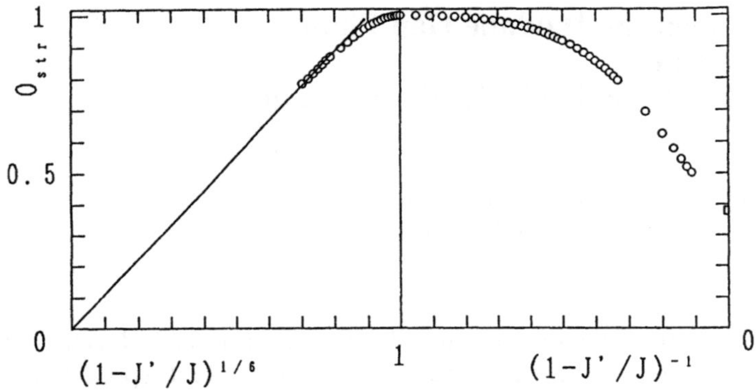

Fig.1 : The string order parameter O_{str} plotted against $(1 - J'/J)^{1/6}$ for $1 > J'/J > 0$, and against $(1 - J'/J)^{-1}$ for $0 > J'/J > -\infty$. The solid line is the guide for eye. The open square is the value for spin-1 AFHC.

$-\infty$, the excited state in the Haldane phase would also keep this character. This picture gives an intuitive explanation of the reversal of anisotropy effect on the excitation spectrum of the Haldane phase[4]. Further detailed results will be published elsewhere[4].

The author is indebted to H. Nishimori for TITPACK ver.2.

References

[1] F.D.M. Haldane: Phys. Lett. **93A** 464 (1983); Phys. Rev. Lett. **50** 1153 (1983).

[2] M. den Nijs and K. Rommelse: Phys. Rev. **B40** 4709 (1989).

[3] H. Tasaki: Phys. Rev. Lett. **66** 798 (1991).

[4] K. Hida: submitted to Phys. Rev (1991).

Ground-State Properties of the Frustrated One-Dimensional S=1 Heisenberg Model

T. Tonegawa[1], *M. Kaburagi*[2], *N. Ichikawa*[1], *and I. Harada*[3]

[1]Department of Physics, Kobe University, Rokkodai, Kobe 657, Japan
[2]College of Liberal Arts, Kobe University, Tsurukabuto, Nada, Kobe 657, Japan
[3]Department of Physics, Okayama University, Tsushima, Okayama 700, Japan

Abstract. Using the method of exact diagonalization of finite-size systems, we have investigated the ground-state properties in the thermodynamic limit of the one-dimensional, $S = 1$, Heisenberg model with antiferromagnetic nearest-neighbor and next-nearest-neighbor interactions. It is found that the competition between both interactions stabilizes the Haldane phase.

The one-dimensional, antiferromagnetic, nearest-neighbor (nn), $S=1$, Heisenberg model has been the subject of numerous theoretical and experimental studies [1]. These are motivated mainly by Haldane's prediction [2] that the ground state of the integer-spin case, in contrast to that of the half-integer-spin case, is in a massive phase characterized by a finite energy gap in the excitation spectrum and by an exponential decay of the two-spin correlation functions. This phase is now called the Haldane phase. Almost all results of the above studies support the prediction, and it is widely agreed that the prediction is correct.

The purpose of this paper is to explore the ground-state properties of the one-dimensional, $S = 1$, Heisenberg model with antiferromagnetic nn and next-nearest-neighbor (nnn) interactions, which is described by the Hamiltonian

$$\mathcal{H} = 2J_1 \sum_{\ell=1}^{N} \vec{S}_\ell \cdot \vec{S}_{\ell+1} + 2J_2 \sum_{\ell=1}^{N} \vec{S}_\ell \cdot \vec{S}_{\ell+2} \tag{1}$$

with periodic boundary conditions $\vec{S}_{\ell+N} = \vec{S}_\ell$. Here $\vec{S}_\ell$ is the $S=1$ spin operator at the ℓth site; $J_1(>0)$ and $J_2(\geq 0)$ are the nn and nnn interaction constants, respectively; N is the number of spins in the system and is assumed to be even. Hereafter, the ratio of J_2 to J_1 is denoted by j, that is, $j=J_2/J_1$.

In order to investigate the ground-state properties of the system in the thermodynamic limit ($N\rightarrow\infty$), we first calculate exactly a variety of physical quantities for finite-size ($N=6, 8, \cdots, 16$) systems by diagonalizing matrices representing the Hamiltonian of Eq. (1) and then extrapolate the results to the limit of $N\rightarrow\infty$. In the calculation we have employed the computer program package coded referring to Nishimori's TITPACK Ver. 2 for the spin-1/2 system [3] and using Nishino's new idea for the two-dimensional search method. Details of the coding technique are discussed in a separate paper in this meeting [4]. According to our calculation, the ground state belongs to the $M=0$ subspace, M being the z-component of the total spin, for any set of the values of j and N. Thus we conclude that the limiting ($N\rightarrow\infty$) ground state is characterized, irrespectively of j, by $M=0$. We denote the ground-state energy and the ground-state eigenfunction for a given value of N by $E_g(N)$ and $\Phi_g(N)$, respectively.

We now turn to the discussion of the results of the calculation. First we discuss the singlet-triplet energy gap $\Delta(N)=E_1(N)-E_g(N)$, where $E_1(N)$ is the lowest eigenvalue of the matrix in the $M=1$ subspace. We have calculated $\Delta(6)$, $\Delta(8)$, $\cdots$, $\Delta(16)$ for various values of j, and have extraporated them by using Shanks' transformation [5] to

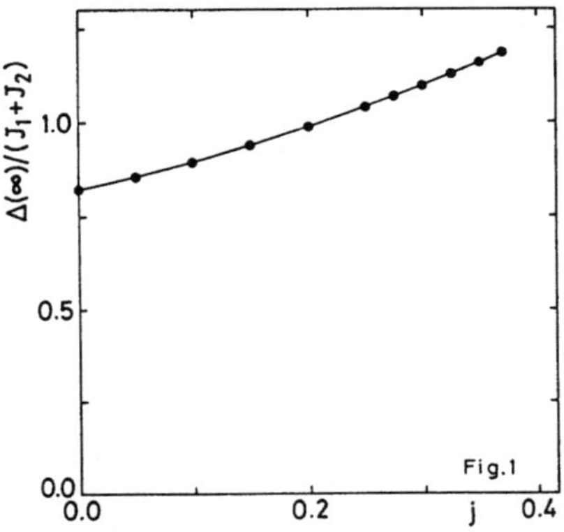

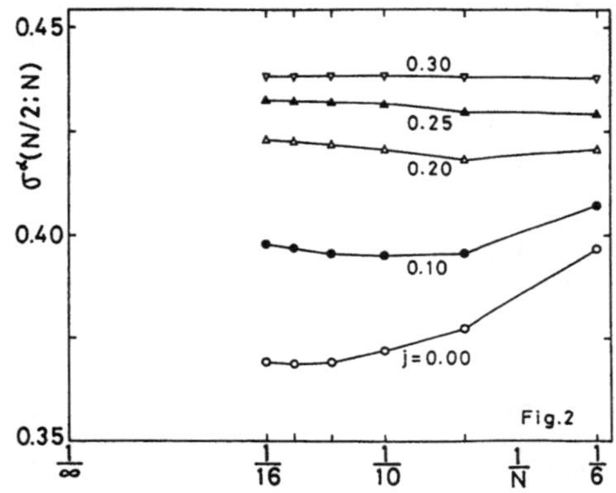

Fig. 1 (left). Plot of $\Delta(\infty)/(J_1 + J_2)$ versus j.
Fig. 2 (right). Plot of $\sigma^\alpha(N/2\,;N)$ versus $1/N$ for various values of j.

estimate $\Delta(\infty)$ which is the quantity corresponding to the Haldane gap. The results are shown in Fig. 1, where $\Delta(\infty)/(J_1 + J_2)$ is plotted as a function of j [6]. Note that this quantity increases with increasing j.

Next we discuss the den Nijs and Rommelse string order parameter [7] defined by $\mathcal{O}^\alpha_{\text{string}} = \lim_{N\to\infty} \sigma^\alpha(N/2\,;N)$ with $\sigma^\alpha(n\,;N) = -\langle\Phi_g(N)|S_1^\alpha \exp(i\pi \sum_{\ell=2}^{n} S_\ell^\alpha) S_{1+n}^\alpha|\Phi_g(N)\rangle$ $(\alpha = x, y, z)$. As has been discussed by Kennedy and Tasaki [8], the fact that $\mathcal{O}^\alpha_{\text{string}}$ is finite is a reflection of the breaking of a hidden $Z_2 \times Z_2$ symmetry, which brings the Haldane gap $\Delta(\infty)$. In Fig. 2 we plot $\sigma^\alpha(N/2\,;N)$ as a function of $1/N$ for various values of j [9]. From this result we see that $\mathcal{O}^\alpha_{\text{string}}$ is finite and increases with increasing j.

In conclusion, we have shown that the competition between the antiferromagnetic nn and nnn interactions stabilizes the Haldane phase.

We wish to thank Professor H. Nishimori for providing us his TITPACK Ver. 2 and to Mr. T. Nishino for his useful comments on coding the program.

References

[1] For a review see I. Affleck, J. Phys. Condens. Matter **1** (1989) 3047.
[2] F. D. M. Haldane, Phys. Lett. **93A** (1983) 464; Phys. Rev. Lett. **50** (1983) 1153.
[3] H. Nishimori, Bussei Kenkyuu **56** (1991) 494 (in Japanese).
[4] M. Kaburagi, T. Tonegawa and T. Nishino, these proceedings.
[5] D. Shanks, J. Math. Phys. **34** (1955) 1.
[6] Our value of $\Delta(\infty)$ for $J_2 = 0$ agrees, of course, with the value reported previously by T. Sakai and M. Takahashi, Phys. Rev. B **42** (1990) 1090.
[7] M. den Nijs and K. Rommelse, Phys. Rev. B **40** (1989) 4709.
[8] T. Kennedy and H. Tasaki, preprints.
[9] The values of $\sigma^\alpha(n\,;N)$ for finite-size systems of up to 14 spins with nn bilinear and biquadratic interactions have been obtained by S. M. Girvin and D. P. Arovas, Physica Scripta **T27** (1989) 156 and by Y. Hatsugai and M. Kohmoto, preprint. We note that the latter authors have also discussed systems with nn bilinear interaction and single-ion-type anisotropy energy.

Numerical Study of Quantum Spin Systems

M. Kaburagi[1], *T. Tonegawa*[2], *and T. Nishino*[3]

[1]College of Liberal Arts, Kobe University, Tsurukabuto, Nada, Kobe 657, Japan
[2]Department of Physics, Kobe University, Rokkodai, Kobe 567, Japan
[3]Department of Physics, Osaka University, Toyonaka, Osaka 560, Japan

Abstract. We have developed a new coding technique of the Lanczös method for exact diagonalization of quantum spin systems. The technique enables us to save the memory-space and CPU time in carrying out numerical calculation for substantially large system size. The algorithm, namely "*subspace coding*", is discussed in detail.

1. Introduction

In many body physics, there are quite few rigorously solvable models and therefore we have to use numerical technique in many cases in studies of quantum systems. For studies of ground state and low-lying excited states, exact diagonalization by the Lanczös method is one of the most useful tools and has played an important role [1]. The purpose of this paper is to explain a new coding technique of the Lanczös method by employing, as a simple model, the spin-S XXZ model on N lattice sites with the symmetry of total S^z conservation. The general form of our Hamiltonian is given as

$$\mathcal{H} = \sum\nolimits_{<ij>} \{ J^x_{ij} (S^+_i S^-_j + S^-_i S^+_j) + 2J^z_{ij} S^z_i S^z_j \} \; . \tag{1}$$

Here S^α_i is the spin-S spin operator on site i ($i=0, \cdots, N-1$); $<ij>$ denotes all interacting pairs of sites; J^α_{ij} is the interaction constant between sites i and j. Although the methodology of our technique is essentially a combination of Lin's two-dimensional search [2] and Nishimori's elements storage method [3], the combination of the two techniques provides us a powerful tool for the numerical studies.

2. Construction of Storage / Lookup Table

With widely used ($2S$+1)-adic notation, we represent the state of the spin configuration by an integer I as $I = \sum_i (S^z_i + S)(2S+1)^i$. In order to employ the two-dimensional search, we divide the lattice into two parts L and H consisting of $N_\mathrm{L} = (N+1)/2$ and $N_\mathrm{H} = N - N_\mathrm{L}$ spins, respectively. The Hamiltonian (1) is then divided into six parts as

$$\mathcal{H} = \mathcal{H}_\mathrm{diag}(\mathrm{LL}) + \mathcal{H}_\mathrm{diag}(\mathrm{HH}) + \mathcal{H}_\mathrm{diag}(\mathrm{LH}) + \mathcal{H}_\mathrm{off}(\mathrm{LL}) + \mathcal{H}_\mathrm{off}(\mathrm{HH}) + \mathcal{H}_\mathrm{off}(\mathrm{LH}) \; , \tag{2}$$

where $\mathcal{H}_\mathrm{diag}(\mathrm{PQ})$ and $\mathcal{H}_\mathrm{off}(\mathrm{PQ})$ represent, respectively, the diagonal and off-diagonal parts containing the interactions between P and Q parts. We note here that in implementing the two-dimensional search algorithm into vectorized code for a certain machine, we have to find the optimal division of the Hamiltonian for the loop unrolling.

We first classify the states I_L (I_H) of L (H) part by total magnetization m_L (m_H) and store them to "storage table" I with number k (j) as $\mathrm{I}(m_\mathrm{L},\mathrm{k}) = I_\mathrm{L}$ ($\mathrm{k}=0, \cdots, \mathrm{n_L}(m_\mathrm{L})-1$) ($\mathrm{I}(m_\mathrm{H},\mathrm{j}) = I_\mathrm{H}$ ($\mathrm{j}=0, \cdots, \mathrm{n_H}(m_\mathrm{H})-1$)), where $\mathrm{n_L}(m_\mathrm{L})$ ($\mathrm{n_H}(m_\mathrm{H})$) is the number of states of L (H) part for a given value of m_L (m_H); the wave function (WF) corresponding to I_L (I_H) is denoted by $\phi_\mathrm{L}(m_\mathrm{L},\mathrm{k})$ ($\phi_\mathrm{H}(m_\mathrm{H},\mathrm{j})$). We also construct "lookup table" J by introducing the inverse function $\mathrm{J}(I_\mathrm{L}) = \mathrm{k}$ ($\mathrm{J}(I_\mathrm{H}) = \mathrm{j}$). The bases $\Phi_\mathrm{K} = \phi_\mathrm{L}(m_\mathrm{L},\mathrm{k})\phi_\mathrm{H}(m_\mathrm{H},\mathrm{j})$ of the eigen WF for a given value of total magnetization $m = m_\mathrm{L} + m_\mathrm{H}$ are classified into several groups by values of m_L and m_H as $\{\phi_\mathrm{L}(m^\mathrm{i}_\mathrm{L},\mathrm{k})\phi_\mathrm{H}(m^\mathrm{i}_\mathrm{H},\mathrm{j}),\ m^\mathrm{i}_\mathrm{L} + m^\mathrm{i}_\mathrm{H} = m\}$.

Springer Proceedings in Physics, Vol. 70 Computational Approaches in Condensed-Matter Physics
Editors: S. Miyashita · M. Imada · H. Takayama

The position number K of the base $\phi_L(m_L^i, k)\phi_H(m_H^i, j)$ in the eigen WF is given as $K = BASE(i) + k + jn(m_L^i)$ ($K = 0, \cdots, n(m) - 1$), where $n(m)$ is the total number of the bases for a given value of m; BASE(i) is the starting position of the i-th group given by $\sum_{m<i} n_L(m_L^m) n_H(m_H^m)$. It should be noted that the maximum length of $n_L(m_L)$ ($n_H(m_H)$) is about $\sqrt{n(m)}$ and much smaller than $n(m)$ and that the "tables" I and J can be used commonly for both L and H parts with replacing m_L (m_H) by $B_L = m_L + N_L S$ ($B_H = m_H + N_H S$). This algorithm of construction of the "storage / lookup table" is therefore a generalization and an improvement of Lin's algorithm. It enables us to save the memory-space, as discussed by Lin [2], and also CPU time significantly in constructing the tables. Another advantage of the present algorithm is that the above construction can be applied to an arbitrary division of the Hamiltonian.

3. Matrix Elements Storage Method

In a usual exact diagonalization for large quantum system by the Lanczös method, because of the limit of memory-space, the matrix elements for the Hamiltonian are calculated repeatedly at each step of the recursions. This is one of the bottle-necks for the Lanczös method. Recently Nishimori devised an excellent storage method of the matrix elements to avoid this bottle-neck [3]. His method is, however, applicable only to the system with relatively small size. In his method, the elements are saved in such a way that after multiplying the $<ij>$-bond Hamiltonian to Φ_K as, for example, $J_{ij}^x S_i^+ S_j^- \Phi_K = A_{ij} \Phi_{K'}$, we store the resultant off-diagonal (diagonal) element A_{ij} (D_{ij}) / (state number K') to array ELM / (LOC) as $ELM_{<ij>}(K) = A_{ij}$ ($ELM_{b+1}(K) = D_{ij}$) / ($LOC_{<ij>}(K) = K'$), where b is the total number of bonds. The length of the arrays required for the double precision of ELM is given by $(3b+1)n(m)$.

In order to reduce the size of the storage array, we combine this method with the expression of the bases given in the previous section. The matrix elements of $\mathcal{H}_{\rm off}(LL)$ in this expression are calculated as

$$\begin{aligned} \langle \phi_L(m_L, k)\phi_H(m_H, j) \mid \mathcal{H}_{\rm off}(LL) \mid \phi_L(m_L', k')\phi_H(m_H', j') \rangle \\ = A_L(k, k')\, \delta(m_L, m_L')\, \delta(m_H, m_H')\, \delta(j, j') , \end{aligned} \tag{3}$$

where $\delta(\cdots, \cdots)$ is the Kronecker delta. That is, the Hamiltonian $\mathcal{H}_{\rm off}(LL)$ is diagonal with respect to ϕ_H. The length of the LL-elements $A_L(k, k')$ is roughly given by $(3b_{LL}+1)\sqrt{n(m)}$, including the diagonal elements. Here b_{LL} is the number of LL-bonds. Similarly the length of the HH-elements is estimated as $(3b_{HH}+1)\sqrt{n(m)}$, where b_{HH} is the number of HH-bonds. As for $\mathcal{H}_{\rm off}(LH)$ and $\mathcal{H}_{\rm diag}(LH)$, we store the elements of boundary spins S_i^α appearing in $\mathcal{H}_{\rm off}(LH)$ and $\mathcal{H}_{\rm diag}(LH)$. It should be noted that for $S_i^+ S_j^-$ in $\mathcal{H}_{\rm off}(LH)$, only the elements between states with $m_L - m_L' = \pm 1$ ($m_H - m_H' = \mp 1$) are nonzero. The total length of these LH-elements is roughly estimated as $3b_{LH}n(m)$. Here b_{LH} is the number of boundary spins and fairly small compared with b_{LL} or b_{HH}.

As shown in the above, the length of the storage array is reduced from $O(n(m))$ to $O\left(\sqrt{n(m)}\right)$ With this reduction, we can apply the storage method to the system with substantially large size. We call this storage method "subspace coding". Details of the coding and its applications [4] will be published elsewhere.

References

[1] See, for example, T. Sakai and M. Takahashi, Phys. Rev. B **43** (1991) 13383.
[2] H. Q. Lin, Phys. Rev. B **42** (1990) 6561.
[3] H. Nishimori, Bussei Kenkyuu **56** (1991) 494 (in Japanese): We thank to Professor H. Nishimori for providing us TITPACK Ver. 2 coded by him.
[4] T. Tonegawa, M. Kaburagi, N. Ichikawa and I. Harada, these proceedings and to be submitted to J. Phys. Soc. Jpn.

The Ground State of Frustrated Quantum Spin Systems and the Marshall Condition

H. Kitatani

Department of Sciences and Mathematics, Nagaoka University of Technology, Kami-Tomioka, Nagaoka 940-21, Japan

Abstract. In some quantum Heisenberg models with competing interactions, the Marshall condition has been found to hold in some finite region of a parameter which indicates the strength of competing interactions, even if the number of spins, N, goes infinity. Moreover, it has been found that the region mentioned above seems to coincide with the region where the Néel order exists.

1. Introduction

An antiferromagnetic quantum Heisenberg model ($S = 1/2$) without competing interactions satisfies the Marshall condition, namely, we can make all the components of the ground state vector into a same sign by the gauge transformation of the spins($S_i^x \to -S_i^x$, $S_i^y \to -S_i^y$, $S_i^z \to S_i^z$) on one sublattice[1]. (We use the bases with $z-$axis quantized.)

Consider the case that we put some competing interactions in a spin system and strengthen these interactions. It is the purpose of this paper to clarify whether the Marshall condition suddenly breaks or remains to hold until some strength of competing interactions with $N \to \infty$ (N is the number of spins).

2. The Marshall Condition

We investigate the ground state of an antiferromagnetic quantum Heisenberg model with nearest and next nearest neighbor interactions on a square lattice, whose Hamiltonian is written as

$$\mathcal{H} = (1-p)J \sum_{n.n} \boldsymbol{S}_i \cdot \boldsymbol{S}_j + pJ \sum_{n.n.n} \boldsymbol{S}_i \cdot \boldsymbol{S}_j,$$

where $J > 0$ and $0 \leq p \leq 1$. At $p = 0$, this model has no competing interaction, so that the Marshall condition holds. At $p > 0$, the model has been studied by several authors[2]. It has been found that the ground state is a Néel state at $0 \leq p < 1/3$, and a collinear state at $1/3 < p$. There is a controversy about the existence of a spin fluid state near $p = 1/3$.

We fix ourselves in a space where $S_{total}^z = 0$. By direct diagonalization using TITPACK ver.2, we get the ground state vector at p with the number of spins, N. Then, we compare each component of the vector with the corresponding

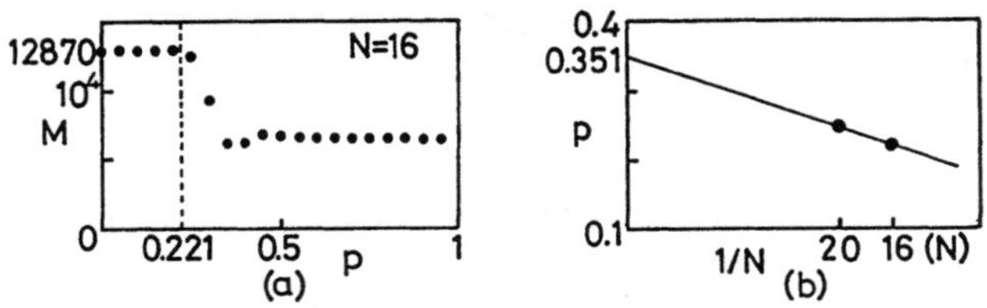

Fig. 1 (a) The plots of M versus p for $N = 16$;(b) the plots of p versus $1/N$.

component of the ground state vector at $p = 0$, and get M, the total number where the two components have a same sign. We make the same procedure with the value of p changed. Figure 1(a) shows the above number, M, versus p. In the region, $0 \leq p \leq 0.221$, we can see that the sign of every component of the ground state vector at p is the same as that of the corresponding component of the ground state vector at $p = 0$. For $N = 20$, the same statement holds in the region, $0 \leq p \leq 0.247$. The rough estimation from $1/N$ plots gives that the Marshall condition holds in a finite region of p, $0 \leq p < 0.351$(Fig. 1(b)). (For some one-dimentional model, several points are, indeed, fairly well arranged on a straight line in the $p - 1/N$ plane.)

We have investigated an antiferrimagnetic quantum Heisenberg model with some competing interactions. We have found that the Marshall condition remains to hold until some strength of competing interactions, though the extrapolation to $N \to \infty$ is rather rough. Moreover, the region that the Marshall condition remains to hold, $0 \leq p < 0.351$, almost coincides with the region that the Néel order exists, $0 \leq p < 1/3$. If we make the extrapolation more precisely, there is a possibility that the two regions become the same. Up to the present, however, the reason why these two regions are similar is not clear, and is to be clarified in the future.

References

1. E. Lieb and D. Mattis:J. Math. Phys. **3** (1962) 749.
2. E. Dagotto and A. Moreo: Phys. Rev. Lett. **63** (1989) 2148.

Ground State of the Two-Dimensional Square-Lattice Antiferromagnet with Dzyaloshinski-Moriya Interaction

Y. Ohta, W. Koshibae, and S. Maekawa

Department of Applied Physics, Nagoya University, Nagoya 464-01, Japan

Abstract. A finite-size exact diagonalization technique is applied for the two-dimensional spin-1/2 Heisenberg Hamiltonian with the Dzyaloshinski-Moriya interaction. The ground-state magnetic structure of the insulating CuO_2 plane is examined for the low-temperature orthorhombic and tetragonal phases of the La_2CuO_4-type crystal.

A small spin-canting in the predominantly antiferromagnetic (AF) CuO_2 plane of the low-temperature orthorhombic (LTO) phase La_2CuO_4 has been discussed in terms of the Dzyaloshinski-Moriya (DM) interaction [1,2]. This interaction should play a role in the low-temperature tetragonal (LTT) phase of the Ba-doped La_2CuO_4 as well. Here we use an exact diagonalization technique to study the spin structure of the insulating CuO_2 plane of the LTO- and LTT-phase La_2CuO_4-type crystal and consider implications of the DM interaction [3] of which not much theoretical work has been made so far.

The model Hamiltonian is composed of the spin-1/2 superexchange (J) plus DM ($\boldsymbol{D}_{ij}$) interaction between the neighboring Cu spins $\boldsymbol{S}_i$ and $\boldsymbol{S}_j$ on the two-dimensional square lattice;

$$H = 2J \sum_{<ij>} \boldsymbol{S}_i \cdot \boldsymbol{S}_j + 4 \sum_{<ij>} \boldsymbol{D}_{ij} \cdot \boldsymbol{S}_i \times \boldsymbol{S}_j .$$

The DM vectors have a spatial structure adapted to the crystal symmetry as shown in Fig. 1. $|\boldsymbol{D}_{ij}|$ is proportional to $(\Delta g/g)J$ where Δg is a deviation in the gyromagnetic ratio g due to the spin-orbit coupling, and is much larger than the exchange anisotrpoy of the higher order in $\Delta g/g$.

We treat the lattices of up to 20 sites with the periodic boundary condition. The ground state is calculated by the (modified) Lanczos methods at the independent $\boldsymbol{q}$-points in the irreducible part of the Brillouin zone. We divide the lattice into two sublattices and calculate the spin correlation function $<S_a^\alpha(-\boldsymbol{q})S_b^\beta(\boldsymbol{q})>$, where $S_a^\alpha(\boldsymbol{q})$ is the Fourier transform of S_i^α, the α-component of the spin operator on the sublattice a.

A number of characteristic spin structures are deduced from the Hamiltonian depending not only on the magnitude of D/J but also on the direction of the DM vector θ in the LTO phase and on the ratio D_2/D_1 in the LTT phase (see Fig. 1).

Springer Proceedings in Physics, Vol. 70 Computational Approaches in Condensed-Matter Physics
Editors: S. Miyashita · M. Imada · H. Takayama

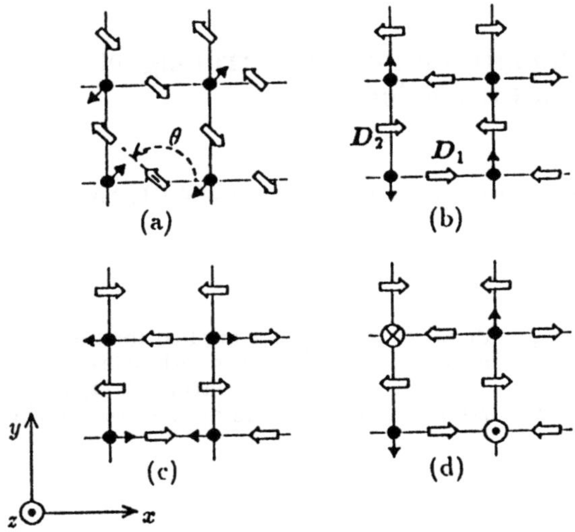

Fig. 1. Schematic representations of the spin structure (black arrow) and direction of the DM vectors (open arrow) in the CuO_2 plane of the La_2CuO_4-type crystal. (a) The LTO phase with the weak ferromagnetism; the DM vector with the direction $\theta=3\pi/4$ is illustrated. (b) The LTT phase with the weak ferromagnetism; the illustrated DM vectors ($D_2/D_1=1$) are appropriate for this state. The spins cant on the DM plane in (a) and (b). (c) The LTT phase with the uniaxially anisotropic AF; ($D_2/D_1=-1$). (d) As in (c) but for the four-sublattice AF.

The results we have obtained are as follows. (i) The weak ferromagnetism induced in the LTO phase is accompanied by a stabilization of the AF with the planar anisotropy. (ii) A fluctuation for the chiral order state is suppressed by the DM interaction of a particular structure that induces the weak ferromagnetism. (iii) The possible magnetic ground state in the LTT phase is either the weak ferromagnetism of the type observed in the LTO phase or the two-sublattice AF with the strong uniaxial anisotropy of the quantum-mechanical origin. A study of the dynamics of doped holes is now under way.

A financial support by a Priority-Areas Grant from the Ministry of Education, Science, and Culture of Japan is acknowledged.

References

[1] I. E. Dzyaloshinski, J. Phys. Chem. Sol. **4** (1958) 241, and T. Moriya, Phys. Rev. Lett. **4** (1960) 51.
[2] D. Coffey, K. S. Bedell, and S. A. Trugman, Phys. Rev. B **42** (1990) 6509.
[3] W. Koshibae, Y. Ohta, and S. Maekawa, *Proc. of M^2S-HTSC Kanazawa* (1991) in press.

Part IV

Critical Phenomena

Numerical CAM Analysis of Critical Phenomena in Spin Systems

M. Suzuki, N. Hatano, and Y. Nonomura

Department of Physics, Faculty of Science, University of Tokyo,
7-3-1 Hongo, Bunkyo-ko, Tokyo 113, Japan

Abstract. The basic idea and some applications of the coherent-anomaly method are briefly reviewed. Numerical analyses on quantum spin systems and spin glasses are surveyed.

Introduction: One of the main subjects of modern studies of phase transitions is to estimate fractional critical exponents. Critical indices obtained by mean-field theories[1], however, always have classical values, which can be also obtained by Landau's phenomenology[2]. In this sense, any extension of mean-field theories has been believed to be useless. Recently one of us (M.S.) proposed the coherent-anomaly method (CAM)[3-50,60-63], with which we can extract the true criticality from a series of mean-field approximations.

Basic Idea of the CAM: Let us consider a phase transition in which a response function diverges in the form $\chi \propto (T - T_c^*)^{-\gamma}$. Mean-field theories give the approximate divergence

$$\chi^{\rm mf}(T) \simeq \overline{\chi} \cdot (T - T_c^{\rm mf})^{-\gamma_0} \tag{1}$$

with the classical exponent γ_0. The mean-field critical coefficient $\overline{\chi}$ of a canonical series may diverge[3,4], as the degree of approximation increases, i.e., as the approximated transition point $T_c^{\rm mf}$ approaches the true critical point T_c^*. This systematic divergence is called "the coherent-anomaly"[3,4]. We assume the canonicality of the series, namely that the asymptotic form of the divergence of the critical coefficient $\overline{\chi}(T_c^{\rm mf})$ near $T_c^{\rm mf} \simeq T_c^*$ is given by[3,4]

$$\overline{\chi}(T_c^{\rm mf}) \sim (T_c^{\rm mf} - T_c^*)^{-\psi} \qquad \text{with} \qquad \gamma = \gamma_0 + \psi. \tag{2}$$

Fitting critical data $\{T_c^{\rm mf}, \overline{\chi}\}$ of several mean-field approximations to the function (2), we obtain T_c^* and γ.

Construction of Canonical Series: The series of the generalized Weiss approximations have been found phenomenologically to be canonical[4].

Springer Proceedings in Physics, Vol. 70 Computational Approaches in Condensed-Matter Physics
Editors: S. Miyashita · M. Imada · H. Takayama

For the time being, we restrict ourselves to ferromagnets with the interaction J. The effective Hamiltonian of a cluster is given by[3,4]

$$\mathcal{H}_{\text{eff}} = \mathcal{H}_\Omega - H \sum_{i\in\Omega} S_i - H_{\text{eff}} \sum_{i\in\partial\Omega} S_i, \tag{3}$$

where $\mathcal{H}_\Omega$ stands for the original Hamiltonian of the cluster Ω. With the expansion of the order-parameter at the center of the cluster, $\langle S_0 \rangle = \chi_\Omega(T)\beta H + \mathcal{F}_\Omega(T)\beta H_{\text{eff}}$, the mean-field self-consistency condition[1,3,4] $J\langle S_0 \rangle = H_{\text{eff}}$ results in the approximated response function[3,4]

$$\chi^{\text{mf}} \equiv \langle S_0 \rangle / H = \chi_\Omega / (1 - \beta J \mathcal{F}_\Omega). \tag{4}$$

This function shows the divergence of the form (1). As we extend the cluster Ω gradually, the approximated transition point T_c is expected to approach the true critical point T_c^*. It is shown[4], using Fisher's asymptotic scaling correlation function[51] $C(R) \sim R^{-(d-2+\eta)} \exp(-R/\xi)$, that the critical coefficient $\overline{\chi}$ of the function (4) shows the coherent anomaly (2) with only a logarithmic correction. A series with the self-consistency condition of the Bethe type[52] $\langle S_0 \rangle = \langle S_{\partial\Omega} \rangle$ is also found to show the coherent anomaly[4].

We take an infinite strip as the cluster Ω. The critical data is obtained with the transfer-matrix method. Hu et al. applied[6] the transfer-matrix CAM theory to the Ising ferromagnet in two dimensions, and obtained the estimate of the exponent $\gamma = 1.749$.

Essential Features of the CAM: The essential difference between the CAM and the finite-size scaling analysis is the disappearance of the system size from the fitting function (2). Thus, we can avoid the ambiguity in the definition of the system size and the irregularity of the shape of the cluster. In most cases, data points of rather small clusters show a good fit to the function (2). Our policy here is that we treat finite-size systems with numerical rigor and analyze them with the CAM. The disappearance of the system size also enables us to use the multi-effective-field theory mentioned in the next section.

Multi-Effective-Field Theory and the CAM: In the above mean-field approximation, the convergence of series of approximations has been achieved through an extension of the cluster. Using the multi-effective-field theory[41,42], we can construct a series of approximation within calculations for a fixed cluster. Instead of the third term of the r.h.s of (3), we consider the multi-body effective fields as follows[41,42];

$$- \sum_{i \in \partial\Omega} J_i^{(1)} S_i - \sum_{i,j \in \partial\Omega} J_{ij}^{(2)} S_i S_j - \cdots - \sum_{i_1,\cdots,i_n \in \partial\Omega} J_{i_1,\cdots,i_n}^{(n)} S_{i_1} \cdots S_{i_n}. \quad (5)$$

Some restricted number of multi-body effective fields can be determined from the self-consistency conditions[41,42] $\langle S_{i_1} S_{i_2} \cdots S_{i_n} \rangle = \langle S_{j_1} S_{j_2} \cdots S_{j_n} \rangle$. The approximation can be rectified by increasing the number of multi-body effective fields. The series thus constructed has been studied by Minami et al.[42] They obtained the exponent estimate $\gamma = 1.7498$ for the Ising ferromagnet in two dimensions.

Double-Cluster Approximation and the CAM: **The cluster-variational approximation motivates us to formulate the double-cluster approximation (DCA)[21,44-50,63], which is based on the consistency condition $\langle S \rangle_A = \langle S \rangle_B$. Here $\langle S \rangle_A$ and $\langle S \rangle_B$ are the order parameters on the cluster Ω_A and Ω_B of different sizes, correspondingly. It has been discussed[44] that the DCA gives better critical data than the single-cluster approximation (3) does. In addition, the DCA shows the coherent-anomaly without any logarithmic correction[44]. The DCA-CAM theory has been applied to the BEG model by Lipowski[45,46], and also to quantum spin systems[47-50] and spin-glasses[63] by ourselves.**

Ground-state Phase Transitions in Quantum Spin Systems: **In quantum spin systems, phase transitions may occur even at $T = 0$, as the strength of quantum fluctuation is varied. The CAM can also be applied[47] to such a problem. To test our theory, we have analyzed[47] the one-dimensional $S = \frac{1}{2}$ XZ model:**

$$\mathcal{H} = -g \sum_i S_i^x S_{i+1}^x - \sum_i S_i^z S_{i+1}^z - H \sum_i S_i^z \quad (6)$$

using the DCA. The order parameter of this model is $\langle S^z \rangle_g$ (the ground-state average of S^z). This model displays critical phenomena in the vicinity of a critical point $g = g_c^*$, and definitions of critical exponents are just the same as those in finite-temperature models (namely, the parameter g plays the role of temperature). Our estimates are in good agreement with the exact solutions[53,54] (see Table 1).

Table 1: The estimates and the exact solutions of the critical point and critical exponents of the one-dimensional $S = \frac{1}{2}$ XZ model.

Quantities	g_c^*	γ	β	δ	α
Our estimates	1.000(8)	1.52(4)	0.264(16)	6.6(1.1)	0(log)
Exact solutions	1	1.5	0.25	7	0(log)

The above method can also be applied to two-dimensional systems[48] and quantum antiferromagnets[49]. Furthermore, it can be coupled[50] with the multi-effective-field theory. This procedure is expected to be powerful in frustrated quantum spin systems, because the enlargement of cluster sizes is very difficult in these systems.

Super-Effective-Field Theory: The exotic phase transitions, e.g., spin-glass transition and chiral transition, have non-local order parameters. To construct a systematic series of approximations also for exotic phase transitions, one of us (M.S.) has proposed the super-effective-field theory[55-59]. We apply, to the boundary of the cluster Ω, the infinitesimal super-effective field Λ conjugate to the non-local order parameter Q. The formulation is quite similar to that for ferromagnets (3)-(4). Thus we can explore the instability of the disordered phase. The theory combined with the CAM has been applied to chiral orders[60,61] and spin-glass transitions[62].

Spin Glasses: As for spin glasses, we consider two real replicas 1 and 2 which have the same bond configurations[55-59,62], and set the order parameter as the product of spins on the replicas $Q = S^{(1)}S^{(2)}$. The super-effective Hamiltonian is now given by[55-59,62]

$$\mathcal{H}_{\text{eff}} = \mathcal{H}_{\Omega}^{(1)} + \mathcal{H}_{\Omega}^{(2)} - \Lambda \sum_{i\in\Omega} S_i^{(1)} S_i^{(2)} - \Lambda_{\text{eff}} \sum_{i\in\partial\Omega} S_i^{(1)} S_i^{(2)}. \tag{7}$$

Spins on the replica 1 fluctuate in favor of spins on the replica 2, and vice versa. This boot-strap mechanism[55-59] equilibrate the whole system. The self-consistency condition[55-59,62] $J[\langle S_0^{(1)} S_0^{(2)}\rangle]_{\text{av}} = \Lambda_{\text{eff}}$ gives the approximate spin-glass susceptibility $\chi_{\text{SG}} \equiv [\langle S_0^{(1)} S_0^{(2)}\rangle]_{\text{av}}/\Lambda$ in the same form as (4). Here the square bracket $[\cdots]_{\text{av}}$ stands for the quenched average. We have analyzed the symmetric $\pm J$ model through the DCA[63], and arrived at the following conclusion: In two dimensions the spin-glass susceptibility diverges at $T = 0$ with the power $\gamma_s = 5.1(1)$. In three dimensions the spin-glass transition take place at $T_{\text{SG}} = 1.18(4)$ with the exponent $\gamma_s = 3.3(4)$. We have also obtained the tentative results in four dimensions, $T_{\text{SG}} = 1.92(7)$.

Summary: Our applications have revealed the good features of the CAM. The CAM is a powerful method for studying systems for which it is difficult to extend the system size, e.g., quantum spin systems and quenched-disordered systems.

References

1. P. Weiss: J. Phys. Radium **6**, 661 (1907)
2. L. D. Landau: Phys. Zhurn. Sowjetunion **11**, 26, 545 (1937)
3. M. Suzuki: J. Phys. Soc. Jpn. **55**, 4205 (1986)
4. M. Suzuki, M. Katori and X. Hu: J. Phys. Soc. Jpn. **56**, 3092 (1987)
5. M. Katori and M. Suzuki: J. Phys. Soc. Jpn. **25**, 3113 (1987)
6. X. Hu, M. Katori and M. Suzuki: J. Phys. Soc. Jpn. **56**, 3865 (1987)
7. X. Hu and M. Suzuki: J. Phys. Soc. Jpn. **57**, 791 (1988)
8. M. Katori and M. Suzuki: J. Phys. Soc. Jpn. **57**, 807 (1988)
9. M. Suzuki: In *Quantum Field Theory*, p.505 (ed. F. Mancini: Elsevier Science, 1986)
10. M. Suzuki: Prog. Theor. Phys. Suppl. **87**, 1 (1986)
11. M. Suzuki: Phys. Lett. **A116**, 375 (1986)
12. M. Suzuki: J. Stat. Phys. **49**, 977 (1987)
13. M. Katori and M. Suzuki: In *Progress in Statistical Mechanics*, p.273 (ed. C. K. Hu: World Scientific, 1988)
14. M. Suzuki: In *Dynamics of Ordering Processes in Condensed Matter*, p.23 (eds. S. Komura and H. Furukawa: Plenum, 1988)
15. M. Suzuki: J. Phys. Soc. Jpn. **56**, 4221 (1987)
16. M. Suzuki: J. Phys. Soc. Jpn. **57**, 1 (1988)
17. M. Suzuki: Sci. Form. **3**, 43 (1988)
18. M. Katori and M. Suzuki: J. Phys. Soc. Jpn. **57**, 3753 (1988)
19. J. L. Monroe: Phys. Lett. **A131**, 427 (1988)
20. N. Ito and M. Suzuki: Int. J. Mod. Phys. **B2**, 1 (1988)
21. T. Oguchi and H. Kitatani: J. Phys. Soc. Jpn. **57**, 3973 (1988)
22. T. Oguchi and H. Kitatani: J. Phys. Soc. Jpn. **58**, 3033 (1989)
23. M. Katori: J. Phys. Soc. Jpn. **57**, 4114 (1988)
24. S. Fujiki: In *Proc. 2nd. YKIS on Cooperative Dynamics in Complex Physical Systems*, p.179 (ed. H. Takayama: Springer, 1988)
25. X. Hu and M. Suzuki: Physica **A150**, 310 (1988)
26. M. Suzuki: Phys. Lett. **127A**, 410 (1988)
27. M. Takayasu and H. Takayasu: Phys. Lett. **A128**, 45 (1988)
28. H. Takayasu, M. Takayasu and T. Nakamura: Phys. Lett. **A132**, 429 (1988)
29. X. Hu and M. Suzuki: In *Proc. 11th Taniguchi Int. Symposium on Space-Time Organization in Macromolecular Fluids*, p.188 (eds. F. Tanaka, M. Doi and T. Ohta: Springer, 1988)
30. M. Suzuki: J. Phys. Soc. Jpn. **58**, 3642 (1989)
31. K. Wada and T. Uchida: J. Phys. Soc. Jpn. **58**, 4358 (1989)
32. N. Konno and M. Katori: J. Phys. Soc. Jpn. **59**, 1581 (1990)
33. N. Kawashima, M. Katori, C. Tsallis and M. Suzuki: Int. J. Mod. Phys. **B4**, 1409 (1990)
34. J. L. Monroe, R. Lucente and P. Hourlland: J. Phys. **A**: Math. Gen. **23**, 2555 (1990)

35. X. Hu and M. Suzuki: J. Phys. **A**: Math. Gen. **23**, 3501 (1990)
36. S. Fujiki, M. Katori and M. Suzuki: J. Phys. Soc. Jpn. **59**, 2683 (1990)
37. Y. Hirata: Prog. Theor. Phys. **82**, 34 (1989)
38. H. S. Mano and K. Nakano: J. Phys. Soc. Jpn. **60**, 548 (1990)
39. A.Patrykiejew and P. Borowski: Phys. Rev. **B42**, 4670 (1990)
40. N. Ito and M. Suzuki: Phys. Rev. **B43**, 3483 (1991)
41. M. Suzuki: In *New Trends in Magnetism* (ed. M. D. Coutino-Filho and S. M. Rezende: World Scientific, Singapore, 1990)
42. K. Minami, Y. Nonomura, M. Katori and M. Suzuki: Physica **A174**, 479 (1991)
43. M. Suzuki: In *Evolutionary Trends in the Physical Sciences* (eds. M. Suzuki and R. Kubo: Springer 1991)
44. M. Suzuki, N. Hatano and Y. Nonomura: J. Phys. Soc. Jpn. **60** (1991).
45. A. Lipowski: Physica **A173**, 293 (1991)
46. A. Lipowski: J. Magn. Mat. **96**, 267 (1991)
47. Y. Nonomura and M. Suzuki: to appear in J. Phys. **A**: Math. Gen.
48. Y. Nonomura and M. Suzuki: to be submitted to J. Phys. **A**: Math. Gen.
49. Y. Nonomura and M. Suzuki: in preparation
50. Y. Nonomura and M. Suzuki: in preparation
51. M. E. Fisher: J. Math. Phys. **5**, 944 (1964)
52. H. A. Bethe: Proc. Roy. Soc. **A150**, 552 (1935)
53. E. Lieb, T. Schultz and D. Mattis: Ann. Phys. **16**, 407 (1961)
54. B. M. McCoy: Phys. Rev. **173**, 531 (1968)
55. M. Suzuki: J. Phys. Soc. Jpn. **57**, 683 (1988)
56. M. Suzuki: J. Phys. Soc. Jpn. **57**, 2310 (1988)
57. M. Suzuki: J. Stat. Phys. **53**, 483 (1988)
58. M. Suzuki: In *Proc. 2nd. YKIS on Cooperative Dynamics in Complex Physical Systems*, p.9 (ed. H. Takayama: Springer, 1988)
59. M. Suzuki: J. de. Phys. (Paris) Colloque **8**, 1519 (1988)
60. N. Kawashima and M. Suzuki: J. Phys. Soc. Jpn. **58**, 3123 (1988)
61. T. Kawarabayashi and M. Suzuki: J. Phys. Soc. Jpn. **60**, 825 (1991)
62. N. Hatano and M.Suzuki: J. Stat. Phys. **63**, 25 (1991)
63. N. Hatano and M.Suzuki: J. Stat. Phys. **66**, Issue 3/4 (1992)

Monte Carlo Study of the Random Potts Model

Y. Okabe[1] *and M. Kikuchi*[2]

[1]Department of Physics, Tohoku University, Sendai 980, Japan
[2]Department of Physics, Osaka University, Toyonaka 560, Japan

Abstract. The three-state Potts model in three dimensions is investigated by use of a Monte Carlo simulation. The order-parameter distribution is examined to clarify the order of the phase transition. The Monte Carlo renormalization group method and the finite-size scaling analysis are used to study the critical properties of the antiferromagnetic Potts model. We also treat the system of a random mixture of the ferromagnetic and antiferromagnetic couplings, and discuss the phase diagram of this model.

1. Introduction

The Potts model has been shown to have interesting properties. The ferromagnetic (F) three-state Potts model is believed to exhibit a first-order transition in three dimensions (3D). On the other hand, the phase transition of the antiferromagnetic (AF) three-state Potts model on a 3D bipartite lattice is of second order, and a recent simulational study has suggested that it belongs to the XY universality class [1].

The random mixture of the F and AF couplings arouses new problems. A zero-temperature transition of the three-state Potts glass in 3D has been inferred by Scheucher et al. [2]. It is interesting to study the overall phase diagram of the random Potts model as a function of the concentration of coupling and the temperature. The rounding of first-order transition due to the randomness is another problem [3, 4]. One may also ask whether the universality of critical exponents holds in random systems.

We present a Monte Carlo study of the random three-state Potts model. The model Hamiltonian is given by

$$\mathcal{H} = \sum_{<i,j>} J_{ij} \delta_{S_i S_j}, \tag{1}$$

where each spin S_i can take three states, a, b and c. We treat the $\pm J$ model with asymmetric probability weight generally. A quenched random variable J_{ij} takes either $+J$ or $-J$, and the concentration of the F coupling is denoted by p.

To study the phase transition and the critical phenomena, we employ the Monte Carlo renormalization group (MCRG) method [5, 6] combined with the finite-size scaling analysis. Different types of the block-spin transformation are required for investigating the F and the AF order. For the F order parameter of the Potts model, we use a simple majority-rule block-spin transformation which again produces a three-state Potts spin. In the case of the AF order, however, we should consider the sublattice structure; we introduce a block-spin transformation which gives the block spin having a six-fold symmetry.

We use the multispin coding algorithm due to Bahnot et al. [7], which was originally used for the Ising model, to perform massive simulation. We assign a single bit to each spin, but we need two words in the case of the three-state Potts model. Writing the code by the logical command, we simulate 32 systems (by a 32-bit machine) simultaneously with a single random-number sequence. It is to be emphasized that the multispin coding technique is also applied to the several types of block-spin transformations.

2. Order-Parameter Distribution of Ferromagnetic Potts Model

First we study the pure F model ($p = 1$), paying attention to the order of the phase transition. We investigate the distribution of the order parameter near the transition point for this purpose [8]. We should be careful about the two-dimensional character of the Potts order parameter. We compare the order-parameter distribution of the 3D and 2D systems. Here only the result will be given. A clear difference has been obtained between these two distributions. The order-parameter distribution for the 3D model has a multi-peak structure, which is characteristic of the first-order transition. On the other hand, the distribution extends over a large region of the order-parameter space for the 2D model. Thus, from the analysis of the distribution of the order parameter, we have clearly shown that the F three-state Potts model exhibits a first-order transition in 3D. It is in contrast to the observation that one cannot determine the order of the transition for the Potts model from the distribution of the *absolute* value of the order parameter [9].

3. MCRG Study of Antiferromagnetic Potts Model

Next, we study the pure AF model ($p = 0$). The two-sublattice structure should be taken into account for the AF order parameter [10]. Three components of the order parameter are defined as follows:

$$m_\sigma \equiv (\sum_{i \in A} \delta_{S_i,\sigma} - \sum_{j \in B} \delta_{S_j,\sigma})/N, \tag{2}$$

where σ denotes one of the three states a, b and c. Any state (m_a, m_b, m_c) can be

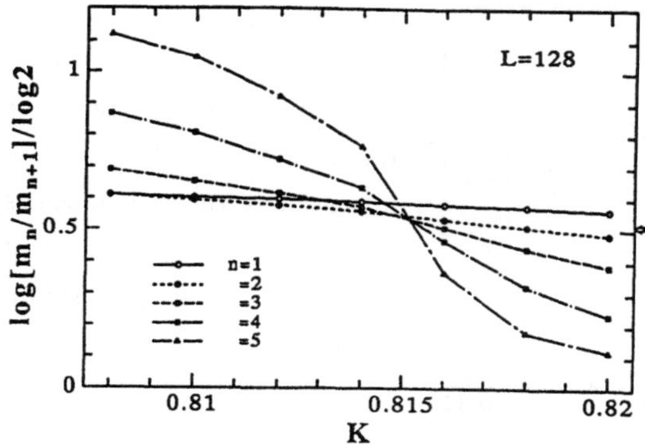

Fig. 1.
Ratio of successive renormalized order parameter for the AF three-state Potts model of $L = 128$. The blocking level is denoted by n. The XY value of the exponent β/ν is shown by an arrow.

expressed in a two-dimensional plane. A detailed study of the order-parameter distribution has shown a six-fold symmetry in the order-parameter space [8]. Moreover, we have confirmed that the ordered state is described by a so-called broken-sublattice order; namely, the order with one sublattice is occupied by one of the three states, while the other sublattice is occupied by the remaining two states at random. Therefore, the block-spin transformation which yields the block spin having the same symmetry as the six-clock model is appropriate for studying the AF order.

We use the MCRG method to study the critical phenomena of the AF three-state Potts model. In Fig. 1, we show the ratio of the renormalized order parameter for the system of $L = 128$ near the critical temperature. The average was taken over 1.6×10^5 sweeps. All the curves are expected to cross at a single point irrespective of the blocking levels at the criticality. The abscissa of the crossing point gives K_c, whereas the ordinate yields the critical exponent β/ν. We carefully examined the size effect, although the system of size $L = 128$ is fairly large. For comparison the critical exponent of the pure 3D XY model is given by an arrow in the figure. We have also used the finite-size scaling analysis of the order parameter, and obtained the result consistent with the MCRG method.

Two critical exponents, that is, the magnetic and the thermal ones, are relevant for ordinary second-order phase transitions. We estimate the other exponent $1/\nu$ by using the direct calculation of the temperature derivative of the order parameter. In general, the temperature derivative of any quantity A is calculated by

$$\frac{d}{d\beta} < A > = < A >< \mathcal{H} > - < A\mathcal{H} > . \tag{3}$$

The finite-size scaling hypothesis gives rise to the relation

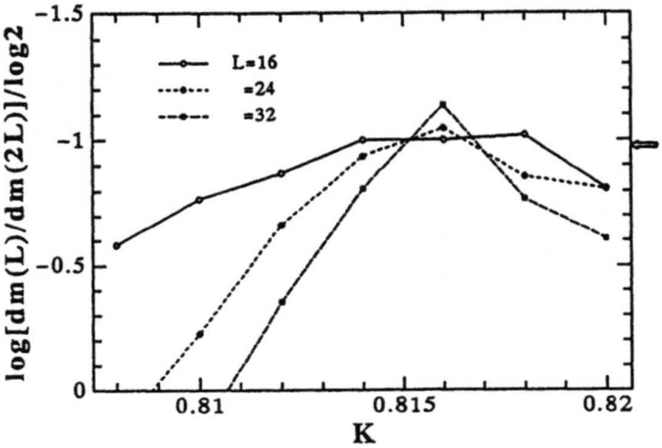

Fig. 2.
Ratio of the temperature derivative of the order parameter for the AF Potts model. The linear dimension of the system size is denoted by L. The XY value of the exponent $(\beta - 1)/\nu$ is shown by an arrow.

$$\frac{dm(L)/d\beta}{dm(L')/d\beta} = (\frac{L}{L'})^{-(\beta-1)/\nu} \tag{4}$$

at $T = T_c$. Thus, we can directly estimate the second critical exponent $1/\nu$ from the analysis of the ratio of temperature derivative of the order parameter of different sizes. Fig. 2 gives this plot for the three-state AF Potts model. For comparison we give the critical exponent of the pure XY model by an arrow. The estimated critical temperature and critical exponents are $K_c = 0.8156 \pm 0.0004$, $\beta/\nu = 0.51 \pm 0.02$ and $1/\nu = 1.51 \pm 0.04$. The present result is consistent with the Ref. [1], and the obtained critical exponents are similar to the 3D XY values within statistical errors. It is *not* in favor of the proposal of new universality class [11].

4. Phase Diagram of the Random Potts Model

We turn to the random mixture of the F and AF couplings. We only show the preliminary result of the calculation in this paper. Using the MCRG method and the finite-size scaling analysis, we have determined the transition temperature. The phase diagram of the $\pm J$ Potts model is shown in Fig. 3. It is to be noted that the phase diagram is not symmetric between the F and AF couplings.

For the weak AF system (say, $p = 0.05, 0.1$, etc.), the phase transition remains to be of second order, and the estimated critical exponents are close to those of the pure AF system ($p = 0$). For the weak F system (say, $p = 0.95, 0.9$, etc.), the phase transition becomes smooth compared to the pure F system ($p = 1$) which shows the first-order transition. In 2D, the rigorous argument has shown the rounding of first-order phase transition due to the quenched disorder [3, 4]. Whether the same argument can be applied to the 3D system awaits for a detailed study of the discontinuity of the energy.

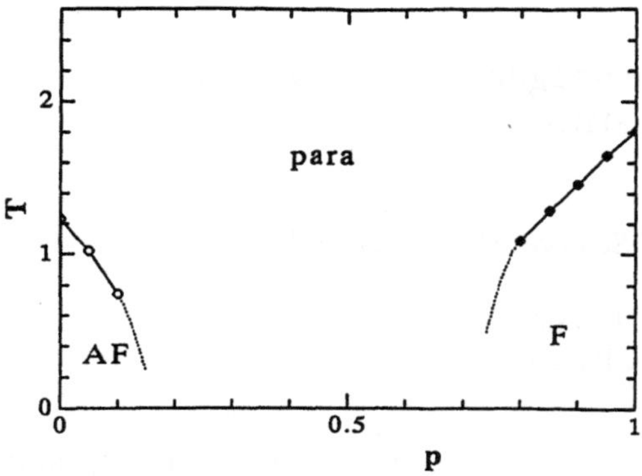

Fig. 3.
Phase diagram of the $\pm J$ Potts model, where p is the concentration of the F coupling. The transition of the AF order is of second order, and the transition of the pure F model ($p = 1$) is of first order.

The Potts glass order has not been studied in detail in this study. To elucidate the phase transition and the critical phenomena of the Potts glass by using the MCRG method will be left to a future study.

References

[1] J.-S. Wang, R.H. Swendsen and R. Kotecký, Phys. Rev. **B42**, 2465 (1990).

[2] M. Scheucher, J.D. Reger, K. Binder and A.P. Young, Phys. Rev. **B42**, 6881 (1990).

[3] M. Aizenman and J. Wehr, Phys. Rev. Lett. **62**, 2503 (1989).

[4] K. Hui and N. Berker, Phys. Rev. Lett. **62**, 2507 (1989).

[5] M. Kikuchi and Y. Okabe, Prog. Theor. Phys. **75**, 192 (1986); **78**, 540 (1987); Phys. Rev. **B35**, 5382 (1987).

[6] M. Kikuchi, Prog. Theor. Phys. Suppl. **101**, 557 (1990).

[7] G. Bahnot, D. Duke and R. Salvador, Phys. Rev. **B33**, 7841 (1986).

[8] M. Kikuchi and Y. Okabe, J. Magn. Magn. Mat., (1991), to be published.

[9] M. Fukugita, H. Mino, M. Okawa and A. Ukawa, J. Phys. **A23**, L561 (1990).

[10] I. Ono, Prog. Theor. Phys. Suppl. **87**, 102 (1986).

[11] Y. Ueno, G. Sun and I. Ono, J. Phys. Soc. Jpn. **58**, 1162 (1989).

Coherent-Anomaly Method in Two-Dimensional Antiferromagnetic Ising Models on Square and Triangular Lattices

A. Lipowski[1,*] *and M. Suzuki*[1]

[1]Department of Physics, Faculty of Science, University of Tokyo, Bunkyo-ku, Tokyo 113, Japan
*Permanent address: Department of Physics, A. Mickiewicz University, Poznań, Poland

In the present paper we have applied the Coherent-Anomaly Method (CAM) [1] proposed by one of the authors (M.S.), to two-dimensional antiferromagnetic Ising models on square and triangular lattices. We determined both critical temperature T_c and critical exponent γ as functions of a uniform external magnetic field H ranging from 0 to H_c. Estimations of T_c and γ were based on the results obtained within the Double-Cluster Approximation (DCA). In that approximation effective fields are obtained from selfconsistency conditions imposed on correlation functions of two different clusters. We have chosen pairs of wider and wider infinite stripes, which seems to be a very appropriate choice for two-dimensional models. A partition function of a stripe can be found as the largest eigenvalue of the corresponding transfer matrix. This version of the CAM has been already applied to ferromagnetic Ising models as well as to Blume-Emery-Griffiths model, and is more thoroughly described in [2,3]. Our results can be summarized as follows.

SQUARE LATTICE: In that case critical temperature vs. magnetic field is in a good agreement with the results obtained by other methods (e.g. finite-size scaling). For $H = 0$ we obtained $T_c = 2.270$, i.e. a temperature which is very close to Onsager's famous result (2.269...). The slope of the curve $T_c(H)$ in the limit $T_c \to 0$ is found to be equal $\eta = -1.481$, which can be compared with the best known result $\eta = -1.499$ obtained by Baxter et al [4]. The lowest approximations of our method lead to strong dependence of γ as a function of H (especially in a high field region). However, higher order approximations reduce that dependence considerably, giving nearly constant values of γ around the expected value $\gamma = 1.75$. It should be emphasized that in the case of the square lattice, the DCA can be applied to stripes with even as well as odd numbers of chains.

Springer Proceedings in Physics, Vol. 70 Computational Approaches in Condensed-Matter Physics
Editors: S. Miyashita · M. Imada · H. Takayama

TRIANGULAR LATTICE: In that case the model is much more complicated and even very sophisticated methods give much less reliable results. Moreover, we have observed that the results obtained by the DCA depend qualitatively on the number of chains in stripes: If wider stripe consists of odd number of chains, then such an approximation will lead in the limit $T_c \to 0$ to wrong values of a critical field. Such approximations were discarded but it resulted in a diminution in our series of approximations. The correct dependence of a critical temperature vs. magnetic field was obtained with the exception of $H = 0$ in which case our method gives finite values of T_c which is in contradiction with Wannier's exact solution [5]. The obtained value $\eta = -0.891$ can be compared with the exact solution $\eta = -0.831$. It is a well known fact that for nonzero values of H, our model belongs to the same universality class as the three-state Pott's model with $\gamma = 13/9$. However, for $H = 0$, it belongs to the universality class of the three-clock model with an infinite value of γ. The values of γ obtained by our method also indicate such crossover between different universality classes. We have currently been applying our method to the three-dimensional Ising antiferromagnet on a s.c. lattice.

One of us (A.L.) would like to thank Inoue Foundation for granting him a scholarship. The numerical calculations were performed on the VAX6440 of the Meson Science Laboratory, Faculty of Science, the University of Tokyo.

REFERENCES

[1] M.Suzuki, J.Phys.Soc. **55** p.4205 (1986).
[2] A.Lipowski, J.Magn.Magn.Mat. **96** p.267 (1991).
[3] A.Lipowski, Physica A **173** p.293 (1991).
[4] R.J.Baxter, I.G.Enting and S.K.Tsang, J. of Stat.Phys. **22** p.465 (1980).
[5] G.H.Wannier, Phys.Rev. **79** p.357 (1950).

Spin Ordering Property of a System with Nearly Degenerate Stable States

O. Nagai[1], *S. Miyashita*[2], *and T. Horiguchi*[3]

[1]Department of Physics, Kobe University, Kobe 657, Japan
[2]Department of Physics, CLAS, Kyoto University, Kyoto 606, Japan
[3]Department of Engineering Science, Tohoku University, Sendai 980, Japan

Abstract. We investigate the magnetic property of antiferromagnetic Ising model of infinite-spin on the triangular lattice by use of Monte Carlo (MC) simulations. We find that there exist two kinds of metastable states in this system and that the Kosterlitz-Thouless (KT) type phase transition occurs.

Very recently, we have investigated the ground state of antiferromagnetic Ising model of general spin, S, on the triangular lattice [1]. We have shown that a long range order exists in the system of infinite-spin, while no long range order exists in the case of S=1/2 [2]. In this paper, we calculate the magnetic properties at finite temperatures for the infinite-spin system by use of the MC simulations. The lattice size and MC steps adopted in our calculation were, respectively, at most $N=360\times 360$ and 10^6 per spin.

Our Hamiltonian is given by $H=J\Sigma_{ij}\sigma_i\sigma_j$, where $\sigma_i=S_i/S$. S_i denotes the Ising spin on the ith lattice site and takes one of the values S, S-1,$\cdots$,-S for the spin S system. $S=\infty$ in our case. Assuming three-sublattice model, we define the sublattice magnetizations by $\mu=\frac{3}{N}\{\Sigma_i\sigma_i,\Sigma_j\sigma_j,\Sigma_n\sigma_n\}$. We summarize our MC result in the following:

(1) The peak of the specific heat (C) is found at the temperature $T\cong 0.51J/k(\equiv T_p)$, k denoting the Boltzmann constant. The data of C do not show any appreciable size-dpendence.

(2) The spin correlation function shows a power law decay $\langle\sigma_i\sigma_j\rangle\propto(1/r_{ij})^{\eta}$. The η value rapidly decreases from 2 to 0.25 when T decreases from T_p to $0.41J/k(\equiv T_{KT})$ and gradually decreases from 0.25 to 0 when T decreases from T_{KT} to 0.

(3) In the low temperature region, $0<T<0.11J/k(\equiv T_0)$, the observed μ is expressed as (M,0,-M) where M is a numerical

Springer Proceedings in Physics, Vol. 70 Computational Approaches in Condensed-Matter Physics
Editors: S. Miyashita · M. Imada · H. Takayama

constant between 0 and 1. We call this spin structure the partial disordered (PD) structure. For $T_0<T<T_p$, $\mu=(0,0,0)$.

(4) C and η do not show any anomalous behavior at $T=T_0$.

From (1), (2), and (4), we think that the phase transition is of the KT-type [3]. Concerning (3), we think that the relaxation time of the PD structure is extremely long and μ would vanish if we made a very long time MC average. However, the PD state may be of the metastable state. In order to see this more explicitly, we consider our system in a small magnetic field. The Zeeman Hamiltonian is written as $-\Sigma_i h_i \sigma_i$ where h_i denotes the magnetic field at the site i. We assumed a staggered field, (h,0,-h), whose symmetry is of the PD structure. Then, the PD structure of $\mu=(M,0,-M)$ is observed in the MC simulations. We note that the M value is large even for small field and high temperature; for example, M=0.77 for h=0.02J and T=0.3J/k. On the other hand, if the system is in a staggered field, (h,-h,-h), whose symmetry is of the ferrimagnetic (FR) structure, then the FR structure of $\mu=(M',-m,-m)$ is observed. We find for example M'=0.88 and m=0.51 for h=0.02J and T=0.3J/k. Thus, we may consider that the FR state is also of the metastable state. There are six equivalent PD states and also six equivalent FR states. The system will successively take these twelve states at finite temperature and then a definite spin structure will not be achieved.

References

1. O. Nagai, S. Miyashita, and T. Horiguchi: submitted
2. G. Wannier: Phys. Rev. 79, 357 (1950)
3. J. M. Kosterlitz, J. Phys. C7, 1046 (1974)

Part V

Spin Glasses

Spin Glasses: Results from Numerical Simulations

P. Young

Physics Department, University of California, Santa Cruz, CA 95064, USA

Spin Glasses are among the simplest examples of statistical mechanics problems with "frustration", i.e. competition between different terms in the Hamiltonian. Monte Carlo simulations are difficult to perform successfully because relaxation times are long, so many Monte Carlo steps are needed. As a result, only rather small sizes can be studied and the techniques of finite-size scaling are used to extrapolate to the thermodynamic limit. This talk will describe results which have been obtained by Monte Carlo simulations on simple spin glass models. Emphasis will be on investigations of the phase transition and the nature of the low temperature phase.

1. Introduction

Spin glasses are magnetic systems which are disordered and "frustrated", by which one means that the interactions between the magnetic moments are in conflict with each other. I do not have room to discuss here the experimental systems in which spin glass behavior arises. Information on this and other aspects of the spin glass problem can be obtained from several reviews[1,2]. In this talk I will emphasize those aspects of the spin glass problem which make it hard. Important question have included whether or not a transition occurs and the nature of the low temperature phase. We shall see that there is a general consensus about the answer to the first question, but that the nature of the low temperature phase remains controversial.

2. Phase Transition in Spin Glasses

No conventional type of long-range order, such as ferromagnetism or antiferromagnetism, can arise in spin glass systems, but the question of whether a sharp phase transition to a new type of "ordered" state can occur has been hotly debated. To see how such a transition may arise it is helpful to make an analogy with ferromagnetism. At a second order phase transition there is a correlation length which diverges. For a ferromagnet this is simply the distance over which the spins have a net parallel alignment. Since correlations decay exponentially, to within a good approximation, we can write

$$<S_i\ S_j> \sim \exp(-R_{ij}/\xi_F) \quad , \tag{1}$$

which defines the ferromagnetic correlation length ξ_F. Here, R_{ij} is the distance between lattice points i and j. For convenience we are assuming Ising spins, where $S_i = \pm 1$, but this is not essential. As the transition temperature, T_c, is approached from above the ferromagnetic correlation length diverges with a power of $t = (T - T_c)/T_c$, i.e.

$$\xi_F \sim t^{-\nu} \quad . \tag{2}$$

While ξ_F can be measured by neutron scattering it is easier to determine the ferromagnetic susceptibility χ_F, which is proportional to the sum of the spin-spin correlation functions over all distances, i.e.

$$\chi_F = \frac{1}{k_B T}\sum_j <S_i\ S_j> \quad . \tag{3}$$

Notice that χ_F is a measure of the number of spins that are ferromagnetically correlated with a given spin and so also diverges at T_c with a critical exponent called γ, i.e.

$$\chi_F \sim t^{-\gamma} \quad . \tag{4}$$

In a spin glass there is no net tendency towards ferromagnetism or antiferromagnetism so ξ_F stays short range. Nonetheless, if one looks at all pairs of spins a given relative distance apart, they will have substantial correlation, but some pairs will be ferromagnetically correlated and others will have an antiferromagnetic correlation, so the *average* is close to zero. This cancellation coming from random signs will not occur if we square the correlation function, so we define the spin-glass correlation length, ξ_{SG} by

$$[<S_i\ S_j>^2]_{av} \sim \exp(-R_{ij}/\xi_{SG}) \quad , \tag{5}$$

where the symbol $[\cdots]_{av}$ denotes an average over different realizations of the random interactions. This is more convenient theoretically than the equivalent procedure of considering a single set of interactions and averaging over all pairs of spins a given distance apart. Just as with ferromagnets, there is a corresponding susceptibility, known here as the spin-glass or Edwards-Anderson[3] susceptibility, obtained by summing the squared correlation function over all distances, i.e.

$$\chi_{SG} = \sum_j [<S_i\ S_j>^2]_{av} \quad . \tag{6}$$

A spin glass transition is characterized by the divergence of ξ_{SG} and χ_{SG} as the temperature is lowered, i.e.

$$\xi_{SG} \sim t^{-\nu} \quad , \tag{7}$$
$$\chi_{SG} \sim t^{-\gamma} \quad ,$$

without a divergence of ξ_F or χ_F. Unfortunately, unlike ξ_F, the spin-glass correlation length cannot be measured by neutron scattering but it turns out that χ_{SG} can be determined by direct magnetic measurements. When a magnetic field, h, is applied one usually determines the *linear* response of the magnetization, controlled by χ_F. This is, however, only the first term of a series, i.e.

$$m = \chi_F h - \chi_{nl} h^3 \cdots . \tag{8}$$

The non-linear susceptibility, χ_{nl}, is the fourth derivative of the free energy with respect to h and so is a four spin cumulant average. Now clearly χ_{SG} is just one piece of the four spin cumulant. It is not difficult to show, for the models we will be discussing below with a symmetric distribution of exchange interactions, that the other terms in the four spin cumulant vanish so

$$\chi_{SG} \propto \chi_{nl} \quad . \tag{9}$$

Even for more realistic systems, where these two susceptibilities are not simply proportional to each other, one argues that the difference comes from the residual ferromagnetic or antiferromagnetic correlations in the system which are of short range, so χ_{SG} and χ_{nl} should still both diverge with the same exponent at T_c. It is, of course, much more difficult to measure the non-linear response because it is generally a small correction to the linear term. Close to T_c it becomes larger but then the higher order terms in eq. (8) become large too because they diverge with larger exponents exponents than γ. Despite these difficulties one can determine χ_{nl} if the experiments are done with sufficient care and most of our knowledge of critical behavior at the spin glass transition has come from such experiments[4-9]. In fact the difficulty of subtracting the linear response can be circumvented by doing an a.c. measurement and looking for the response at the third harmonic[8,9].

Below the transition one expects an order parameter to develop. For ferromagnetism this is, of course, the magnetization

$$m = \langle S_i \rangle \quad . \tag{10}$$

For spin glasses the analogous quantity would vanish because the spins will freeze in directions which appear random (though they are of course determined by the random interactions through the requirement that the free energy be minimized). In fact the expectation value of the spin at any wavevector will be zero (more precisely of order $N^{-1/2}$, where N is the number of spins) but, once again, one circumvents the cancellations occurring from random signs by squaring the thermal average. This leads to the Edwards-Anderson[3] order parameter

$$q_{EA} = [<S_i>^2]_{av} \quad . \tag{11}$$

Historically, the first measurement which indicated the possibility of a phase transition was the a-c susceptibility experiment of Cannella and Mydosh[10]. This showed a simple Curie, or Curie-Weiss, law at high temperatures, but, at a fairly well defined freezing temperature, the curve showed a cusp and the susceptibility dropped on lowering the temperature further. By itself, though, this did not show whether a sharp transition occurs at the cusp temperature, in which case the susceptibility would be expected to drop at lower temperatures as in an antiferromagnet, or whether the cusp simply shows that relaxation times become very long and the system drops out of equilibrium on time scales of the measurement. The difference is that there is a divergent length, ξ_{SG} in the first scenario while, in the second picture, the correlation length stays fairly small but relaxation times become very large, though strictly not infinite. From the above discussion, it is clear that this issue can be resolved by looking at the non-linear susceptibility. Careful measurements[4-9] show that this becomes enormous and an analysis of the temperature dependence indicates fairly conclusively that it really diverges at a temperature close to the freezing temperature. While the earlier measurements[4-6] gave a value for γ near 3, more recent experiments[7-9] which were able to probe temperatures closer to T_c, give a value near 2.

3. The Model

In order to study this problem theoretically it is useful to have a relatively simple model which incorporates the essential physical ingredients of randomness and frustration. The most commonly studied model is that of Edwards and Anderson[3] in which the Hamiltonian, for Ising spins, is

$$H = -\sum_{<i,j>} J_{i,j} S_i S_j \quad , \tag{12}$$

where the spins lie on a regular lattice and the interactions, assumed nearest neighbor here, are independent random variables with a probability distribution $P(J_{ij})$. We will consider the "pure spin glass case" where the mean of the distribution is zero, and will also set the standard deviation to unity to define the temperature scale. Two distributions are especially popular, the $\pm J$ distribution, in which

$$P(J_{ij}) = \frac{1}{2}[\delta(J_{ij} - 1) + \delta(J_{ij} + 1)] \tag{13}$$

and the Gaussian distribution:

$$P(J_{ij}) = \frac{1}{\sqrt{2\pi}} e^{-J_{ij}^2/2} \quad . \tag{14}$$

One interesting question is whether systems with different bond distributions lie in the same universality class.

4. Computing Aspects

Several theoretical techniques have been used to study the spin glass problem including, high-temperature series expansions, domain wall renormalization group methods and Monte Carlo simulations. Here I will discuss the application of the Monte Carlo technique, combined with finite-size scaling, which I have been working on in collaboration with Ravin Bhatt[11-13]. Unfortunately, one immediately runs into several difficulties when applying Monte Carlo techniques to spin glasses.

(a) Although the usual ideas of "self-averaging" indicate that one sample should give the same answer, in the thermodynamic limit, as an average over all samples with interactions chosen from the given distribution, this is not the case for the finite systems studied in simulations and one typically finds that an average of several hundred or several thousand samples must be simulated to do the configuration average. In fact, self-averaging may not even apply below the spin glass transition temperature (so many samples would need to be averaged over even in the thermodynamic limit) since it is known not to hold[14-16] for the infinite range model[17]. Since many samples must be averaged over the simulations require substantial computer time.

(b) Another reason that the simulations require a lot of CPU time is that, just as in the actual experiments, relaxation times become very long as the transition temperature is approached. It is therefore imperative to have a sound criterion for deciding if the simulation ran for enough sweeps for the system to be in equilibrium. The necessary number of sweeps will, of course, vary with temperature and system size.

(c) Since the spin glass correlation length diverges at T_c, the results close to the transition will be in the regime $\xi_{SG} > L$ where L is the linear dimension of the system. These will inevitably be affected by finite size effects.

Because simulations of spin glasses require a lot of computer power, use has been made of special purpose computers and parallel processors, as well as vector pipeline machines such as the Cray. Great care has also been taken to ensure that the system being simulated is in thermal equilibrium. These technical aspects have been discussed elsewhere[11-13]. Here will concentrate on the the contribution that simulations have made to understanding spin-glass physics.

5. Finite Size Scaling

Because the relaxation times are long in the vicinity of the transition (as well as everywhere in the low temperature phase), only rather small sizes can generally be studied. This means that one needs to use the technique of finite-size scaling to extrapolate to extrapolate to the thermodynamic limit. One convenient quantity to study by finite-size scaling is the order parameter distribution, $P(q)$ given by

$$P(q) = [<\delta(q - q_{12})>]_{av} \quad , \tag{15}$$

where $q_{12} = N^{-1}\sum_i S_i^{(1)} S_i^{(2)}$, and the labels (1) and (2) refer to two identical, independent copies of the system with the same set of bonds. Clearly q_{EA} and χ_{SG} are related to the first two moments of $P(q)$, i.e.

$$q_{EA} = <q> \quad , \tag{16}$$

$$\chi_{SG} = N<q^2> \quad ,$$

where the brackets here mean an average over $P(q)$.

Since q is the order parameter of the problem, its distribution should satisfy the finite-size scaling ansatz

$$P(q) = L^{\beta/\nu}\overline{P}(L^{\beta/\nu}q, L^{1/\nu}t) \quad , \tag{17}$$

where β is the order parameter exponent and $\overline{P}$ is a scaling function independent of L and $T - T_c$. The basic assumption in deriving eq. (17) is that finite size effects are controlled entirely by the ratio of the size of the system to the correlation length of the infinite system. The number of parameters involved can be reduced following Binder[18] by looking at ratios of moments such as

$$g = \frac{1}{2}\left[3 - \frac{<q^4>}{<q^2>^2}\right] . \tag{18}$$

It follows from eq. (17) that g satisfies the finite size scaling form

$$g = \overline{g}(L^{1/\nu}(T - T_c)) \quad , \tag{19}$$

where $\overline{g}$ is a scaling function. Note that g is dimensionless (since the powers of q in the numerator and denominator of eq. (18) are equal) so there is no power of L multiplying the scaling function in eq. (19). This makes g a particularly convenient quantity to study, as we shall see. In the paramagnetic phase for $T > T_c$ and for sizes $L \gg \xi$, $P(q)$ tends to a Gaussian centered around $q = 0$ of width $N^{-1/2} \sim L^{-d/2}$, so $g \to 0$. On the other hand, for $T < T_c$, where the infinite system develops a non- zero order parameter, $g \to 1$ as $L \to \infty$. From eq. (19) one sees that the range of temperature in which the value of g changes from 0 to 1 varies with L as $L^{-1/\nu}$. Eq. (19) also shows that g is independent of size at T_c

so curves of $g(T)$ for different sizes must all cross at T_c, with g being a decreasing function of L above T_c and an increasing function of L below T_c. This should provide a very convenient criterion for locating T_c. In addition, the exponent ν may be obtained from determining for which value the data are best fit by the scaling form, eq. (19). The second independent exponent, η, can be obtained from the finite size scaling form for the spin glass susceptibility

$$\chi_{SG} = L^{2-\eta}\, \overline{\chi}(L^{1/\nu}(T - T_c)) \tag{20}$$

and the exponent γ is then given by the scaling law

$$\gamma = (2-\eta)\nu \quad . \tag{21}$$

All other exponents can be obtained from ν and η using scaling laws.

6. Results.

These finite size scaling ideas all work out very cleanly in 4-dimensions[13]. We show, in fig. 1, data for g obtained on a hypercubic lattice with a Gaussian distribution of bonds for $2 \le L \le 6$, where each point is an average of 200 - 800 samples. The curves do intersect at (or at least close to) a single temperature, from which we estimate that $T_c = 1.75 \pm 0.05$, below the mean field estimate of $T_c^{MF} = \sqrt{8} = 2.83$. Scaling the data to the expected finite size forms, as discussed above, leads to the following estimates of the exponents: $\nu = 0.8 \pm 0.15$ and $\eta = -0.3 \pm 0.15$, which, combined with the scaling law, eq. (21), gives

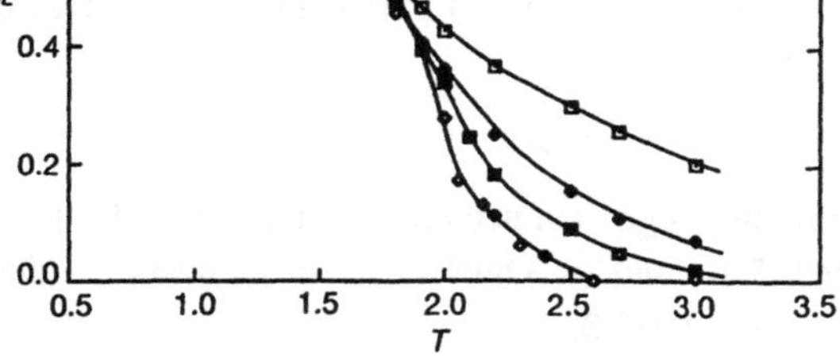

Fig. 1. Results for g against T in 4-dimensions for the Gaussian distribution. T_c is estimated to be 1.75 ± 0.05 from the intersection of the curves. The curves for different sizes splay out below T_c showing that the spin glass order parameter is non zero. The lines are just guides to the eye.

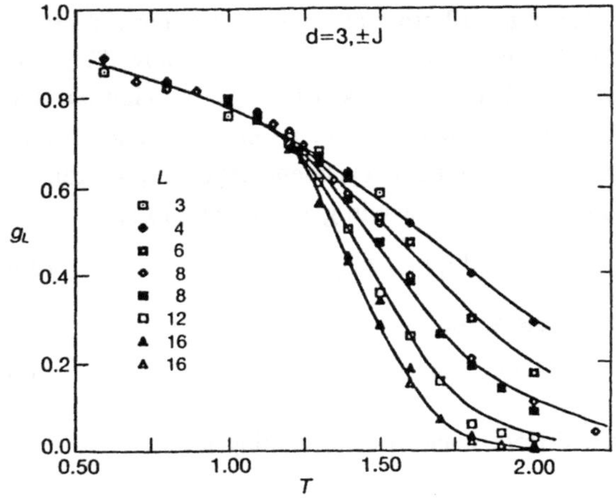

Fig. 2. Plot of g against T in 3-dimensions for the $\pm J$ distribution. The lines are just guides to the eye. The data for the open diamonds $L = 8$ and the filled triangles $L = 16$ are from A. T. Ogielski (private communication).

$$\gamma = 1.8 \pm 0.4 \quad (d = 4) \quad . \tag{22}$$

This value for γ agrees well with the estimate of $\gamma = 2.0 \pm 0.4$ obtained for a $\pm J$ distribution by high temperature series expansions[19]. This provides some evidence for the Gaussian and $\pm J$ models lying in the same universality class, though the error bars are too large to make a firm statement on this.

Unfortunately, the results in three dimensions[11-13,20,21] did not work out quite so cleanly. Data for g for the $\pm J$ distribution is shown in fig. 2. While the data at higher temperatures show the expected behavior and the curves come together at about $T = 1.2$, which is therefore the estimate of T_c, the curves for different sizes do not splay out again below T_c. The data above T_c scale quite nicely and from the fit one gets an estimate of $\nu = 1.3 \pm 0.3$. In addition, results for χ_{SG} [11,13] give $\eta = -0.3 \pm 0.2$ and hence

$$\gamma = 3.0 \pm 0.9 \quad (d = 3) \quad . \tag{23}$$

These values agree well with the high temperature series results[19] of $T_c = 1.2 \pm 0.1$ and $\gamma = 2.9 \pm 0.5$ and Ogielski's[20] result that $T_c = 1.175 \pm 0.025$, $\nu = 1.3 \pm 0.1$ and $\gamma = 2.9 \pm 0.3$.

The γ values also agree fairly well with earlier experimental values, which used data rather far from T_c, but is rather higher than the most recent experimental values of around 2 [7,9]. which used data obtained closer to T_c. It is therefore quite possible that the numerical results are still rather far from the

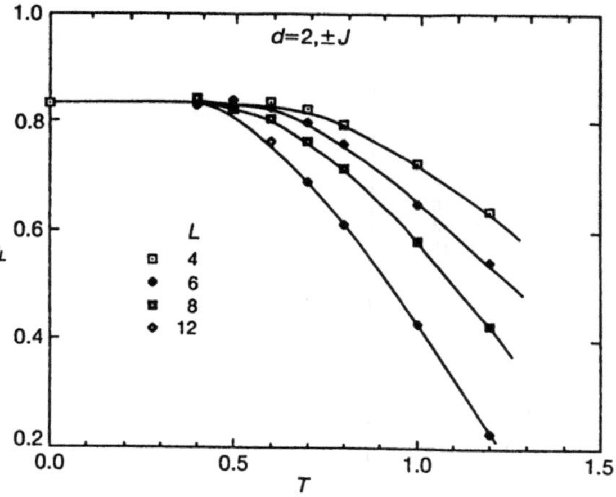

Fig. 3. Plot of g against T for various sizes in $d = 2$ for the $\pm J$ distribution. The lines are just guides to the eye.

asymptotic limit, $L \to \infty$, $T - T_c \to 0$, in which scaling is supposed to hold, and that the quoted results are "effective exponents" rather than the true asymptotic exponents. We suspect that corrections to scaling are also responsible for the apparent violation of scaling below T_c. Note, however, that we cannot rule out the possibility that T_c may be somewhat lower than estimated. It does not, however, seem possible for T_c to be any higher than around 1.2.

While the theory definitely predicts a transition in $d = 4$ and almost certainly one in $d = 3$, we shall now see that, by contrast, only a zero temperature transition is predicted in $d = 2$ [12,13,19,22-27]. Results for g for the $\pm J$ model are plotted in fig. 3. In addition to the Monte Carlo results, we also show a point at $T = 0$ for $L = 4$ obtained by direct enumeration of all the states. Notice that the temperature at which the curves merge gets lower as the system size increases and that that g has already reached its zero temperature value when the merging occurs. This is different from the behavior observed in three dimensions. Fig. 4 shows that scaling works well with $T_c = 0$ and an analysis gives $\nu = 2.6 \pm 0.4$ as the range of fits. Similar results for χ_{SG} gives $\eta = 0.2 \pm 0.05$ and so the estimate for gamma is

$$\gamma = 4.6 \pm 0.5 \quad (d = 2) \quad , \tag{24}$$

which is somewhat smaller than the value of $\gamma = 5.3 \pm 0.3$, obtained from high temperature series expansions[19], though the values do overlap. Neither of these calculations definitely rule out a transition at extremely low temperature but, bearing in mind that the mean field transition temperature is $T_c^{MF} = 2$, and that our data certainly rules out a transition with $T_c > 0.3$, we feel that a finite T_c is very unlikely.

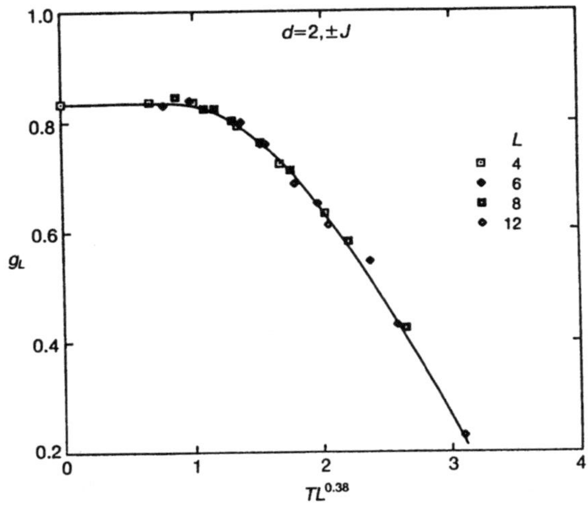

Fig.4 A fit of the data in fig. 3 to the finite size scaling formula, eq. (19), with $T_c = 0$ and $1/\nu = 0.38$ for the $\pm J$ distribution in $d = 2$.

Since the calculations give very different predictions for 2 and 3-dimensional systems, it would obviously be desirable to find an experimental system which is a 2-dimensional spin glass. Recently this has been achieved by the beautiful work of Dekker et al.[28-30] on $Rb_2Cu_{1-x}Co_xF_4$. The crystal structure is the K_2NiF_4 type in which the magnetic ions are situated on a simple square lattice which are widely separated from each other. The magnetic interactions are predominantly nearest neighbors in the plane. The strong Ising anisotropy on the Co sites presumably makes the whole system Ising-like, so this material is indeed an almost perfect representation of the 2-d Ising $\pm J$ spin glass. Dekker et al. find $T_c = 0$ and exponent values $\gamma = 4.5 \pm 0.2$ and $\nu = 2.4 \pm 0.3$, which are in excellent agreement with the Monte Carlo predictions. They also look at the dynamics and show that a characteristic relaxation time, τ, varies exponentially as $T \to 0$, also as expected from theory[31-34]. Essentially the theory predicts that relaxation proceeds by thermal activation over barriers whose characteristic height, ΔE, grows with some power of the correlation length, i.e.

$$\Delta E \sim \xi^{\psi} \sim T^{-\psi\nu} \quad , \tag{25}$$

which defines a "barrier height" exponent, ψ. Hence the characteristic relaxation time τ varies as

$$\tau \sim \exp(C/T^{1+\psi\nu}) \quad . \tag{26}$$

Dekker at al. find

$$\psi\nu = 2.0 \pm 0.3 \quad , \tag{27}$$

which is again in excellent agreement with their value of ν and a recent prediction of Cieplak (unpublished) that $\psi = 1$. Thus, at least in 2-d, a really detailed comparison between experiment and theory has been obtained.

While $Rb_2Cu_{1-x}Co_xF_4$ is very anisotropic, and so is well described by an Ising model, most of the three-dimensional systems studied, such as *CuMn*, have relatively little anisotropy, so one might expect a Heisenberg model to be more appropriate. Surprisingly, several calculations[35-38] on 3-d isotropic Heisenberg spin glass models indicate that $T_c = 0$ and that the lower critical dimension , d_l, (the dimension below which $T_c = 0$) is close to 4. Thus the Ising model gives a better agreement with experiment than the Heisenberg model even for spin glasses with relatively small anisotropy. Presumably then, the anisotropy cannot be ignored and must be included to explain the experiments in detail. Why a small anisotropy should have such a big effect is not entirely clear. One possible explanation is that the commonly studied spin glasses have the RKKY interaction, which falls of as r^{-3}, whereas the models studied by simulation have short range couplings. It has been argued[39,40] that RKKY Heisenberg systems are *at* their lower critical dimension, as opposed to short range systems which are below d_l. This means that any small anisotropy *will* have a large effect, the transition temperature, for example, increasing from zero with the log of the anisotropy, D, whereas, for systems below d_l, T_c increases with a power of D. However a detailed numerical study including both anisotropy and RKKY interaction has not yet been attempted.

7. Low Temperature Phase

Two possible models for the low temperature state below T_c have been proposed. The first assumes that the short range systems are similar to the infinite range Sherrington Kirkpatrick[17] model, which appears to have been solved by Parisi[41]. Below T_c, any given realization of the SK model has many thermodynamic states[42,14,15], unrelated by symmetry, which differ in the (total) free energy from the ground state by only an amount of order unity. This leads to a non-trivial distribution, $P(q)$, below T_c with a peak at finite q and a tail with a finite weight at $q = 0$. The alternate scenario is the "droplet model" [31-34], according to which there is just a single thermodynamic state and $P(q)$ for an infinite system is just a delta function. Model Carlo simulations[43] of the $d = 4$ Ising model below T_c for sizes $2 \le L \le 6$ find behavior similar to that of the SK model. This is consistent with recent simulations[44] which find evidence for a transition in magnetic field. This is expected in the SK model[45] but not in the droplet picture. However, the range of sizes and temperatures studied in both these simulations was rather limited, so the nature of the low temperature phase is not yet definitely settled.

8. Conclusions

The question of a spin glass transition, which had remained controversial for so long, seems now to be settled in favor of a finite temperature transition. It is particularly gratifying to see how well the simulations agree with experiments on two-dimensional systems. Agreement between theory and experiment in 3-d is perhaps reasonable, but not very precise. Presumably the discrepancy arises from the simulation (and possibly the experiment) not being in the asymptotic scaling region and from the various crossover effects mentioned above. The numerical methods also work very well in 4-d, where we will presumably not see any real experiments. It is frustrating but perhaps typical that the 3-d case is the hardest to solve. Finally, the nature of the low temperature phase remains controversial.

Acknowledgements

Much of my own work reported here was done is collaboration with R. N. Bhatt and J. D. Reger. I would like to thank them as well as S. Jain, J. A. Olive and D. Sherrington for many stimulating discussions. This work is supported in part by NSF grant DMR 91-11576.

References

1. Binder K. and Young A. P., Rev. Mod. Phys. **58**, 801 (1986).
2. Van Hemmen, J. L. and Morgenstern, I. (ed.), *Heidelberg Colloquium on Glassy Dynamics*, Lecture Notes in Physics, Vol 275, Springer (1987).
3. Edwards S. F. and Anderson P. W., J. Phys. F, **5**, 965 (1975).
4. Malozemoff A. P., Imry Y. and Barbara B. J., Appl. Phys. **53**, 7672 (1982).
5. Omari R., Prejean J. J. and Souletie J., J. Phys. (Paris) **44**, 1079 (1983).
6. Beauvillain P., Chappert C. and Renard J. P., J. Phys. (Paris) Lett. **44**, L665 (1984).
7. Bouchiat H., J. Phys. (Paris) **47**, 71 (1986).
8. Levy L., Phys. Rev. B**38**, 4963 (1988).
9. Levy L. and Ogielski A. T., Phys. Rev. Lett. **57**, 3288 (1986).
10. Cannella V. and Mydosh J. A., Phys. Rev. **B6**, 4220 (1972).
11. Bhatt R. N. and Young A. P., Phys. Rev. Lett. **58**, 924 (1985).
12. Bhatt R. N. and Young A. P., in *Heidelberg Colloquium on Glassy Dynamics*, edited by J. L. Van Hemmen and I. Morgenstern, (Springer, Berlin,), p. 215 (1987).
13. Bhatt R. N. and Young A. P., Phys. Rev. **B37**, 5606 (1988).

14. Mézard M., Parisi G., Sourlas N., Toulouse G. and Virasoro M., Phys. Rev. Lett. **52**, 1156 (1984).

15. Mézard M., Parisi G., Sourlas N., Toulouse G. and Virasoro M., J. Phys. (Paris) **45**, 843 (1984).

16. Young A. P., Bray A. J. and Moore M. A., J. Phys. C **17**, L149 (1984).

17. Sherrington D. and Kirkpatrick K., Phys. Rev. Lett. **35**, 1972 (1975).

18. Binder K., Z. Phys. **B48**, 319 (1982).

19. Singh R. R. P. and Chakravarty S., Phys. Rev. Lett. **57**, 245 (1986).

20. Ogielski A. T., Phys. Rev. **B32**, 7384 (1985).

21. Ogielski A. T. and Morgenstern I., Phys. Rev. Lett. **54**, 928 (1985).

22. Morgenstern I. and Binder K., Phys. Rev. Lett, **43** 1615 (1979).

23. Cheung H. F. and McMillan W. L., J. Phys. C **16**, 7027 (1983).

24. Cieplak, M. and Banavar, J., Phys. Rev. **B27**, 293 (1983).

25. McMillan W. L., Phys. Rev. B.**28, 5216 (1983).**

26. Bray A. J. and Moore M. A., J. Phys. C **17**, L463 (1984).

27. Huse D. and Morgenstern I., Phys. Rev. **b32**, 3032 (1985).

28. Dekker C., Arts A. F. M. and de Wijn H. W. J., Appl. Phys. **63**, 4334 (1988).

29. Dekker C., Arts A. F. M. and de Wijn H. W. J., Appl. Phys. **63**, Phys. Rev. **B38**, 8985 (1988).

30. Dekker C., Arts A. F. M. H., de Wijn H. W. J., van Duyneveldt A. J. and Mydosh J. A. (1989) Phys. Rev. B. (to be published).

31. Fisher D. S. and Huse D. A., Phys. Rev. Lett. **56**, 1601 (1986).

32. Fisher D. S. and Huse D. A., Phys. Rev. **B36**, 8937 (1987).

33. Bray A. J. and M. A. Moore M. A., in *Heidelberg Colloquium on Glassy Dynamics*, edited by J. L. Van Hemmen and I. Morgenstern, Lecture Notes in Physics, Vol. 275 (Springer, Heidelberg), p. 121 (1987).

34. Bray A. J., Comments Cond. Mat. Phys. **14**, 21 (1988).

35. Banavar J. R. and Cieplak M., Phys. Rev. Lett. **48**, 832 (1982).

36. McMillan W. L., Phys. Rev. **31**, 342 (1985).

37. Morris B. W., Colborne S. G., Moore M. A., Bray A. J. and Canisius J., J. Phys. C **19**, 1157 (1986).

38. Olive J. A., Young A. P. and Sherrington D, Phys. Rev. **B34**, 6344 (1985).

39. Bray A. J. Moore M. A. and Young A. P., Phys. Rev. Lett. **56** 2641 (1986).

40. Reger J. D. and Young A. P., Phys. Rev. **B37**, 5493 (1987).

41. Parisi, G., J. Phys. A, **13**, 1101 (1980).

42. Parisi, G., Phys. Rev. Lett., **50**, 1946 (1983).

43. Reger, J. D., Bhatt, R. N. and Young, A. P., Phys. Rev. Lett. **64**, 1859 (1990).

44. Grannan, E. R. and Hetzel, R. E., Phys. Rev. Lett., **67**, 907 (1991).

45. de Almeida, J. R. L. and D. J. Thouless, J. Phys. A, **11**, 983 (1978).

A Hybrid Monte-Carlo Spin-Dynamics Method and Its Applications to the $\pm J$ Heisenberg Models in Three Dimensions

F. Matsubara, T. Iyota, and S. Inawashiro

Department of Applied Physics, Tohoku University, Sendai 980, Japan

An efficient method of simulating the Heisenberg model is proposed taking into account the spin dynamics. Using the method, we obtain the phase diagram of the short-range $\pm J$ Heisenberg model. We determine the critical concentration of $p_c = 0.79 \pm 0.01$. We also study effects of an anisotropy and find that the anisotropy induces the spin glass phase transition.

1. Introduction

The Ising spin glass (SG) has been studied extensively by means of computer simulations[1-3]. The relaxation time in these simulations increases very rapidly as the temperature is decreased, limiting the system size and the temperature range that can be studied. However, one can use the discreteness of the Ising variables to write very efficient codes[4,5], thus keeping the required computer time within reach. This is not so for models where the variables take continuous values. For this reason, the Heisenberg spin glasses have not yet been very well studied[6-8], although many SG materials are Heisenberg-like. Many interesting problems remain such as what effects anisotropy has on the SG phase transition[7,9], and why the reentrant mixed phase[10] occurs in many SG materials. Here, we propose a simple but powerful method of studying Heisenberg SG models[11]. By the use of the method, we study the phase diagram of the short-range $\pm J$ model[12] and effects of anisotropy[13]. We find that the SG phase occurs only when the anisotropy is present.

2. Method

We start with the classical $\pm J$ Heisenberg model described by the Hamiltonian

$$H = \sum_{\langle ij \rangle} J_{ij} S_i S_j \quad , \tag{1}$$

where $|S_i| = 1$, $J_{ij} = J$ or $-J$ with probabilities of 1/2 and $\langle ij \rangle$ runs over all nearest-neighbor pairs on a simple cubic lattice of $L \times L \times (L+1) \equiv N$. The spin dynamics of the model is described by the following equation of motion,

$$\hbar \dot{S}_i = [\, S_i \times H_i \,] \ , \tag{2}$$

where $H_i = \Sigma_j J_{ij} S_j$. In this dynamics, the system never reaches any equilibrium state because the energy and total magnetization $M = \Sigma_i S_i$ are conserved. To get an equilibrium spin configuration, we combine the Monte-Carlo method with the spin dynamics. Hereafter we call this method *hybrid Monte-Carlo spin-dynamics* (*HMCSD*) method.

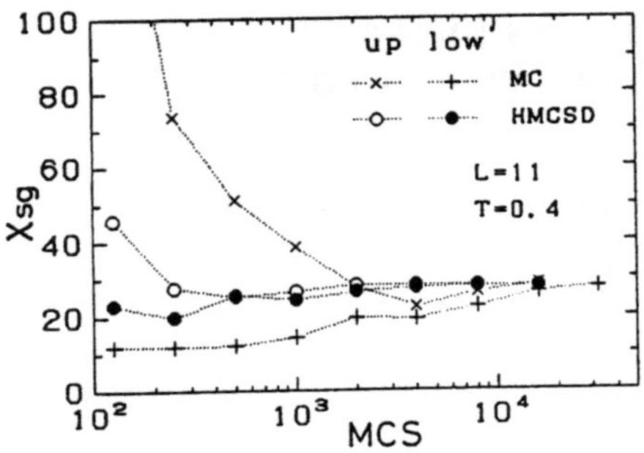

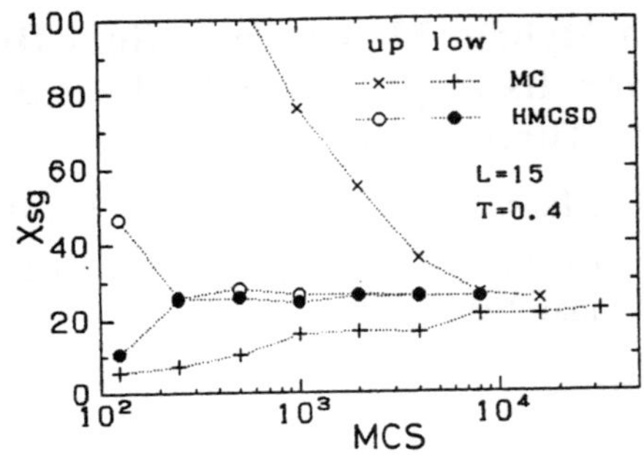

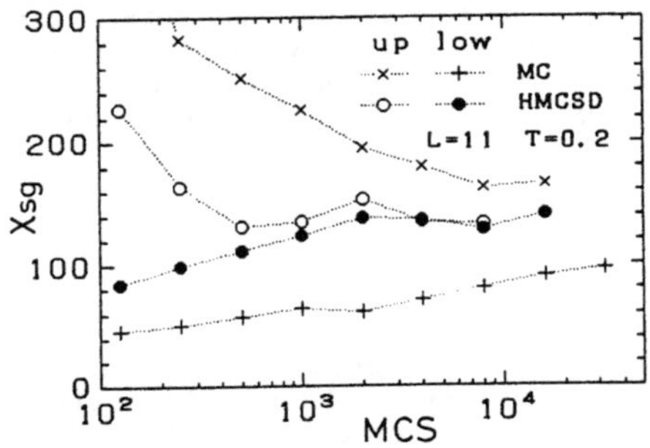

Fig.1 The upper and lower bounds of χ_{sg}. These are calculated starting with a paramagnetic spin configuration.

By the use of the HMCSD method, we calculate various physical quantites and find that the method reproduces all results of the standard MC method[6,8] *within much smaller MC steps.* In Fig.1, we present the upper and lower bounds of the SG susceptibility χ_{sg} obtained by a similar method of Bhatt and Young[1]. The equilibrium state is attained when the two bounds agree with each other. We see the HMCSD method reduces drastically the MC steps K needed to reach the equilibrium state. The HMCSD method becomes much more efficient as the size of the lattice is increased and/or the temperature is decreased. We also calculate SG order parameter for various MC steps K, $Q(K)$, and find that $Q(K)$ rapidly decreases with increasing K.

3. Phase diagram of the $\pm J$ Heisenberg model

We study the phase diagram of the $\pm J$ Heisenberg model using the HMCSD method. We study the same model as (1) with the concentrations of ferromagnetic and antiferromagnetic bonds being p and $1-p$, respectively. We first calculate the magnetization, $\langle M \rangle$, and the absolute magnetization, $\langle |M| \rangle$, to discuss whether or not the ferromagnetic phase occurs. Here, M is the magnetization per spin and $\langle \ldots \rangle$ means the thermal average. To examine whether the random spin freezing occurs or not, we also calculate the SG order parameter. Since $\langle M \rangle$ does not vanish in the ferromagnetic region of p, we define the SG order parameter, $Q(K)$, for a finite step K as

$$Q(K) = \frac{1}{N}\sum_{i}^{N} \langle \, (\, \langle S_i \rangle - \langle M \rangle \,)^2 \, \rangle_J \, , \qquad (3)$$

where $\langle \ldots \rangle_J$ means the sample average. Even if M is conserved, when the spin structure changes from a stage to another stage, $Q(K)$ decreases as

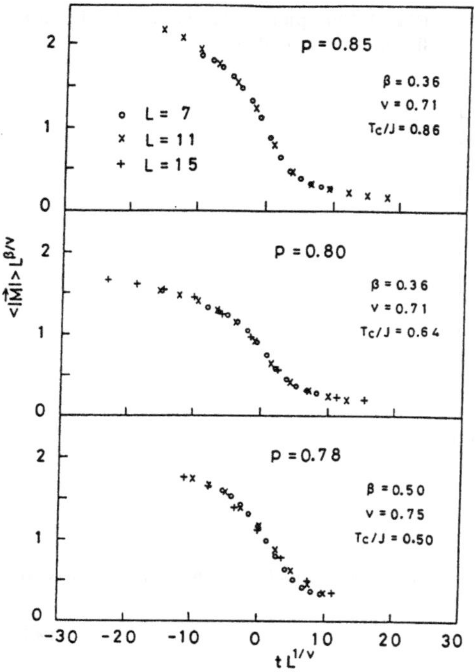

Fig.2 Finite size scaling plots of $<|M|>$.

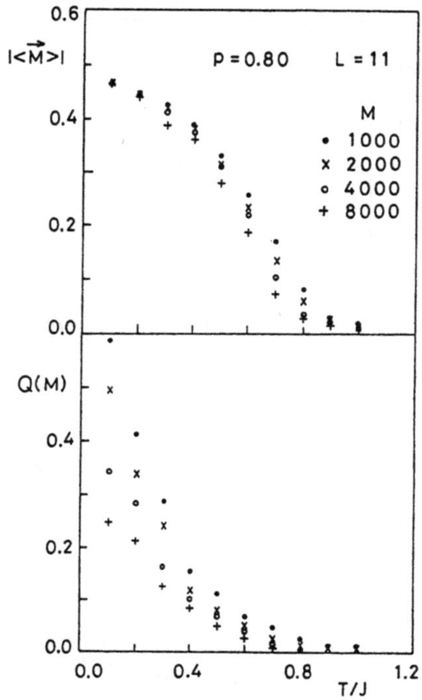

Fig.3 The magnetization $<M>$ and the SG order parameter $Q(K)$ for different MC steps K.

K increases. Only when some components of the spin randomly freeze, $Q(K)$ has a certain non-vanishing value.

We perform detailed simulations for p = 0.85, 0.80 and 0.78. We calculate $<|M|>$ for different sizes of the lattices and make finite size scaling plots[14]. We find that data for p = 0.85 and 0.80 are scaled well using values of β = 0.36 and ν = 0.71 of the Heisenberg ferromagnet[15], where β is the exponent of the magnetization, and ν the exponent of the correlation length. However, we cannot scale data for p = 0.78 even when different values of β and ν are used. These are shown in Fig. 2. We conclude, hence, that the critical concentration is $p_c = 0.79 \pm 0.01$ and that the phase transition for $p > p_c$ belongs to the same universality class as that of the pure Heisenberg ferromagnet.

To examine whether the random spin freezing occurs or not, we calculate $Q(K)$ for different K and find that it decreases rapidly as K increases(see Fig.3). This K-dependence of $Q(K)$ is not due to the uniform rotation of all the spins, because $<M>$ at low temperatures does not exhibit any considerable K-dependence. We conclude, hence, that the random spin freezing does not occur. That is, no mixed phase occurs in our model.

The phase diagram thus obtained is presented in Fig. 4. The phase boundary between the ferromagnetic and the paramagnetic phases will be vertical at low temperatures, because the ferromagnetic phase transition disappears abruptly in the range of $0.78 < p < 0.80$ and no evidence of another phase transition is seen at lower temperatures. However, further studies are necessary to confirm this point.

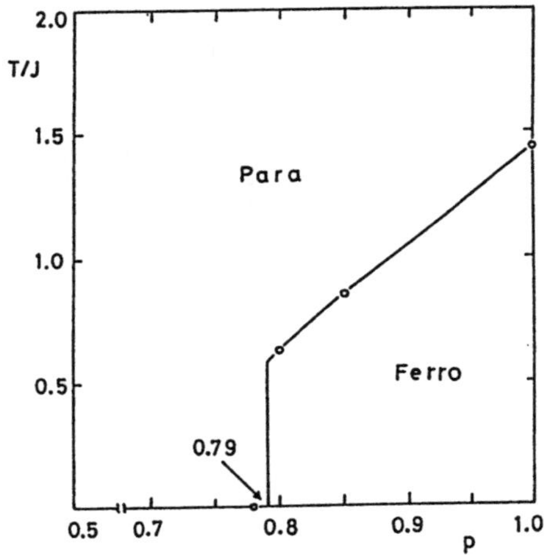

Fig.4 The phase diagram of the $\pm J$ Heisenberg model.

4. Effects of anisotropy

Now we discuss effects of anisotropy. Here we consider the anisotropy of the form:

$$H = \sum_{<ij>} \sum_{\alpha \neq \beta} D_{ij}^{\alpha\beta} S_i^{\alpha} S_j^{\beta} \quad , \tag{4}$$

where $D_{ij}^{\alpha\beta}$ are uniform random values between $-D$ and D. Adding this term to eq.(1), we make a similar simulation in the case of $D/J = 0.1$.

We find that the susceptibility χ exhibits a cusp-like peak at $T/J \sim 0.35$ suggesting the occurrence of the spin glass phase. To confirm this, we calculate the spin glass susceptibility χ_{sg}'s for different sizes of the lattice and make a finite size scaling[14]. As shown in Fig.5, all the data can be scaled well using values of $T_c/J = 0.30^{+0.02}_{-0.05}$, $\eta = 0.7^{+0.2}_{-0.4}$ and $\nu = 1.15 \pm 0.10$, where T_c is the transition temperature and η is the exponent which describes the decay of the correlation at $T = T_c$. The value of T_c obtained here is in agreement with that estimated from the

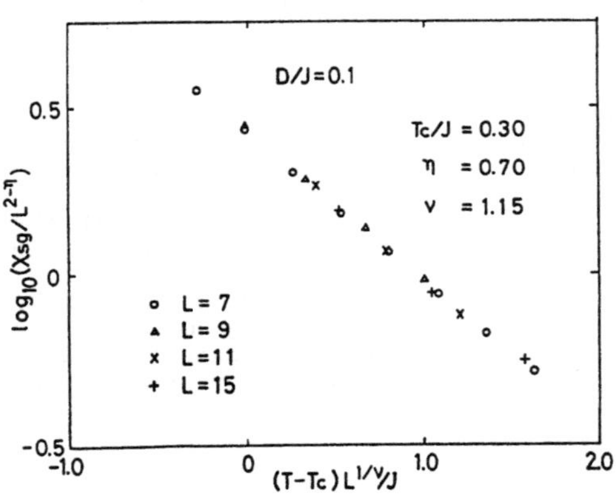

Fig.5 A finite size scaling plot of χ_{sg}

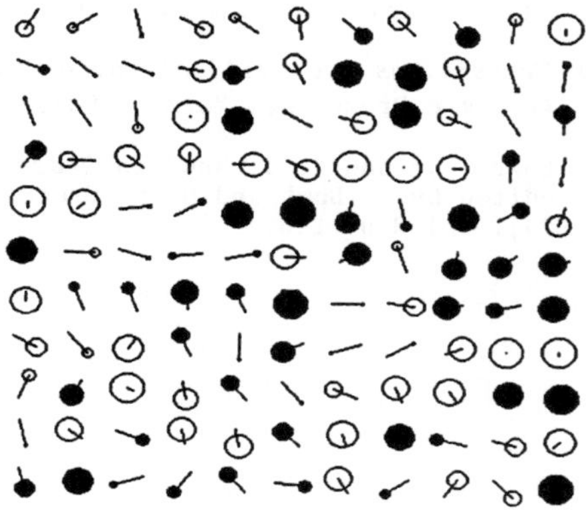

Fig.6 A snapshot of an equilibrium spin structure at T/J=0.2 projected on the xy-plane. The z-component of the spin is represented by open (S^z>0) or closed(S^z<0) circle.

cusp of χ. From these results, we conclude that the anisotropy induces the SG phase transition at a finite temperature. The exponent $\eta \sim 0.6$ in our model is much larger than η = -0.3 $\sim$ -0.2 predicted in the Ising SG[1,2]. This will come from the difference in the spin structure between the two models. A typical example of the snapshot of the spin structure at an equilibrium state is presented in Fig.6. The difference in the value of η together with the difference in the spin structure strongly suggest that the Heisenberg spin glass with the random anisotropy belongs to another universality class.

5.Conclusion

We have proposed a powerful method for simulating the Heisenberg model taking into account its intrinsic spin dynamics. We have applied the method to the Heisenberg spin glasses and solved various problems which have been desired to be studied numerically.

The authors wish to thank Dr. T. Shirakura for valuable discussions, and Mr. M. Iguchi for helping with numerical calculations.

References

1. R. N. Bhatt and A. P. Young: Phys. Rev. Lett. **54**(1985) 924.
2. A.T. Ogielski and I. Morgenstern: Phys. Rev. Lett. **54**(1985)928.
3. Y. Ozeki and H. Nishimori: J. Phys. Soc. Jpn **56**(1987) 1568,
4. M. Kikuchi and Y. Okabe: Phys. Rev. **B35**(1987) 5382, and references therein.
5. N. Ito and Y. Kanada: Supercomputer **25**(1988) 31, and references therein.
6. A. Ghazali, P. Lallemand and H.T. Diep: Physica **134A**(1986) 628.
7. B. W. Morris, S. G. Colborne, M. A. Moore, A. J. Bray and J. Canisius: J. Phys. C: Solid State Phys. **19**(1986) 1157.
8. J. A. Olive, A. P. Young and D. Sherrington: Phys. Rev. **B34**(1986) 6341.
9. A. J. Bray and M. A. Moore: Phys. Rev. **B34**(1986) 6561.
10. M.Gabay and G.Thoulouse, Phys. Rev. Lett. **47** (1981) 201.
11. F. Matsubara, T. Iyota and S. Inawashiro: J. Phys. Soc. Jpn. **60**(1991) 41.
12. F. Matsubara, T. Iyota and S. Inawashiro: J. Phys. Soc. Jpn.in press.

13. F. Matsubara, T. Iyota and S. Inawashiro: Phys. Rev. Lett.67 (1991) 1458.
14. See, e.g., M. N. Barber: in *Phase Transitions and Critical Phenomena*, edited by C. Domb and J. Lebowitz (Academic Press, New York, 1983), Vol.8, p. 146.
15. See, e.g., G. S. Rushbrooke, G. A. Baker: Jr and P. J. Wood,in *Phase Transition and Critical Phenomena*, edited by C. Domb and M. S. Green (Academic Press, London New York, 1974), Vol.3 p. 245.

Chiral Order in Vector Spin Glasses

H. Kawamura

Department of Physics, College of General Education,
Osaka University, Toyonaka 560, Japan

Abstract. The nearest-neighbor $XY(n = 2)$ and Heisenberg ($n = 3$) spin glasses with $\pm J$ distribution of interactions are studied in both two and three dimensions by use of the domain-wall renormalization-group method at zero temperature, with a focus on the chiral ordering.

1. Introduction

Owing to the recent extensive studies, it now seems well-established that the lower critical dimension (LCD) of an Ising ($n = 1$) spin glass with short-ranged interactions is between $d = 2$ and 3, while LCD of vector spin glasses ($n \geq 2$) is greater than $d = 3$, probably around $d = 4$. In other words, at $d = 3$ an Ising spin glass exhibits an equilibrium phase transition at a finite temperature, whereas vector spin glasses like XY ($n = 2$) and Heisenberg ($n = 3$) spin glasses exhibit only a zero-temperature transition.

Meanwhile, Villain pointed out in 1977 that XY spin glasses possess a twofold Ising-like degeneracy, often called "chirality", in addition to a continuous degeneracy associated with the original XY-spin symmetries [1]. By analyzing the properties of an XY spin glass, Villain pointed out the importance of this chiral degree of freedom for the spin-glass behavior. It should be noticed that, since the chirality is a multispin variable as will be shown below, the usual criterion of a spin-glass transition does not apply to the possible chiral ordering. Thus, the above-mentioned property that the LCD of vector spin glasses is greater than 3 does not necessarily exclude the possibility of a finite-temperature *chiral ordering* at $d = 3$.

Studies on the chiral components of vector spin glasses have been scarce. In 1985, Kawamura and Tanemura performed the first numerical study on the chiral ordering in vector spin glasses for the case of a two-dimensional $\pm J$ XY model [2]. It was then found that the ordering tendency of chirality was much enhanced as compared with that of the XY spins: In fact, the chiral and XY-spin orientational degrees of freedom ordered in a markedly different way, namely, while the standard spin-glass susceptibility exhibited a power-law divergence at zero temperature with the exponent $\gamma = 1.9 \pm 0.1$, the chiral susceptibility could be mapped onto the spin-glass susceptibility of a pure *Ising* spin glass rather accurately. Although chiral ordering appeared to take place at zero temperature, the associated chiral-susceptibility exponent, $\gamma_\kappa \gtrsim 4.5$, was significantly larger than the corresponding susceptibility exponent $\gamma \simeq 1.9$. Similarities to the Ising spin glass were also found in the dynamics of the chirality: Indeed, Batrouni and Dagotto found by numerical simulation of the two-dimensional XY spin glass that the dynamics of the chiral variables is of thermal-activation type, in sharp contrast to the dynamics of the XY spins [3].

A preliminary Monte Carlo study on the chirality in a $d = 3$ Heisenberg spin glass was also made by Olive, Young and Sherrington, although no definite conclusion was obtained as to its ordering properties [4].

In the present article, I report on the results of the recent calculations on the chiral ordering of the random $\pm J$ XY ($n = 2$) and Heisenberg ($n = 3$) models at zero temperature both in two and three dimensions, based on the domain-wall renormalization-group (DWRG) method combined with a spin-quench algorithm. The purpose of the calculation is to clarify whether the chiral spin-glass ordering occurs either at a finite temperature or at zero temperature, and also to clarify the relation to the conventional spin-glass ordering. Due to the length limitation, only the results will be presented in the following sections. For details of the calculation, readers are referred to the cited references.

2. Chirality

The models studied are the random $\pm J$ XY ($n = 2$) and classical Heisenberg ($n = 3$) models on the d=2 dimensional square and d=3 dimensional simple cubic lattices. The Hamiltonian is given by

$$\mathcal{H} = -\sum_{<ij>} J_{ij}\mathbf{S}_i \cdot \mathbf{S}_j, \tag{2.1}$$

where $\mathbf{S}_i$ is a classical unit vector with $\mathbf{S}_i = (S_i^x, S_i^y)$ or $\mathbf{S}_i = (S_i^x, S_i^y, S_i^z)$ for the $n = 2$ and 3 cases, respectively. The sum runs over all nearest-neighbor pairs $< ij >$, while the J_{ij} are random variables taking the values $+J$ and $-J$ with equal probability.

Let us analyze the global symmetry of the model. If all spins in a ground state are rotated equally, one obtains another ground state. In unfrustrated systems with collinear spin ordering, this operation generates all possible ground states. In a frustrated system with noncollinear spin ordering this is no longer so, since the state with opposite "handedness" or "chirality" can never be reached by mere rotations from the original state. In the XY case, such a ground state with opposite chirality can only be reached by a global spin *reflection* with respect to an arbitrary axis in the spin space. In the Heisenberg case, a ground state with opposite chirality can be reached by a global reflection with respect to an arbitrary chosen plane, (or by inversion of spins which can also be viewed as a combination of reflection and rotation). Anyway, in either case of XY and Heisenberg spins, the full set of ground-state manifolds consist of at least two disconnected manifolds characterized by opposite chiralities, which are axial scalar variables representing the handedness of the noncollinear spin ordering.

More specifically, in case of XY spins a local chirality may be defined for *two* spins by

$$\kappa_{ij} = [\mathbf{S}_i \times \mathbf{S}_j]_z, \tag{2.2}$$

whereas in case of Heisenberg spins it may be defined for *three* spins as

$$\kappa_{ijk} = \mathbf{S}_i \cdot \mathbf{S}_j \times \mathbf{S}_k. \tag{2.3}$$

Note that the chirality defined above for Heisenberg spins is a *scalar* quantity and is distinct from the *vector* chirality defined by $\mathbf{S}_i \times \mathbf{S}_j$: For any noncollinear but *planar* spin configurations, the former quantity vanishes while the latter quantity can take a finite value. In both cases of XY and Heisenberg spins, the chirality defined by (2.2) or (2.3) is invariant under any global rotation of the spins whereas it changes sign under any global reflection of the spins. Evidently, chiral order can be regarded as a manifestation of the breaking of the Z_2 symmetry associated with a global spin reflection.

3. XY spin glasses

In this section, the results of recent DWRG calculations on XY spin glasses are presented [5]. The standard DWRG method, in which the domain wall energy is defined as a difference between the ground-state energies for periodic and anti-periodic boundary conditions, was applied to the XY spin glass with a Gaussian distribution of nearest-neighbor interactions by Morris, Colborne, Moore, Bray and Canisius [6]. They found that the spin-glass phase transition occurred only at zero temperature with the standard spin-glass correlation-length exponents $\nu = 1.09 \pm 0.05$ for $d = 2$ and $\nu = 2.20 \pm 0.05$ for $d = 3$. In the type of boundary conditions adopted by Morris *et al*, however, only the contribution of the spin domain wall, which was associated with the continuous deformation of spins, was included whereas the contribution of the chiral domain wall was *not* included. This is because anti-periodic boundary conditions are equivalent to making a π rotation on boundary spins and do not cause of the flipping of chirality.

Thus, by use of the DWRG calculations employing *reflection boundary conditions* which cause the flipping of chirality and introduce the chiral domain wall into the sample, Kawamura and Tanemura found the followings [5]: (I) In $d = 2$ dimensions, both the spin and chirality exhibit a zero-temperature transition characterized by mutually different correlation-length exponents, $\nu = 1.2 \pm 0.15$ and $\nu_\kappa = 2.6 \pm 0.3$, respectively. The estimate for the usual spin-glass correlation-length exponent is consistent with the previous works [6],[7], whereas the estimate for the chiral correlation-length exponent is new and is markedly larger than that for ν. However, the estimated ν_κ is rather close to the standard correlation-length exponent of the $d = 2$ *Ising* spin glass, supporting the chirality–Ising spin glass analogy. (II) In $d = 3$ dimensions, the chiral component iterated towards strong coupling, suggesting the occurrence of a finite-temperature chiral spin-glass transition. The low-temperature phase is characterized by a broken reflection symmetry with orientational symmetry

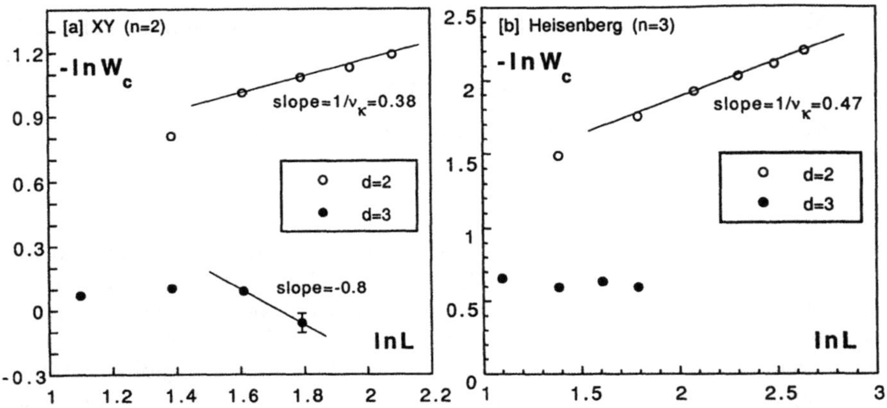

L-dependence of of the chiral domain-wall energy, $W_c(L)$, on a log-log plot for the $\pm J$ XY [Fig.a] and Heisenberg [Fig.b] models on $L\times L$ square and $L\times L\times L$ simple-cubic lattices. The statistical error on each point is smaller than the size of the point unless otherwise indicated.

preserved. By contrast, XY-spin orientational degrees of freedom were found to order only at zero temperature with the correlation-length exponent $\nu \simeq 2.4$, consistent with the previous works [6],[7]. The size dependence of the calculated chiral domain-wall energy, $W_c(L)$, is displayed in Fig.1.

4. Heisenberg spin glasses

The standard type of DWRG calculations employing periodic and anti-periodic boundary conditions were performed by McMillan for the Heisenberg spin glass with a Gaussian distribution of nearest-neighbor interactions [8]. He found that the spin-glass phase transition occurred only at zero temperature in both cases of $d = 2$ and 3 with the exponents $\nu = 0.714\pm0.015$ for $d = 2$ and $\nu = 1.54\pm0.19$ for $d = 3$. In the Heisenberg case, the application of anti-periodic boundary conditions causes the flipping of chirality and introduces the chiral domain wall, in sharp contrast to the XY case. Thus, even the standard domain-wall energy includes the contribution of the chiral domain wall in addition to that of the usual spin domain wall. The observation of a zero-temperature transition in ref.[8] seems to exclude the possibility of a finite-temperature chiral ordering in Heisenberg spin glasses either in two or three dimensions. However, since the investigated lattice sizes were rather small, there remains a possibility that the contribution of the chiral domain wall was masked by that of the spin domain wall in ref.[8].

In order to detect the contribution of the chiral domain wall directly, the present author recently performed a preliminary DWRG calculation employing a different type of domain-wall energy from the standard one, in which the contribution of the chiral domain wall is included but the contribution of the spin domain wall is suppressed: It is defined as a difference between the ground-state energies for *reflection boundary condition*, in which boundary spins are reflected with respect to a plane, and for *rotation boundary conditions* in which boundary spins are rotated by 180° around an axis. The size dependence of the calculated chiral domain-wall energy, $W_c(L)$, is displayed in Fig.2. The obtained results can be summarized as follows: (I) In $d = 2$ dimensions, both the spin and chirality exhibit a zero-temperature transition characterized by mutually different correlation-length exponents, $\nu = 1.2 \pm 0.2$ and $\nu_\kappa = 2.1 \pm 0.4$, respectively. As in the XY case, the estimate for ν_κ seems to be considerably larger than that for ν. (II) In $d = 3$ dimensions, the chiral component exhibits a marginal behavior. At the present stage, it seems impossible to say definitely whether a Heisenberg spin glass can sustain a chiral state at finite temperatures. In any case, the result seems to suggest that $d = 3$ is close to the LCD of the chiral components.

References

[1] J.Villain, J. Phys. **C10**, 4793 (1977).
[2] H. Kawamura and M. Tanemura, J. Phys. Soc. Jpn. **54**, 4479 (1985); Phys. Rev. **B36**, 7177 (1987).
[3] G.G. Batrouni and E. Dagotto, Phys. Rev. **B37**, 9875 (1988).
[4] J.A. Olive, A.P. Young and D. Sherrington, Phys. Rev. **B34**, 6341 (1986).
[5] H. Kawamura and M. Tanemura, J. Phys. Soc. Jpn. **60**, 608 (1991).
[6] B.W. Morris, S.G. Colborne, M.A. Moore, A.J. Bray and C. Canisius, J. Phys. **C19**, 1157 (1986).
[7] S. Jain and A.P. Young, J. Phys. **C19**, 3913 (1986).
[8] W.L. McMillan, Phys. Rev. **B31**, 342 (1985).

Simulation of Models for Isotropic and Anisotropic Orientational Glasses

K. Binder

Institut für Physik, Johannes-Gutenberg-Universität Mainz,
W-6500 Mainz, Staudinger Weg 7, Fed. Rep. of Germany

Abstract:"Orientational glass" behavior is found when molecular crystals are randomly diluted, and quadrupole moments get frozen by random alignment of the molecules, similar to "spin glass" behavior of randomly diluted magnets. Monte Carlo simulation of lattice models where quadrupole moments interact with nearest neighbor Gaussian coupling is a unique tool to study this behavior. The time-dependent glass order parameter exhibits anomalously slow relaxation, compatible with the Kohlrausch–Williams–Watts (KWW) stretched exponential function. Both isotropic and anisotropic models exhibit in d=2 and d=3 spatial dimensions glass transitions at zero temperature only. While the glass correlation length and glass susceptibility in the isotropic case exhibit power law divergences as $T \to 0$, exponential divergences are found for the p=3-state Potts glass in d=3, suggesting that the latter system is at its lower critical dimension.

1. Introduction

The glass transition by which a supercooled fluid freezes into a amorphous solid still is a great puzzle of condensed matter physics [1,2]. One hopes, however, that spin glasses [3] or quadrupolar glasses [4–7] are simpler understandable prototypes of glassy materials:

(i) there is a clear-cut distinction between quenched disorder (random occupation of lattice sites by atoms without spin or quadrupole moments, Fig.1 and thermal disorder (in the alignment of spins or quadrupole moments)

(ii) Order parameters and correlation functions that measure glass-like order can be identified and calculated [3–7].

Here we focus on two "random-bond" type models [3–7]:

(i) the fully isotropic quadrupolar glass [8,9]

$$\mathcal{H}^{\text{isotropic quadrupolar}} = - \sum_{<i,j>} J_{ij} \left[(\vec{S}_i \cdot \vec{S}_j)^2 - 1/3\right], \tag{1}$$

$\vec{S}_i$ being a unit vector along the axis of the uniaxial molecule at lattice site i, $<i,j>$ denotes a sum over nearest neighbor pairs, and J_{ij} is drawn from a Gaussian distribution of zero mean and width unity.

(ii) The p-state Potts glass [10–13] with p=3,

$$\mathcal{H}^{\text{potts}} = - \sum_{<i,j>} J_{ij}\, \delta_{n_i n_j}, \quad n_i = 1,2,\dots,p. \tag{2}$$

Springer Proceedings in Physics, Vol. 70 **Computational Approaches in Condensed-Matter Physics**
Editors: S. Miyashita · M. Imada · H. Takayama

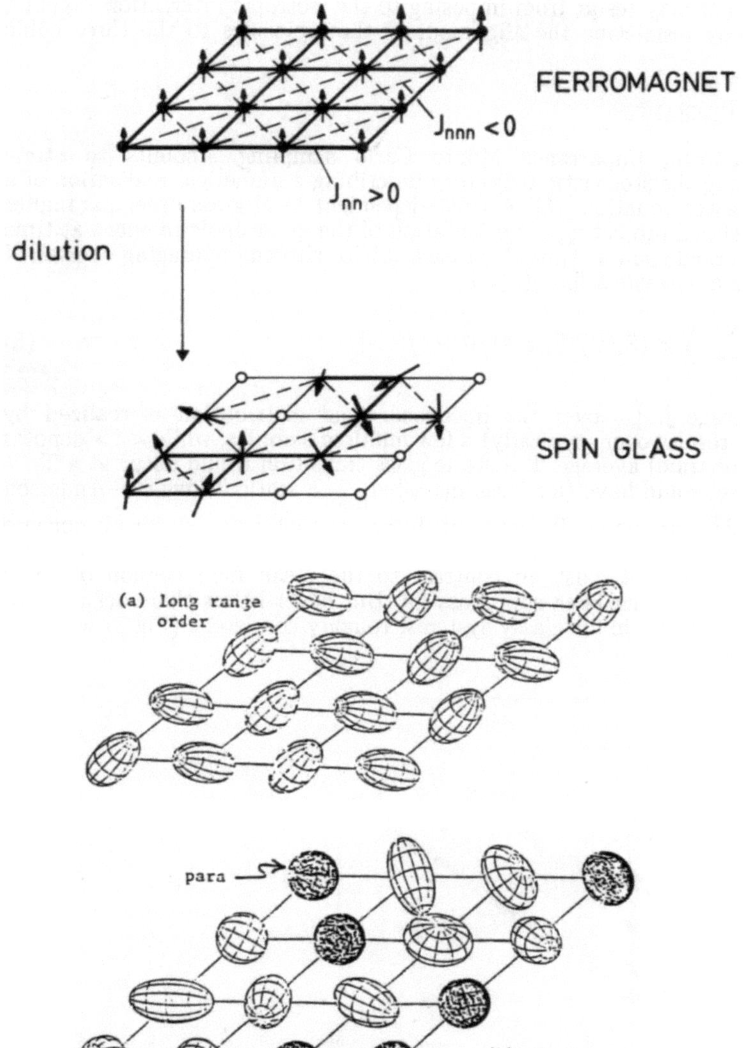

Fig. 1 A spin glass results from random dilution of a ferromagnet (magnetic atoms are shown by black dots) with nonmagnetic atoms (open circles), if there is competing exchange ($J_{nn}>0$, $J_{nnn}<0$; upper part). A quadrupolar glass results from random dilution of a molecular crystal (e.g. oH_2, lower part (a)) with spherical atoms (e.g. pH_2, lower part (b)).

Note that Eq.(2) may result from imposing to the isotropic interaction Eq.(1) a cubic anisotropy restricting the alignment of the molecules to the three cubic axes [14].

2. Dynamic Properties

As is well known, importance Monte Carlo sampling amounts to "time averaging" along the stochastic trajectory describing a numerical realisation of a Markovian master equation [15]. A time-dependent local glass order parameter q(t) is then defined projecting the orientation of the quadrupole moment at time t'+t onto its orientation at time t' for each lattice site and averaging over all N sites, choosing a normalization q(0)=1

$$q(t) \equiv \Big[\frac{3N}{2} \sum_i < (\vec{S}_i(t') \cdot \vec{S}_i(t'+t))^2 - 1/3 >\Big]_{av}. \tag{3}$$

Here the average $[...]_{av}$ over the quenched bond distribution is realized by averaging the results over (typically) a few hundred samples, while <..> denotes the thermal (or time) average. If a static glass transition would occur at a $T_f > 0$, for $T<T_f$ we would have (for $N \to \infty$) $q(t \to \infty) \to q_{EA}$, a static "Edwards–Anderson [16]–type" order parameter. It turns out, however, that $q(t \to \infty) \to 0$ for all nonzero temperatures for both models Eq.(1), (2), i.e. in thermal equilibrium a glassy ordered phase does not exist, in contrast to the mean field version of these models [8,10,11] and the Ising spin glass [3]. But nevertheless there occurs slow relaxation, characteristic for glassy systems, roughly consistent (Fig.2) with the KWW function,

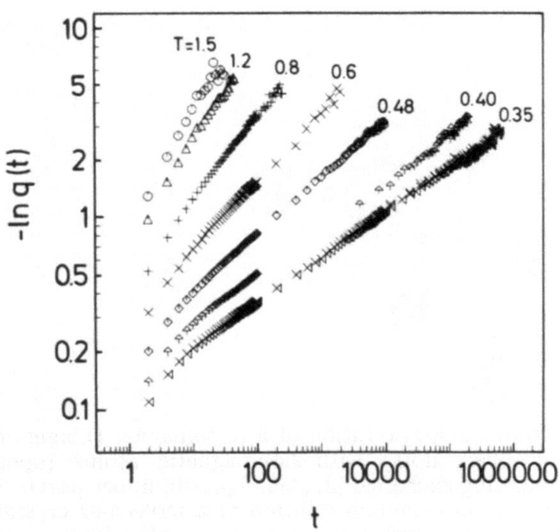

Fig. 2 Log–log plot of {–lnq(t)} vs. t, for the three-dimensional orientational glass, simulating 12x12x12 simple cubic lattices with periodic boundary conditions. Various symbols denote different temperatures T as indicated on the plot. Note that for C=1 Eq.(4) in this plot corresponds to a straight line with slope y. From Hammes et al.[9].

$$q(t) = C \exp[-(t/\tau)]^y, \tag{4}$$

C being a constant, τ a characteristic average relaxation time, and y a parameter characterizing the width of the relaxation spectrum. From Fig.2 it can be inferred that y decreases as T is lowered. Data for the model in d=2 and the Potts glass in d=3 look similar [9,12,17]. One distinction, however, is the behavior of the relaxation time τ: for the d=3 Potts glass, it diverges faster than any Arrhenius law $\tau \sim \exp(\Delta E/T)$, implying that diverging energy barriers ΔE develop in phase space as $T \to 0$ [12]. For the isotropic orientational glasses, the data [9] are compatible with simple power laws, $\tau \sim T^{-\nu z}$, with $\nu z \approx 4.3$ (d=2) and $\nu z \approx 6.8$ (d=3), respectively. Of course, these exponents may be only "effective exponents" since it is not obvious that the asymptotic critical region of the zero temperature phase transition actually has been reached.

3. Static Properties

In all the models studied, the dramatic slowing down at low temperatures can be linked to an onset of static correlations $g_G(\vec{R})$, $\vec{R}$ being the distance between sites i and j. This glass correlation function is defined as [7,9]

$$g_G(\vec{R}) = [\langle (\vec{S}_i \cdot \vec{S}_j)^2 - 1/3 \rangle_T^2]_{av}, \tag{5}$$

and the data (Fig.3) show that it is compatible with simple exponential decay, $g_G(\vec{R}) \sim \exp(-R/\xi_G)$: $1/\xi_G(T)$ can be found from fitting straight lines to the data of Fig.3. Both $\xi_G(T)$ and the associated susceptibility $\chi_G(T) = \sum_{\vec{R}} g_G(\vec{R})$

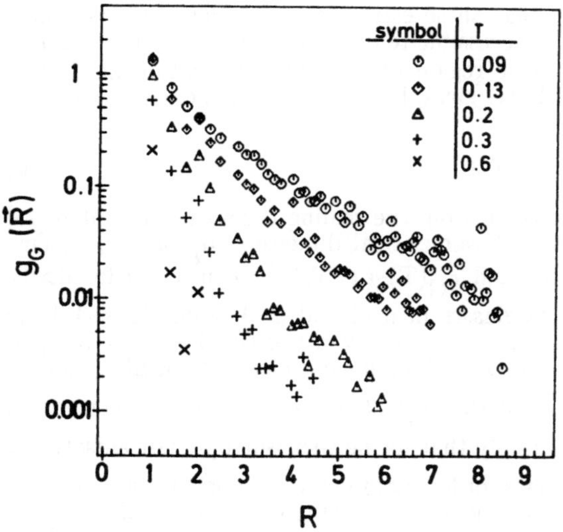

Fig. 3 Semilog plot of the glass correlation function $g_G(\vec{R})$ vs.distance, for the d=3 isotropic model, Eq.(1). Five temperatures are shown as indicated. Lattice sizes were 12x12x12 for the four higher temperatures and 18x18x18 for T=0.08, respectively. From Hammes et al.[9].

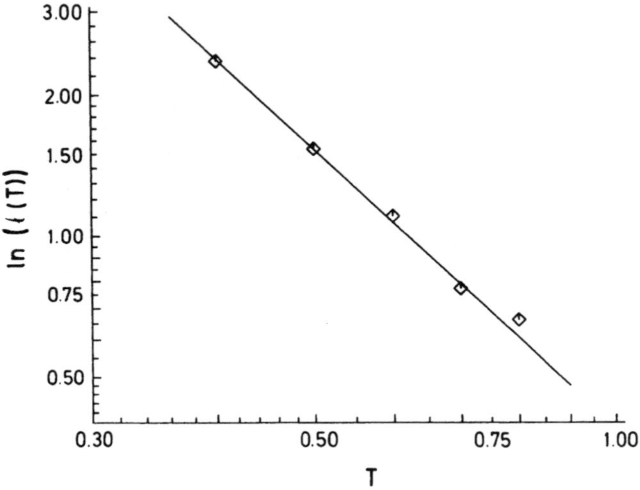

Fig. 4 Log–log plot of ln ℓ(T) vs.T, for the three–state Potts glass in d=3 dimensions. Here ℓ(T) is a characteristic length extracted from a finite size scaling analysis and is proportional to ξ_G(T). From Scheucher et al. [13].

diverge as T→0. For the isotropic orientational glass power laws are found, $\xi_G(T) \sim T^{-\nu}$, $\chi_G(T) \sim T^{-\gamma}$, with $\nu \approx 0.63$, $\gamma \approx 1.35$ (d=2) and $\nu \approx 1.02$, $\gamma \approx 2.7$ (d=3). Again the caveat must be made that these estimates are "effective exponents" characterizing the behavior for the accessible temperature range, since it is not clear to what extent the asymptotic region of the phase transition at T=0 is reached. For the case of the d=3 Potts glass, however, an exponential divergence is found,

$$\xi_G(T) = \exp(\text{const } T^{-\sigma}), \ \sigma \approx 2, \ T \to 0 \tag{6}$$

which is exactly the behavior expected from the scaling theory of McMillan [18] for a system that is precisely at its lower critical dimension d_ℓ, see Fig.4. One expects that $\xi_G(T) \sim T^{-\nu}$ for $d < d_\ell$ and $\xi_G(T) \sim (T-T_c)^{-\nu}$ with $T_c > 0$ for $d > d_\ell$. Monte Carlo simulations for Potts glasses in d=2 and d=4 dimensions [19] are consistent with these predictions and hence confirm the conclusion of Ref.[13] that d=3 is either exactly equal to d_ℓ or at least very close. Remember that in the Ising spin glass [3] it is believed that $d_\ell \approx 2.5$ while in isotropic vector spin glasses $d_\ell = 4$ [3]. Thus our results imply that for neither isotropic nor anisotropic orientational glasses a static glass order parameter is nonzero at physical dimensionalities, and hence in these models the freezing transition is of a gradual nature and a nonequilibrium phenomenon only, depending on the time scale of observations [3]. However, more work is still needed in order to explain the dynamics of these models and to find out whether Eq.(4) is more than a phenomenological function for an approximate fit of both experimental and simulational data.

Acknowledgements

This brief review is based on research done with H.–O. Carmesin, D. Hammes, J.D. Reger, M. Scheucher and A.P. Young; it is a pleasure to thank them for a fruitful collaboration. Support from the Deutsche Forschungsgemeinschaft (SFB 262) is gratefully acknowledged, as well as grants of computer time at VP 100 (Kaiserslautern) and CRAY–XMP (HLRZ Jülich) supercomputers.

References

1. J. Jäckle, Rep. Progr. Phys.49, 171 (1986)
2. J.P. Hansen, D. Levesque and J. Zinn–Justin (eds) Liquids, Freezing and the Glass Transition (North–Holland, Amsterdam 1990)
3. K. Binder and A.P. Young, Rev. Mod. Phys.58, 801 (1986)
4. K. Knorr, Phys. Scripta T 19, 531 (1987)
5. A. Loidl, Ann. Rev. Phys. Chem.40, 29 (1989)
6. U.T. Höchli, K. Knorr and A. Loidl, Advanc. Phys. 39, 405 (1990)
7. K. Binder and J.D. Reger, Advanc. Phys.(in press)
8. P. Goldbart and D. Sherrington, J. Phys. C18, 1923 (1985)
9. D. Hammes, H.–O. Carmesin and K. Binder, Z. Phys. B76, 115 (1989)
10. D. Elderfield and D. Sherrington, J. Phys. C16, L497, L971, L1169 (1983)
11. D.J. Gross, I. Kanter and H. Sherrington, Phys. Rev. Lett.55, 304 (1985)
12. H.–O. Carmesin and K. Binder, J. Phys. A21, 4035 (1988)
13. M. Scheucher, J.D. Reger, K. Binder and A.P. Young, Phys. Rev. B42, 6881 (1990)
14. H.–O. Carmesin, J. Phys. A22, 297 (1989)
15. K. Binder (ed.) Monte Carlo Methods in Statistical Physics, 2nd ed.(Springer, Berlin 1986)
16. S.F. Edwards and P.W. Anderson, J. Phys. F5, 965 (1975)
17. K. Binder, J. Non–Cryst. Solid 131–133, 262 (1991)
18. W.L. McMillan, J. Phys. C17, 3179 (1984)
19. M. Scheucher and J.D. Reger, to be published.

The Phase Diagram of the Three-Dimensional $\pm J$-Ising Spin Glass Model

N. Ito

Computing and Information Systems Center, Japan Atomic Energy Research Institute, Tokai-mura, Naka-gun, Ibaraki 319-11, Japan

Abstract. The location of tricritical point of paramagnetic, ferromagnetic and spin glass phase of the $\pm J$-Ising spin glass model on the cubic lattice is studied numerically. The spontaneous magnetization is estimated along the Nishimori line and the critical density of ferromagnetic bond is estimated to be 0.7715(30).

The behavior of the SK model for spin glass system is now well-established. The SK model is an infinite dimensional and infinite range interaction model. The behavior of finite dimensional model is, however, not fully understood yet. The nature of the spin glass phase in three-dimensional lattice is important because the real material has this dimensionality. Therefore the theoretical understanding of the spin glass model in three-dimensional lattice is highly expected[1,2].

The model Hamiltonian is the form of

$$H = -\sum_{<i,j>} J_{ij}\sigma_i\sigma_j \quad (\sigma_i = \pm 1, \quad J_{ij} = \pm J, \quad J > 0),$$

where the summation runs over all nearest-neighbor pairs of lattice points on the cubic lattice. The bond variables J_{ij} are treated as quenched and the values are determined randomly so that the density of ferromagnetic bonds ($+J$-bonds) will be p.

The p-T phase diagram of this model was studied firstly by Ozeki and Nishimori[3] and the location of the tricritical point was $p_c^{\text{tri.}} = 0.767(4)$ using the Monte Carlo renormalization group method. Singh[4] estimated $T_c^{\text{tri.}} = 1.690(16)$ using the high-temperature expansion method and this critical temperature correspond to $p_c^{\text{tri.}} = 0.7656(20)$ because the tricritical point will be on the Nishimori line[5,6].

Recently the magnetization curve of the three-dimensional ferromagnetic Ising model was studied numerically[7,8] and the critical point and exponent β were estimated. It is shown that this method works efficiently to estimate the critical behavior based on the Monte Carlo study of the finite lattices. The values of spontaneous magnetization m_s are obtained from the thermal expectation values of squared magnetization $< M^2 >$ and the extrapolation formula

$$\frac{< M^2 >}{N^2} = m_s^2 + \frac{kT\chi}{N} + \text{(less significant correction)},$$

where N, k, T and χ denote the number of spins in the lattice, the Boltzmann's constant, the absolute temperature and the magnetic susceptibility, respectively.

In this paper, this method is applied to the study of tricritical point $p_c^{\text{tri.}}$ of spin glass model on three-dimensional lattice. The values of spontaneous magnetization are estimated along the Nishimori line which is defined by the relation $p = 1/(1+\exp(-2K))$, where K denotes J/kT. The lattice sizes, numbers of Monte Carlo steps and numbers of bond samples are summarized in Table 1.

Springer Proceedings in Physics, Vol. 70 Computational Approaches in Condensed-Matter Physics
Editors: S. Miyashita · M. Imada · H. Takayama

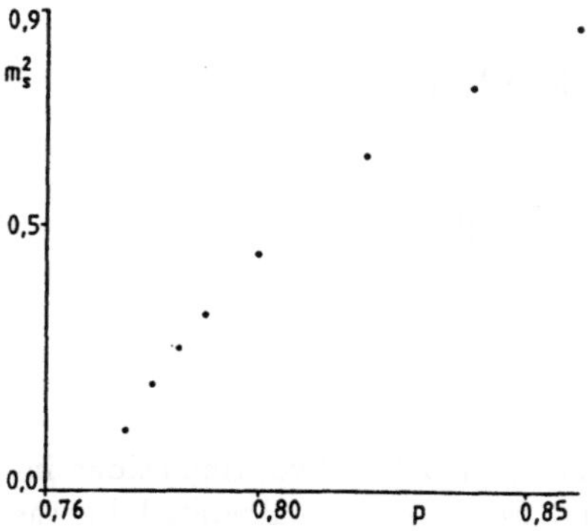

Figure 1: The estimated values of m_s^2 along the Nishimori line are shown.

Table 1: The Monte Carlo data are listed. The numbers in column "sample"denote the number of different bond configurations. The numbers, $11^2 \cdot 12 \sim 41^2 \cdot 42$, denote the lattice sizes. The numbers 10^7 and 10^8 mean the numbers of Monte Carlo sweep for the systems.

p	sample	$11^2 \cdot 12$	$15^2 \cdot 16$	$21^2 \cdot 22$	$31^2 \cdot 32$	$41^2 \cdot 42$
0.75	32	10^7	10^7	10^7	–	–
0.76	32	10^8	10^8	10^8	–	–
0.775, 0.780, 0.785, 0.790	8	10^7	–	10^7	10^7	10^7
0.800, 0.820, 0.840, 0.860	8	10^7	–	10^7	10^7	–

At $p = 0.75$ and 0.76, the values of m_s are estimated to be almost 0. Therefore the system is in paramagnetic phase at these values of p and the value of $p_c^{\text{tri.}}$ is clearly larger than 0.76. Non-zero values of m_s are obtained at p larger than or equal to 0.775. They are shown in Fig. 1.

The χ^2-fitting and maximum entropy method(AIC analysis) are used for estimating the value of $p_c^{\text{tri.}}$ based on the obtained values of m_s^2. The model functions for the χ^2-fittings were assumed to be simple-pole type, that is, $m_s^2(p) = [\sum_{k=1}^{N} a_k(p-p_c^{\text{tri.}})^k]^{2\beta}$ $(N = 2, 3, 4)$. And the estimation $p_c^{\text{tri.}} = 0.7715(30)$ is obtained. The error ± 0.0030 denotes the two-standard-deviation region. This value is consistent with the previous estimations by Ozeki and Nishimori and Singh.

The FACOM VP2600/10 in Japan Atomic Energy Research Institute was used for the Monte Carlo simulation.

References

[1]R. N. Bhatt and A. P. Young, Phys. Rev. Lett. **54** (1985) 924.

[2]A. T. Ogielski and I. Morgenstern, Phys. Rev. Lett. **54** (1985) 928.

[3]Y. Ozeki and H. Nishimori, J. Phys. Soc. Jpn. **56** (1987) 1568.

[4]R. R. P. Singh, Phys. Rev. Lett. **67** 1991) 899.

[5]H. Nishimori, Prog. Theor. Phys. **66** (1981) 1169.

[6]H. Nishimori, J. Phys. Soc. Jpn. **55** (1986) 3305.

[7]N. Ito and M. Suzuki, J. Phys. Soc. Jpn. **60** (1991) 1978.

[8]N. Ito, to be published in the proceedings of "Computer Aided Studies in Statistical Physics"(Taipei, 1991 June).

Phase Diagram and Ordered Phase of the $\pm J$ Model by the Interfacial Method

Y. Ozeki and *Y. Ueno*

Department of Physics, Tokyo Institute of Technology, Oh-okayama, Meguro-ku, Tokyo 152, Japan

Abstract. We study the asymmetric $\pm J$ Ising model in two dimensions investigating the size dependence of interfacial free energies calculated by the transfer matrix method.

There have been found some interesting properties[1-5] for the asymmetric distributions of the $\pm J$ Ising models. One is the vertical phase boundary along the temperature(T)-axis of the ferromagnetic phase (FP). Another is the random antiphase state (RAS) which exists adjacent to the FP. Ground state properties of the RAS in 2d have been studied intensively[1,5], but they have not yet been confirmed. The interface approach has been applied to the SG problem by several authors[7-9]. We make use of concentration- (of ferromagnetic bonds; p) and temperature-dependences of two stiffness exponents (defined later) relevant to the FP and the random phase (RP; which we generically call the ordered phase without uniform magnetization such as SG or RAS), and obtain the phase diagram of the RAS clearly.

In our calculations, the periodic boundary condition is imposed in one direction. The spins at boundaries in the other direction are fixed either all up or up at one boundary and down at the other. We calculate the interfacial free energy $\Delta F_L\{J\}$ for a bond configuration $\{J\}$ in the system of linear size L, from the difference of free energies between these two boundary conditions. Although we call ΔF the interfacial free energy, it is not certain whether or not ΔF always reflects a single interface as in the pure Ising ferromagnets. In random systems, in general, both boundary conditions yield some distortions or defects. However, it still remains unchanged that ΔF is a response to the strong force applied through the boundaries; it tells one whether a long range order exists or not, by the property how it goes to infinity or vanishes as $L \to \infty$. Thus it is natural to consider the mean and mean deviation of ΔF [6]:

$$\begin{aligned} W_L(p,T) &= \langle \Delta F_L\{J\}\rangle_J, \\ \tilde{W}_L(p,T) &= [\langle \Delta F_L\{J\}^2\rangle_J - \langle \Delta F_L\{J\}\rangle_J^2]^{\frac{1}{2}}, \end{aligned}$$

where $\langle\cdots\rangle_J$ denotes the bond average. Let us define stiffness exponents from the asymptotic forms, $W_L(p,T) \sim L^{a(p,T)}$ and $\tilde{W}_L(p,T) \sim L^{\tilde{a}(p,T)}$. It is clear that W_L is relevant to the FP in such a way that W_L vanishes at the phase boundary. Since $\tilde{W}_L$ arises owing to the randomness, we naturally assume that it is relevant to the random phase.

Springer Proceedings in Physics, Vol. 70 Computational Approaches in Condensed-Matter Physics
Editors: S. Miyashita · M. Imada · H. Takayama

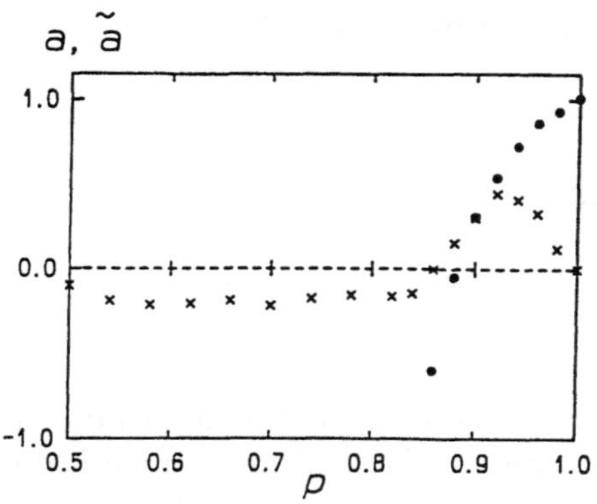

Fig. 1. Stiffness exponens[6] a and $\tilde{a}$ vs. concentration p at T=0.15 with $a(\bullet)$ and $\tilde{a}(\times)$.

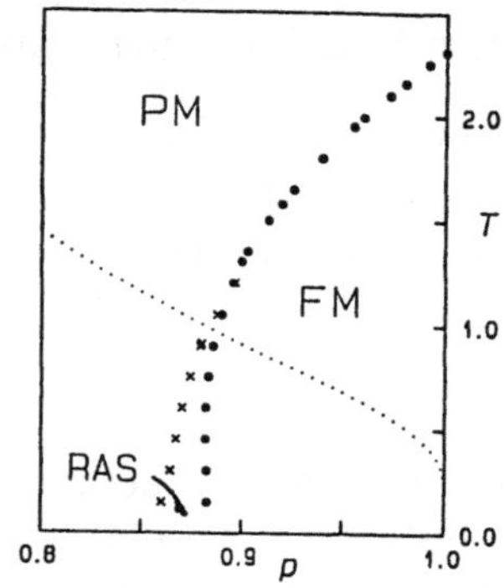

Fig. 2. Calculated phase diagram[6]. Closed circles and crosses are obtained by a=0 and $\tilde{a}$=0, respepctively. The dotted line is the Nishimori-line[2].

As it arises also in the FP, one cannot easily separate the contributions from the FP and the RP to see if they coexist or not. So one gets the phase boundary of the RP by $\tilde{a} \to 0$ only in the case where the FP is absent. It is natural to expect the following properties of stiffness exponents at phase boundaries among FP, RP and the disordered phase (DP): a=0 and $\tilde{a}$=0 at the boundary of FP↔DP, $a < 0$ and $\tilde{a}$=0 at the boundary of RP↔DP, a=0 and $\tilde{a} > 0$ at the boundary of FP↔RP.

Calculations are carried out for the square lattice of sizes L=6, 8, 10, 12 at almost 100 points in the $p - T$ plane located at intervals of 0.01 in p and 0.15 in T. At each point, we estimate W_L and $\tilde{W}_L$ by the averages in randomly chosen 2000 samples of bond configurations. A typical example of the concentration-dependence of stiffness exponents at T=0.15 are shown in Fig.1. One can easily see that there exist two phase-boundaries, FP↔RAS and RAS↔DP, in this temperature range. From the results like in Fig.1, we obtain the phase diagram in Fig.2, which is consistent with those obtained before[1-6].

References

1. R. Maynard and R. Rammal, J. Phys. Lett. (Paris) **43**,L347 (1982)
2. H. Nishimori, J. phys. Soc. Jpn. **55**,3305 (1986).
3. Y. Ozeki and H. Nishimori, J. Phys. Soc. Jpn. **56**,3265 (1987).
4. J. D. Regar and A. P. Young, J. Phys.: Condes. Matter **1**,915 (1989).
5. Y. Ozeki, J. Phys. Soc. Jpn. **59**,3531 (1990).
6. Y. Ueno and Y. Ozeki, J. Stat. Phys. **64**,227 (1991)
7. W. L. Mcmillan, Phys. Rev. B**29**,4026 (1984); B**31**,340 (1985).
8. A. J. Bray and M. A. Moore, J.Phys. C**17**,L463 (1984)
9. J. R. Banavar and M. Cieplak, Phys. Rev. Lett. **48**,832 (1982)

Replica Optimization Method for Ground-State Search of Random Spin-Systems

N. Kawashima and *M. Suzuki*

Department of Physics, Faculty of Science, University of Tokyo, Hongo, Tokyo 113, Japan

Abstract. A new method to find ground states is proposed for random spin systems. The efficiency of this method is confirmed numerically in the case of the two-dimensional Ising spin glass with the gaussian bond-distribution in a uniform field. Increase of computational time with respect to system size is moderate and well fitted by a power-law up to $L = 32$. Magnetizations are calculated for various magnetic fields using the new method. Size dependence of the susceptibility is found to be $\chi(L) \propto L^x$ with $x = 0.476(5)$. This value of the exponent x is somewhat larger than the predicted value by the droplet argument.

1 Replica Method

For the last 20 years, random spin systems have been attracting interests of many physicists. Validity of theories in this area has to be confirmed by numerical calculations. As for the investigation of the critical behavior of the two-dimensional random spin-systems, one has to find ground states or low-lying excited states by the numerical calculations. The most common stochastic method for the ground state search is so-called simulated annealing. Recently a new algorithm for the Monte Carlo simulation of the random spin-systems at a finite temperature is proposed[1]. In the present paper, we report a result of an approach which is along the same line but much more effective for the present purpose.

Suppose that the initial conditions are given by a certain configurations of Ising spins $\{S^{(\mu)}(\boldsymbol{R})\}$. First, calculate the following integer:

$$\psi(\boldsymbol{R}) \equiv f(S^{(1)}(\boldsymbol{R}), S^{(2)}(\boldsymbol{R}), \ldots, S^{(n)}(\boldsymbol{R})) \equiv \sum_{\mu=2}^{n} 2^{\mu-3}(1 - S^{(\mu)}(\boldsymbol{R})). \quad (1)$$

Then, divide the whole system into clusters using this integer. Here a cluster is a group of sites which have the same value of $\psi(\boldsymbol{R})$ and connected with each other by interaction bonds. Next examine whether the total energy of the system is lowered by simultaneous flipping all spins in a cluster. If so, flip all of those spins. Do this for all the clusters one by one. After examining all the clusters, calculate the integer $\psi(\boldsymbol{R})$ again. This time, the resulting clusters must be larger than the first ones. Then,

Springer Proceedings in Physics, Vol. 70 **Computational Approaches in Condensed-Matter Physics**
Editors: S. Miyashita · M. Imada · H. Takayama

examine all the new clusters one by one. Repeat all of these procedures until there is no cluster to be flipped.

We have applied this operation repeatedly to a set of systems until true ground states of almost all the systems are reached. For large systems, the necessary computational time τ per system is smaller than that for other methods. It is remarkable that τ's obtained in the present method are well fitted by the power law $\tau \propto N^{\lambda}$ ($\lambda \sim 2.6$) where N is the total number of sites. This increase of computational time is moderate compared to other methods.

2 Magnetization at $T = 0$

We have calculated magnetizations of the two-dimensional EA models with the Gaussian bond-distribution with various system sizes and in the presence of magnetic fields of various strength. The least square analysis to the present data for susceptibilities results in

$$\chi(L) = aL^{x} \qquad x = 0.476(5). \tag{2}$$

On the other hand, according to the domain-wall renormalization-group (DWRG) argument, the above exponent x is directly related to the stiffness exponent θ by $x = -\theta$. Using the transfer matrix method at $T = 0$ the stiffness exponent θ has been estimated as $\theta = -0.291(2)$ [2]. Our estimation (2) is clearly larger than the prediction by the DWRG argument beyond the statistical error.

To summarize, we have presented a new method to calculate the exact ground states for the two-dimensional systems in several conditions which are not tractable by the ordinary methods. Our result about the exponent x suggests that the DWRG argument with the numerical estimations of the stiffness exponent fails to predict the value of the exponent x precisely. It seems that further consideration is necessary for the complete understanding of the critical behavior of the present model.

References

[1] R.H.Swendsen and J.-S.Wang: *Phys.Rev.Lett.* **57**2607(1986).

[2] A.J.Bray and M.A.Moore: *J.Phys.C* **17**L463(1984).

Anomalous Phase in the 2D EA Ising Spin Glass

K. Nemoto, K. Hukushima, and H. Takayama
Institute of Physics, University of Tsukuba, Tsukuba 305, Japan

Abstract. We examine the possibility of the anomalous phase in the 2D $\pm J$ EA Ising spin glass, which corresponds to the Griffiths phase in the diluted ferromagnetic system. By Monte Carlo simulation, we find that the density of eigenvalues of the spatial correlation matrix has a weak anomaly around T_c, the transition temperature for the pure ferromagnetic system, and that the autocorrelation function is scaled to a function of the power of time, the exponent of which deviates from 1 below T_c.

It has been suggested that there exists a kind of the Griffiths phase in the Edwards-Anderson (EA) Ising spin glass (SG) systems[1,2]. Ogielski observed by his enormous simulation for the 3D $\pm J$ SG system that the autocorrelation function $q(t) \equiv \sum_i \langle\langle S_i(0)S_i(t)\rangle_T\rangle_J$ follows the stretched exponential decay between the SG transition temperature, T_g, and the transition temperature of the pure ferromagnetic system, T_c[2]. If the anomalous *phase* corresponds to the Griffiths phase in the diluted ferromagnetic system, the same phase should also exist in the 2D system, where the spin glass transition does not occur.

In order to investigate the phase in the 2D $\pm J$ SG system, we performed Monte Carlo simulation to observe the two-point correlation function matrix $C_{ij} \equiv \langle S_i S_j \rangle_T$. Then we evaluated the density of eigenvalues of the inverse of the matrix, $\rho_J(\lambda)$. For a transition to occur, the lower bound of $\rho_J(\lambda)$, $\lambda_{\min}$, reaches to 0 and the corresponding eigenstate percolates. We analyzed the behavior near the edge of $\rho(\lambda) \equiv \langle \rho_J(\lambda) \rangle_J \sim (\lambda - \lambda_{\min})^x$ and found that the exponent x, which is an increasing function of T, has a knee-like anomaly at T_c. Further, $\Lambda \equiv \lambda_{\min}^x$ is constant above T_c and increases with decreasing temperature below T_c. Note that the linear susceptibility χ_0 can be evaluated as $\chi_0 = \int \lambda^{-1}\rho(\lambda)\mathrm{d}\lambda$. Therefore the increasing Λ means the increasing contribution of the lower states to χ_0 since $\Lambda \sim \int \lambda^{-1}(\lambda - \lambda_{\min})^x \mathrm{d}\lambda$. This behavior is a signal to the singularity in the external field dependence of the free energy, which characterizes the Griffiths phase in the diluted system.

Springer Proceedings in Physics, Vol. 70 **Computational Approaches in Condensed-Matter Physics**
Editors: S. Miyashita · M. Imada · H. Takayama

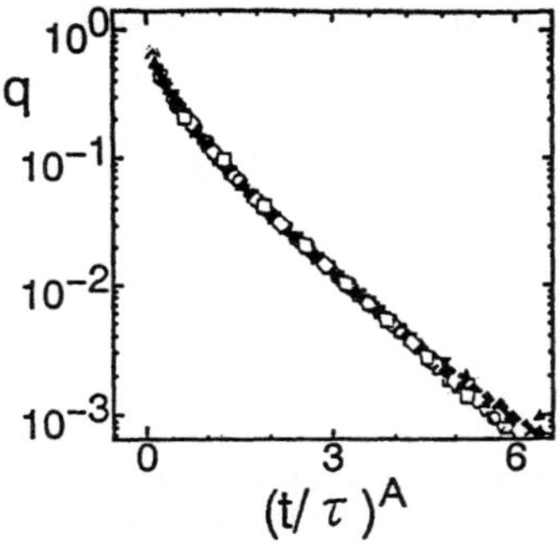

Fig. 1. Scaling plot of $q(t)$ of various T

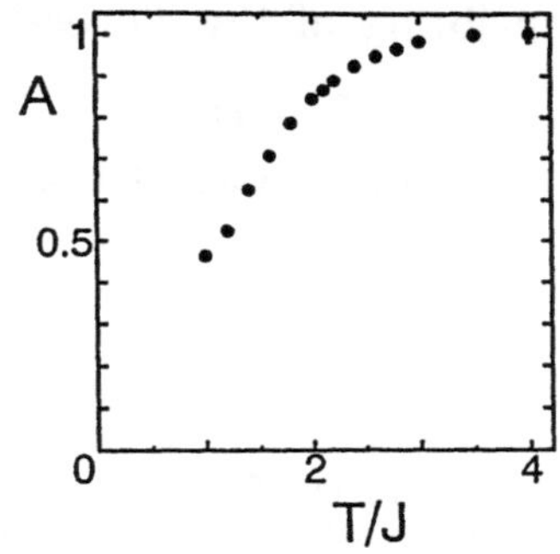

Fig. 2. T dependence of the exponent A

The above anomaly is seen more moderately in the autocorrelation function $q(t)$. We observed $q(t)$ at temperatures around T_c ($J < T < 4J$) and found that it can be scaled as $q(t) = Q(x)$ where $x = (t/\tau)^A$ as shown in Fig. 1. Note that $q(t)$ obeys the stretched exponential law if $Q(x) \sim \exp(-x)$. From the scaling we can extract $A(T)$ and $\tau(T)$ except for the common multiplicative factors, which are determined from the high temperature region where $q(t)$ seems to be a simple exponential function. As shown in Fig. 2, $A(T)$ deviates from 1 below T_c. The result resembles that of Ogielski for the 3D system[2]. We expect that the change of $A(T)$ near T_c becomes the sharper if the smaller value of $q(t)$ than 10^{-3} can be used for the scaling. Another interesting feature is that $A(T)$ depends linearly on T below T_c. This dependence is widely seen in the various systems which show the stretched exponential decay of relaxation. This can be explained, for example, by a phenomenological theory of diffusion process on a hierarchical tree.

In conclusion, our results indicate the existence of anomalous phase in the 2D $\pm J$ SG, which corresponds to the Griffiths phase in the diluted system.

[1] M. Randeria, J.P. Sethna and R.G. Palmer: Phys. Rev. Lett., **54** (1985) 1321
[2] A.T. Ogielski: Phys. Rev., **B32** (1985) 7384

AC Susceptibility of the Three-Dimensional $\pm J$ Ising Model

T. Shirakura and S. Inawashiro

Department of Applied Physics, Tohoku University, Sendai 980, Japan

Recently AC susceptibilities $\chi(\omega)$ and relaxation time distribution $g(\tau)$ in the $\pm J$ Ising spin glasses(SG) were calculated by Monte Carlo simulations, and characteristic differences were found in the two-dimentional(2D) and three-dimensional(3D) systems by Suzuki et al.[1]. We showed that these behaviors can be reproduced by a model based on the picture of thermally activated hopping of state in an ultrametric phase space [2]. The work in Ref.[1] is restricted to the case of p=0.5, where p is the probability that J_{ij} is +J. Here we study AC susceptibilities in the 3D $\pm J$ Ising model on a simple cubic lattice for some p's in the range of $0.5<p<1$, using the MC simulation. The details of the model and the method are referred to Ref.[1].

We concentrate on the following two points: (i) in the case of p=0.5, the imaginary parts of AC susceptibilities $Im\chi(\omega)$'s show no frequency dependence at low temperatures for the calculated frequencies $\omega=2\pi/n_{cycl}$, $n_{cycl}=20\sim6000$ MC steps per spin (MCS)[1]. This suggests by the fluctuation dissipation theorem that the fluctuation spectrum of the magnetization shows 1/f-noise behavior at low temperatures. We would like to know that whether or not this 1/f-noise behavior holds for $p\neq0.5$. (ii) Ozeki and Nishimori [3] estimated the multicritical point in this model as $(p_c,T_c/J)=(0.767\pm0.004,1.68\pm0.025)$. Through the behavior of $Im\chi(\omega)$, we might see the AT transition[4]-like behavior at a ferromagnetic concentration p near p_c. It was observed for the pure ferromagnet [5] that the magnitude of $Im\chi(\omega)$ becomes small rapidly below the para-ferro transition temperature. On the other hand, for SG (p=0.5), $Im\chi(\omega)$ holds a finite magnitude even at low temperatures [1]. Therefore we might expect a double-peak structure for $Im\chi(\omega)$ versus T/J at a ferromagnetic concentration p near p_c: one peak near a para-ferro transition temperature and the other peak near the ferro-mixed phase (AT) transition temperature (which might be expected only on a finite time scale).

The parameters used in the MC simulation are almost the same as those in Ref.[1], except for the field amplitude δh. Here we chose $\delta h=0.05J$ and $0.025J$. In the ferromagnetic region, we want to choose δh as small as possible, but owing to the finiteness of the number of spins $N=14\times14\times14$, we must choose $\delta h>J/\sqrt{N}\fallingdotseq0.02J$ to obtain the data whose scatter is small in the SG region. A staionarity of the measured values of $Im\chi(\omega)$ was confirmed by the same method as in Ref.[1].

We show the results in Figs.1(a)~(d). In the case of p=0.6 (Fig.1(a)), $Im\chi(\omega)$ exhibits no frequency dependence and the 1/f-noise behavior appears at low temperatures, similarly to the case of p=0.5. For p=0.7, we observe a deviation from the 1/f-noise behavior in the range of the measured cycles $n_{cycl}=200\sim6000$ MCS. In the cases of p=0.8 and 0.82 (Figs.1(b) and (c)), $Im\chi(\omega)$ shows a sharp peak which is strongly frequency-dependent near the para-ferro transition temperature. At low temperatures, $Im\chi(\omega)$ has a finite magnitude and exhibits no frequency-dependence. The temperature below which $Im\chi(\omega)$ becomes frequency-independent decreases more in p=0.82 than in p=0.8. For p=0.85 (Fig.1(d)), we observe a similar behavior of $Im\chi(\omega)$ to that in the pure ferromagnet (p=1) [5]. These behaviors in the cases of p=0.8 and 0.82 suggest the AT transition on a finite time scale.

Next we would like to investigate that whether or not there is the AT transition on an infinite time scale, i.e., in equilibrium in this model. In order to study this problem, we investigate the order-parameter distribution P(q) at p=0.8 by means of the method of Bhatt and Young [6]. We chose the number of spins $N=6\times6\times6$ and $10\times10\times10$, and the number of samples $N_s=1280$.

Springer Proceedings in Physics, Vol. 70 Computational Approaches in Condensed-Matter Physics
Editors: S. Miyashita · M. Imada · H. Takayama

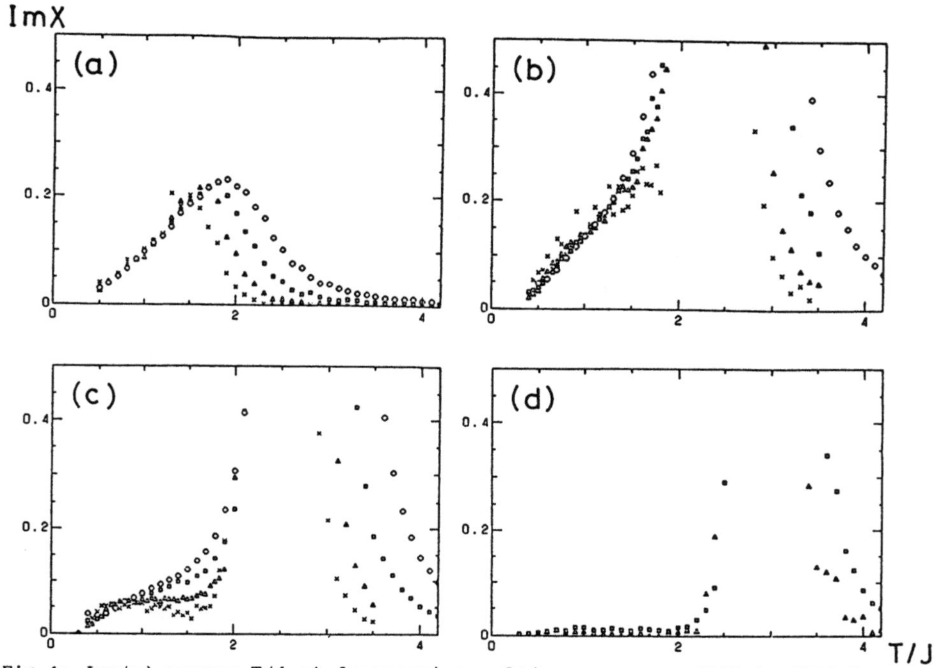

Fig.1: $Im\chi(\omega)$ versus T/J at frequencies $\omega=2\pi/n_{cycl}$, $n_{cycl}=200(o)$, 600(□), 2000(Δ) and 6000(x). (a) p=0.6, (b) p=0.8, (c) p=0.82 and (d) p=0.85.

In the case of p=0.8 with 100000 MCS, our measurements for T≧0.8J satisfy the equilibration condition. At low temperatures, we could not see any characteristic change of P(q) from its ferromagnetic behavior at least for T≧0.8J. The details will be published elsewhere.

In conclusion, we have obtained the characteristic behaviors of $Im\chi(\omega)$ which suggest the AT transition on a finite time scale in the 3D ±J model. However we could not find any indication that there exists the AT transition as an equilibrium phase transition at least for T≧0.8J and p=0.8.

This work was financially supported by a Grant-in-Aid for Scientific Research from the Ministry of Education, Science and Culture of Japan.

1. M. Suzuki, T. Shirakura and S. Inawashiro: J. Phys.: Condens. Matter **3**, 3139 (1991)
2. T. Shirakura and S. Inawashiro: J. Phys.: Condens. Matter **3**, 3785 (1991)
3. Y. Ozeki and H. Nishimori: J. Phys. Soc. Jpn. **56**, 1568 (1987)
4. J. R. L. de Almeida and D. J. Thouless: J. Phys. A11, 983 (1978)
5. T. Shirakura, H. Kajitani and S. Inawashiro: J. Phys. **C20**, 6061 (1987)
6. R. N. Bhatt and A. P. Young: Phys. Rev. Lett. **54**, 924 (1985)

Part VI

Other Complex Systems

Monte Carlo Simulation of Polymeric Materials – Still a Challenge?

K. Binder

Institut für Physik, Johannes-Gutenberg-Universität Mainz, W-6500 Mainz, Staudinger Weg 7, Fed. Rep. of Germany

Monte Carlo simulation of polymeric materials is difficult, since they exhibit nontrivial structure over many different length scales, from the bond length (~1Å) to the radius of the random coil (~10^2Å) and still larger collective length scales, and similarly, motions occur on very different time scales. Hence it is a nontrivial problem to devise suitable coarse–grained models which capture the essential physics and are accessible to simulation.

As an example for a useful and versatile lattice model, the bond fluctuation model will be introduced and applications will be discussed to model the dynamics of dense solutions and melts, critical phenomena in polymer mixtures, and the glass transition.

It will be argued that the huge demands in computing power needed for the simulation of such problems can either be handled by massive effort on supercomputers or by parallel computers such as the multitransputer facility at the University of Mainz. It will be shown that these simulations constitute a link between analytic theory and experiment, and improve significantly our understanding of polymeric materials.

1. Introduction

Macromolecular materials have an enormous technologocal impact but pose at the same time very interesting fundamental problems in the context of the statistical thermodynamics of condensed matter [1–3]. Long flexible linear polymers exist in the solution, melt, and amorphous solid in a randomly coiled state (Fig.1). Computer simulation of polymers is a problem, due to the many length scales involved: already a single chain exhibits structure from the scale of a single chemical bond (≈1Å) to the persistence length (≈10Å) to the coil radius (≈100Å).

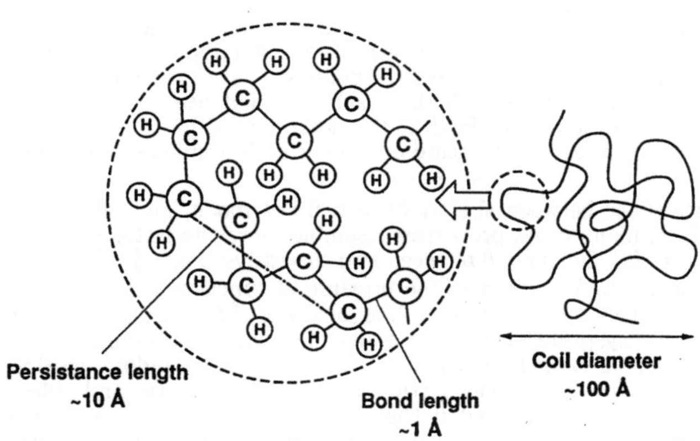

Fig.1: Length scales characterizing the structure of a long polymer (example: polyethylene)

Still larger lengths arise from collective phenomena near phase transitions: near the critical point of unmixing of a binary (AB) polymer mixture we expect correlation lengths of concentration of the order of 1000Å [4–7]; similar large length scales occur in layers of endgrafted chains at walls [8–11], in ordered mesophases of block copolymers [12–14], etc. At the same time, intermediate length scales (such as the "blob size" over which excluded volume interactions are screened out in semidilute solutions [2]) are important for many of these problems [8,11,15,16].

The existence of many length scales makes the simulation of macromolecular matter much more demanding than other simulation problems such as simple fluids, phase transitions of models for magnets, alloys, adsorbed layers etc. [17,18]. But the situation even is worse for the second factor which controls the feasibility of simulations, namely time scales: motions such as bond length or bond angle vibrations occur on a time scale $\tau_{vib} \approx 10^{-13}$sec; reorientation jumps of bonds in their torsional potential need about a time $W^{-1} \approx 10^{-11}$sec in a polymer melt; and the time needed to renew the global configuration of a coil (Fig.1) in an entangled melt is still much larger: reptation theory [2,3] yields $\tau_N = W^{-1}(N/N_e)N^2$, N being the number of monomers forming the chain, N_e the number of monomers between entanglements (typically $N_e \approx 50$ [19]). For N=500 one gets $\tau_N \approx 10^{-5}$sec, 8 orders of magnitude more than τ_{vib}. Again collective phenomena are still slower, due to the critical slowing down at phase transitions.

Hence a straightforward simulation of macromolecular matter in full atomistic detail is neither possible by molecular dynamics nor by Monte Carlo methods [20–22]. The strategy thus is to eliminate degrees of freedom that are less important for the problem that shall be simulated and invent suitable simplified models. This approach, of course, has a long tradition in polymer science [23–25], but the models that have been devised are not unique and often have many limitations [25]. In this brief review we shall only focus on the recently suggested bond fluctuation model [26,27]. We describe the motivation for it and its properties in the next section, and then (Sec.3) turn to studies of the dynamics of polymer solutions and melts [28,29], polymer mixtures [30–33] (Sec.4), and the glass transition [34,35] (Sec.5).

2. A Coarse–Grained Model for Polymers: The Bond Fluctuation Model

For simulating collective phenomena in dense polymer systems details on very small length scales and associated fast motions must be omitted, we only wish to model the important long wavelength degrees of freedom and their slow dynamics. This is achieved (Fig.2) by integrating n chemical bonds along the backbone of a chain into one "effective bond" of the lattice model. Since the actual length of the end–to–end vector of a group of n chemical bonds fluctuates, due to the bond length and bond angle vibrations and (most importantly) due to the torsional degree of freedom [36], it is natural to allow also some fluctuation in the length of the effective bond. While in the traditional polymer lattice models [23–25] polymers are represented by selfavoiding walks (SAW's) on the lattice such that each occupied site is an effective monomer and each effective bond length is fixed at the nearest neighbor distance, the bond fluctuation model implies a lattice structure on a finer scale than the size of the effective monomer (Fig.2).

This model thus is intermediate between the traditional SAW model and off–lattice (continuum) models. In fact, many of its properties resemble more the latter than the former – here are 108 choices for the angle θ between two successive bonds [30] rather than only ±90°, 180° as for the SAW, and also the pressure as function of the volume fraction ϕ of occupied sites has a continuum – like behavior [37]. Both the pair correlation [30] and the density profile near hard walls [11] show continuum – like properties also. A particular advantage is that a single type of random motion {monomers attempt random hops by one lattice spacing in ±x, ±y and (d=3 dimensions) ±z direction, which are accepted if the move does not lead to a double occupancy of lattice sites} suffices to yield Rouse–like dynamics (Rouse model=harmonic chain in a heat bath [3]) for isolated chains both in d=2 and in d=3. Suitable restriction of the set of allowed bond lengths ($2 \leq \ell < \sqrt{14}$ in d=2, $2 \leq \ell \leq \sqrt{10}$ in

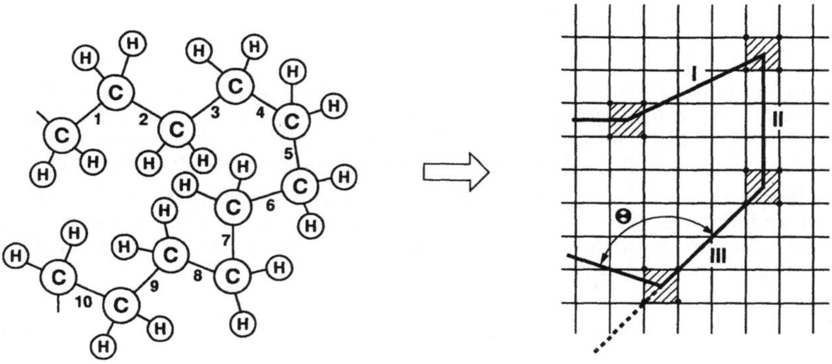

Fig.2: Approximate mapping of a chemically realistic polymer (e.g.polyethylene) - to the bond fluctuation model on the simple cubic lattice (for simplicity the square lattice is shown here). In this coarse–graining we combine n successive chemical bonds (e.g.n=3 here) into one "effective bond" between "effective monomers" which then block 8 adjacent sites of an elementary cube on the simple cubic lattice (or 4 sites along a plaquette of the square lattice) from occupation by other monomers. The chemical bonds 1,2,3 correspond to effective bond I, bonds 4,5,6 to effective bond II, etc. From Ref. [36].

d=3 with the additional exclusion of $l=\sqrt{8}$) automatically ensures "entanglement constraints", i.e. chains cannot intersect themselves or other chains in the course of their random motions. Very efficient simulation codes exist, with a speed of 2–3 millions of attempted monomer moves per second on a CRAY–YMP processor [27], and a similar performance on parallel computers [28]. This efficiency allows to study the relaxation over sufficiently large time scales (up to $3 \cdot 10^7$ attempted moves per monomer [29], which means up to about 10^{-6}sec in real time) and with high statistical accuracy (128000 {132000} monomers are averaged over in d=2 {d=3, respectively}). Such precise data over a wide time range are indispensable for a meaningful analysis of polymer dynamics, as will be seen below.

While the model described so far is athermal and a reasonable description for universal properties of polymers in good solvent, it is easily possible to include intermolecular potentials, intramolecular potentials, or both [30–35]. In fact, it is natural to assume that the potentials controlling the lengths of the chemical bond, the bond angles and torsional angles are "translated" into effective potentials $U_{eff}(\ell)$ for the length ℓ of an effective bond and $V_{eff}(\theta)$ for the angle θ (Fig.2) between two successive effective bonds [35,36]. So far, it has been assumed [35] that these potentials have a (quasi–) harmonic form, $U_{eff}(\ell)=U_0(\ell-\ell_0)^2$, $V_{eff}(\ell)=V_0(\cos\theta-\cos\theta_0)^2$. These assumptions receive some justification from a calculation of the distribution functions $P_n(\ell)$, $P_n(\theta)$ of the length of effective monomers formed from n successive chemical bonds [36], and of the associated bond angle θ, from microscopically realistic continuum models of single chains. The mapping indicated in Fig.2 is then effected by adjusting the parameters U_0, ℓ_0, V_0, $\cos\theta_0$ such that the distribution function of isolated chains in the bond fluctuation model $\{P_{eff}(\ell) \propto \exp[-U_{eff}(\ell)/k_B T_f]$, $P_{eff}(\theta) \propto \exp[-V_{eff}(\theta)/k_B T_f]$, T_f being some reference temperature$\}$ matches the chemically realistic distributions $P_n(\ell)$, $P_n(\theta)$ as closely as possible [35]. Alternatively, one can work with ad–hoc model potentials [30–34,38]: thus the model is very suitable for a study of various phase transition phenomena in polymeric materials. However, first we shall study the fully athermal case, all potentials being set to zero, only excluded–volume and entanglement constraints being taken into account.

3. Crossover from Polymer Solutions to Melts

Polymer coils in good solvents are "swollen" as long as the solution is dilute: mean square gyration radii vary with chain length N as [2] $<R_g^2> \propto N^{2\nu}$ with [39] $\nu \approx 0.59$ in d=3 dimensions. In a melt, excluded volume forces are screened out and the chains behave like ideal random walks, $<R_g^2> \propto N$. With increasing volume fraction ϕ the coils overlap: excluded volume can then be neglected on length scales exceeding the screening length $\xi(\phi) \sim \phi^{-\nu/(3\nu-1)}$; on scales smaller than ξ the coils still are "swollen". For $\phi << 1$ and N large $\xi(\phi)$ is larger than the persistence length (Fig.1) and in such "semidilute" solution the crossover from the "swollen" to the "ideal" gaussian behavior is predicted to be universal [2],

$$<R_G^2>_\phi \; / \; <R_G^2>_{\phi=0} = f_G\{N/N_{blob}(\phi)\}, \; \xi(\phi) \sim [N_{blob}(\phi)]^\nu. \tag{1}$$

Here $f_G(\xi)$ is a scaling function with $f_G(\xi >> 1) \sim \xi^{-(2-1/\nu)}$ to ensure gaussian behavior for large N, and N is rescaled by the number N_{blob} of monomers inside a "blob" of size ξ.

Fig.3 shows that this crossover scaling concept indeed works for a wide range of ϕ and N (data [28] are included from N=20 to N=200, $0.025 \lesssim \phi \lesssim 0.5$).

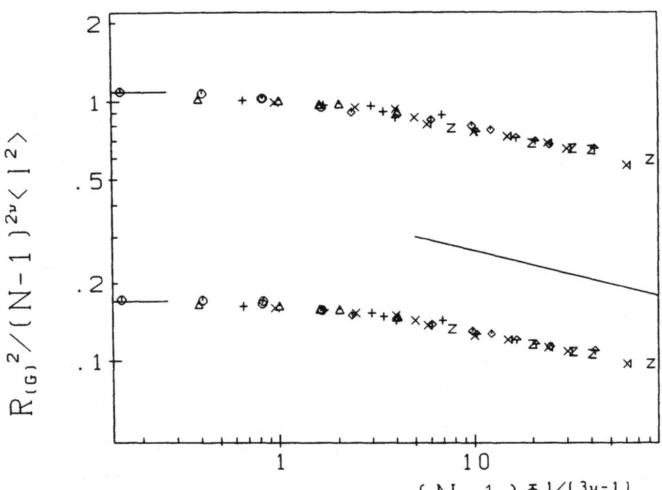

Fig.3: Log–log plot of the scaling function f_G(lower part) and f_R (for the mean square end–to–end distance $<R^2>$, upper part) versus scaling variable $(N-1)(\phi<\ell^2>^{3/2})^{1/(3\nu-1)}$. This variable is essentially the same as N/N_{blob}. Both $<R^2>$ and $<R_G^2>$ are normalized by their asymptotic behavior in dilute solution, $<R^2>_0 \sim <\ell^2>(N-1)^{2\nu}$, ℓ being the length of a single effective bond, and N–1 their number (N being the number of effective monomers). The indicated straight line shows the slope of the scaling function, $-(2-1/\nu) \approx -0.3$. Different symbols indicate the volume fractions: ϕ=0.025 (circles), 0.05 (triangles), 0.075 (+), 0.1 (x), 0.2 (diamonds), 0.3 (↑), 0.4 (x|) and 0.5 (Z), respectively. From Paul et al.[28].

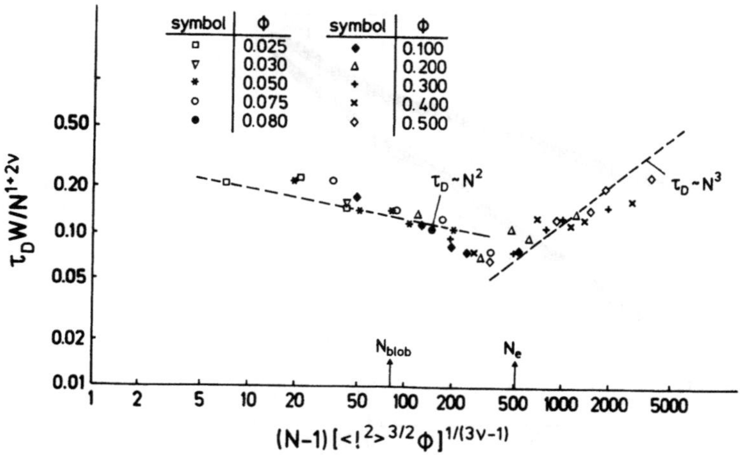

Fig.4: Log–log plot of the normalized relaxation time τ_D for the center of mass diffusion versus the scaling variable $(N-1)[<\ell^2>^{3/2}\phi]^{1/(3\nu-1)}$. Different symbols show various volume fractions, as indicated in the figure. Arrows indicate $N=N_{blob}$ and $N=N_e$, respectively. From Paul et al. [29].

This concept can be carried over to dynamics as well: in the dilute solution the relaxation time behaves as $\tau(\phi=0)\sim W^{-1} N^{1+2\nu}$ [2]. Thus the extension of Eq.(1) to dynamics is

$$\tau(\phi)/\tau(\phi=0) = \tau(\phi)\, W/N^{1+2\nu} = f_\tau\,\{N/N_{blob}(\phi)\}. \tag{2}$$

Eq.(2) is tested in Fig.4. The scaled relaxation time starts to decrease for $N \geq N_{blob}(\phi)$, reflecting Rouse–like behavior of short chains in a melt $(\tau \approx W^{-1}N^2)$, but then it increases again $\{$for $N \gtrsim N_e(\phi)\}$, due to the onset of entangled behavior.

For dense systems ($\phi=0.5$) we estimate $N_e \approx 30$ and thus the longest chain (N=20) clearly should show reptation [3]. A study of mean–square displacements $g_1(t) \equiv <[\vec{r}_i(t)-\vec{r}_i(0)]^2>$ of inner monomers and of the center of gravity displacement $g_3(t) \equiv <[\vec{r}_{CG}(t)-\vec{r}_{CG}(0)]^2>$ (Fig.5) indeed shows the predicted crossover from Rouse behavior $\{g_1(t) \propto t^{1/2}\}$ to reptation $\{g_1(t) \propto t^{1/4}\}$ at τ_e and a second crossover $\{g_1(t) \propto t^{1/2}$ ideally, while the data rather suggest $t^{0.62}\}$ at τ_R {Fig.5}. Consistent with the reptation prediction, $g_3(t)$ exhibits an intermediate $t^{0.5}$ regime. Together with corresponding molecular dynamics (MD) work [40], this is the most complete evidence for the reptation model so far [28,29]. However, there are slight problems – $g_1(t)$ behaves as $t^{0.3}$ rather than $t^{0.25}$ but this may be a crossover problem, N being still somewhat short. Also $g_3(t)$ should behave as $g_3(t) \sim t$ rather than $t^{0.8}$ in the Rouse regime (the $t^{0.8}$ power is also observed for short non–entangled chains [29]. Actually the time where finally $g_3(t) \sim t$ sets in was used as estimate for the diffusion time τ_D in Fig.4).

The crossover from Rouse to reptation is also clearly recognized when one plots the data for $\phi=0.5$ as D/D_{Rouse} versus N/N_e, Fig.6 [28,29,41]. This also shows that

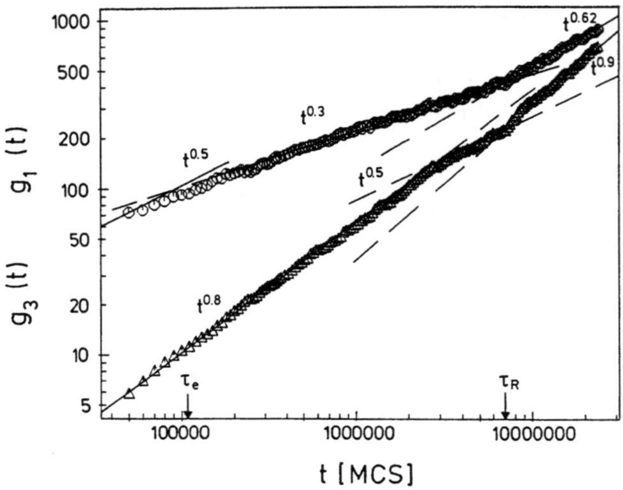

Fig.5: Log–log plot of $g_1(t)$ [circles] and $g_3(t)$ [triangles] vs. time, for $\phi=0.5$ and N=200. Arrows indicate the time τ_e where reptation sets in and Rouse time τ_R where the chain is equilibrated in its tube. Straight lines show various power laws as indicated in the figure. From Paul et al. [29].

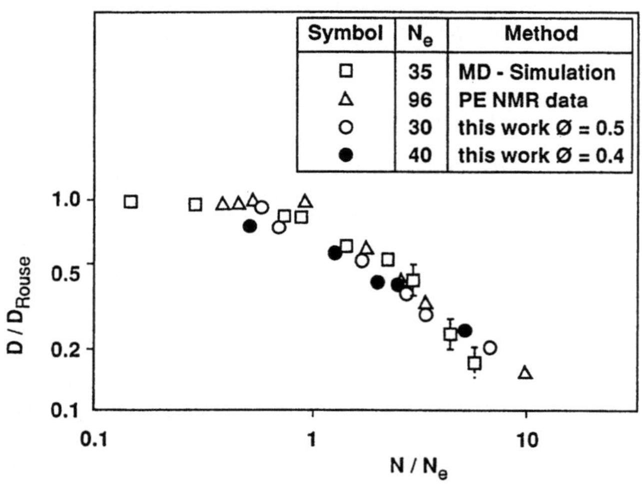

Fig.6: Log–log plot of the self–diffusion constant D of polymer melts vs. chain length. D is normalized by the diffusion constant of the Rouse limit, D_{Rouse}, which is reached for short chain lengths. N is normalized by N_e; experimental data for polyethylene (PE) [41] and MD results [40] from a bead–spring model are included. From Paul et al. [28].

results of simulations of coarse–grained models can be compared directly to experimental data, if the parameters specific for the material are absorbed in scale factors, such as D_{Rouse} and N_e.

Often Monte Carlo (MC) methods are considered less suitable for simulations of dynamic processes than MD. This is true for quick processes, such as vibrations, which are not tractable by MC at all. However, the opposite is true for very slow processes, such as slow diffusive motions in polymer mixtures – such processes are hardly accessible with MD at all. For selfdiffusion in polymer melts (Fig.5,6) MC is certainly competitive with MD, and even reaches somewhat longer times: $3 \cdot 10^7$ attempted moves per monomer in Fig.5 correspond roughly to 10^{-6}sec in real time: this has taken an effort of about 10 weeks CPU time at the multitransputer parallel computer at the Condensed Matter Theory Group at Mainz.

4. Critical Phenomena in Polymer Mixtures

Since 50 years [42] Flory–Huggins theory has been the accepted basis for interpreting compatibility and unmixing in polymer mixtures. This theory actually is a very crude mean field approximation to a lattice model, where polymers are modelled by SAW's and energy parameters ϵ_{AA}, ϵ_{AB} and ϵ_{BB} are introduced to model the enthalpy between nearest neighbor AA, AB and BB pairs of effective monomers on the lattice. To make the problem tractable, many approximations are made – in counting the configurational entropy it is ignored that the chains must not intersect each other; considering the probability that a monomer of type A has a neighbor of type B, correlations in the occupancy of lattice sites are completely ignored. In counting of how many B–monomer neighbors an A monomer may have, a crude overestimation is made due to the neglect of excluded volume effects on local scales and self–contacts (which should not be included in the count). The prediction for the critical point of a symmetric ($N_A=N_B=N$) mixture then becomes

$$k_BT_c/\epsilon = zN(1-\phi_v), \qquad 2\epsilon = \epsilon_{AB} - (\epsilon_{AA} + \epsilon_{BB})/2, \tag{3}$$

z being the coordination number, ϕ_v the volume fraction of vacant sites.

First Monte Carlo studies of this problem [43,44] showed that indeed this theory is rather inaccurate, typically T_c is overestimated by a factor of three by Eq.(3). It was suggested that due to a small part this failure is due to the neglect of statistical fluctuations (the latter also lead to critical behavior of Ising type, while Flory Huggins theory implies the mean field exponents), the major part being attributed to a reduction of the effective coordination numbers by the effect quoted above. Now it has been suggested by an integral equation theory [45,46] that due to a correlation effect the reduction of the prefactor in Eq.(3) is by a factor $1/\sqrt{N}$, which would imply $T_c \sim \sqrt{N}$ instead of $T_c \sim N$. It was claimed [46] that the chain length employed in [43,44] were too small to see this effect.

Now the Monte Carlo study of critical phenomena in polymer mixtures is difficult [43,44]: the diverging correlation length requires a careful finite size scaling [47,48] analysis, and although for $N_A=N_B$ one can work in the grand–canonical ensemble of the mixture, critical slowing down is a problem. Only by combining the very fast bond fluctuation method [30] with efficient histogram techniques [49] for making optimal use of the generated configurations has it become possible [31–33] to obtain precise estimates for T_c in the range $16 \lesssim N \lesssim 256$. Fig.7 shows that the data are well fitted by a linear relation, $k_BT_c/\epsilon \approx 2.15N+1.35$. Since the effective coordination number of the model is [30] $z \approx 14$, $\phi_v=0.5$, again the coefficient of N is about a factor of three smaller than expected from Eq.(3). Apart from this discrepancy, the result agrees qualitatively with Eq.(3) but not with Refs.45,46. The additive term (1.35) is understood as a correction due to the free ends.

The simulations [30–33, 43, 44, 50] have yielded many other results useful both for interpreting experiments and elucidating interesting theoretical questions:

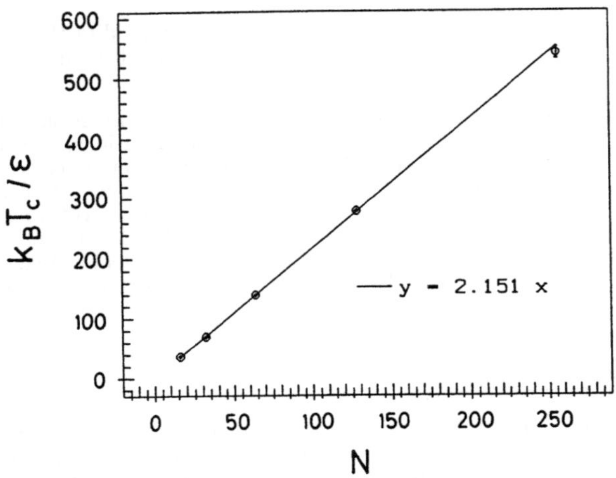

Fig.7: Plot of k_BT_c/ϵ vs.N. For $N \lesssim 128$ the errors are smaller than the size of the symbol, while for N=256 an error bar is given. From [31].

(i) The coexistence curve between the unmixed phases does not have a parabolic shape as predicted [1,42] (mean–field critical exponent $\beta=1/2$), but is much flatter ($\beta\approx0.32$ as in [39] the Ising model).

(ii) The inverse critical scattering intensity does not vanish linearly as T_c is approached (critical exponent $\gamma=1$ in Flory–Huggins theory [1,42]), but according to $(T-T_c)^\gamma$ with $\gamma\approx1.24$ as in the Ising model [39], which has been established experimentally [6,7].

(iii) Chain radii decrease with increasing interaction strength [44], which is a hint that the de Gennes [2] "random phase approximation" is inaccurate (see also [14]). Such a decrease occurs also during the initial stages of phase separation kinetics [50], at least for deep quenches.

(iv) If one fits the Flory Huggins theory to the simulation data as experimentalists do with real data, a spurious U–shaped concentration dependence of the Flory Huggins enthalpy parameter results, again qualitatively similar to real symmetric mixtures [51]. For details, we refer to recent reviews [21,52].

5.Modelling the Glass Transition

We now choose (for d=2) the intramolecular potential for the bond length $\{U_{eff}(\ell)=U_0(\ell-\ell_0)^2\}$ with $\ell_0=\sqrt{10}$: then 20% of all lattice sites cannot be occupied in the ground state [34]. When the volume fraction is of order $\phi=0.5$ or larger, configurational entropy favors that many bonds have $\ell=2$, to allow tight packing of monomers: this choice of ℓ_0 thus yields a conflict between bond energetics and the tendency towards close packing. This "frustration effect" is thought [34] responsible for a glass transition in this model.

Fig.8 shows the selfdiffusion constant found when the polymer melt is cooled from $1/T=0$ to a trial temperature T_f as $1/T(t)=(1/T_f)\Gamma t$, with $\Gamma=4.10^{-7}$. Units of T are fixed by $k_B\equiv1$ and the choice $U_{eff}(2)=1$. Near $T_g\sim0.2$ practically D(T) vanishes (for $T\lesssim0.3$ one can no longer estimate D(T) reliably but only obtains an upper bound). Since the acceptance rate of moves near $T_g\sim0.2$ is still distinctly nonzero, local motions

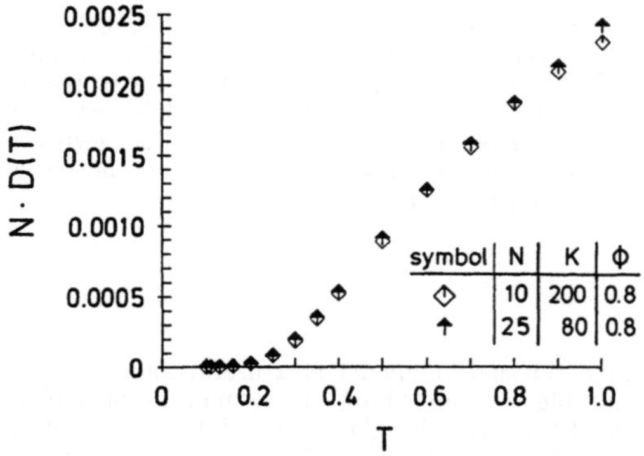

Fig.8: Normalized selfdiffusion constant ND(T) plotted vs.T for $\phi=0.8$, obtained from an average over 64 lattices of size 100x100 with periodic boundary conditions run in parallel, each lattice containing K chains as shown. From Wittmann et al. [34]

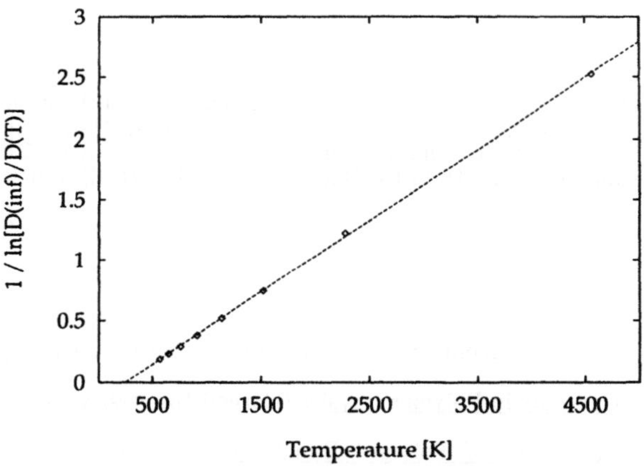

Fig.9: Plot of $1/\ln(D_0/D_\infty)$ vs. T for N=20, $D_0=D(T)$, $D_\infty=D(T=\infty)$. Straight line indicates the Vogel–Fulcher fit, intercept being T_0. From Paul et al. [35]

still occur – mobile monomers being "arrested in cages" formed by their environment. The significance of the observed T_g also shows up in rounded kinks of the temperature dependence of compressibility, free volume, etc. [34]. But the relaxation times do not diverge at T_g, rather they behaves as $\ln \tau \propto T^{-2}$, implying a zero temperature phase transition with a divergence stronger than an Arrhenius law. At $T_g \approx 0.2$ the system has only a "kinetic freezing" by falling out of equilibrium, for the chosen cooling rate. In contrast, a related calculation in d=3, where $U_{eff}(\ell)/T_f=a(\ell-\ell_0)^2$ and $V_{eff}(\ell)/T_f=b(\cos\theta-\cos\theta_0)^2$ with a=4, b=5, $\theta_0=90^0$, $\ell_0=3$ for $T_f=570$K was chosen [35] yielded a decrease of the selfdiffusion constant consistent with the Vogel–Fulcher law [33], $D(T)/D(T=\infty)\propto\exp\{-\text{const}/(T-T_0)\}$, with a Vogel–Fulcher temperature T_0 of about 250K (Fig.9). Here $\phi=0$ was chosen and the above potentials represent a choice appropriate for the polymer bisphenole–A–polycarbonate.

Both in d=2 and in d=3 occurs anomalous slow relaxation, compatible with Kohlrausch–Williams–Watts [54] relaxation functions; evidence for the time–temperature superposition principle [55] is found [34,35], and thus many features of the observed glass transitions [55] are reproduced by the present models. At this point, however, it is still unclear whether the freezing in d=3 is fully gradual (with a static glass transition occurring at T=0 only, as is probably true in d=2), or whether a finite temperature thermodynamic glass transition occurs.

6.Concluding remarks

In this talk it was shown that the bond fluctuation model is a useful and versatile coarse–grained model of polymers, suitable for describing the dynamics of melts, mixing thermodynamics, freeze in phenomena near the glass transition, etc. Clearly, some of the results are somewhat qualitative, and more effort is clearly required. Since the results shown are based on large scale computations already (e.g.Figs.5,9 have taken a CPU time of 10 weeks each at the multitransputer facility at the university of Mainz, Figs.7,8 each have needed several hundred hours CPU time at CRAY–YMP processors), more powerful computers clearly would have a large impact for the simulation of polymeric materials.

Acknowledgements

The material presented in this lecture is taken from collaborative work with H.P.Deutsch, D.W.Heermann, K.Kremer, W.Paul and H.P.Wittmann (Refs 27–35). It is a pleasure to thank them for a very fruitful interaction. Support from the BAYER AG and the BMFT (grant N°03M4028), the DFG (SFB262/D2) and the HLRZ Jülich is gratefully acknowledged.

References

1. P.J.Flory, Principles of Polymer Chemistry (Cornell University Press, Ithaca, New York 1953)
2. P.G.de Gennes, Scaling Concepts in Polymer Physics (Cornell University Press, Ithaca, New York 1979)
3. M.Doi and S.F.Edwards, Theory of Polymer Dynamics, (Clarendon Press, Oxford 1986)
4. K.Binder, J.Chem.Phys.79, 6387 (1983)
5. K.Binder, Colloid & Polymer Sci.265, 273 (1987)
6. I.Stepanek, T.Lodge, Kodronski, and F.S.Bates, J.Chem.Phys.94, 8289 (1991)
7. G.Meier, B.Momper and E.W.Fischer, preprint; D.Schwahn, K.Mortensen, and H.Yee–Madeira, Phys.Rev.Lett.58, 7544 (1987)
8. A.Halperin, M.Tirrell and T.P.Lodge, Adv.in Polymer Science 100 (1991, in press)
9. H.J.Taunton, C.Toprakcioglu, L.J.Fetters, and J.Klein, Macromolecules 23, 571 (1990)
10. A.Murat and G.S.Grest, Macromolecules 22, 4054 (1989)
11. P.–Y.Lai and K.Binder, J.Chem.Phys.(1991, in press)
12. F.S.Bates and G.H.Fredrickson, Ann.Rev.Chem.41, 525 (1990)
13. L.Leibler, Macromolecules 13, 1602 (1980)
14. H.Fried and K.Binder, J.Chem.Phys.94, 8349 (1991); Europhys.Lett.(1991, in press)
15. D.Broseta, J.F.Joanny and L.Leibler, Macromolecules 20, 1935 (1987)
16. K.Binder, in Computer Simulation Studies in Condensed Matter Physics, Recent Developments, ed.by D.P.Landau, K.K.Mon, and H.–B.Schüttler (Springer, Berlin 1988) p.84.

17. K.Binder (ed.) Monte Carlo Methods in Statistical Physics (Springer, Berlin 1979)
18. K.Binder (ed.) Applications of the Monte Carlo Method in Statistical Physics (Springer, Berlin 1984)
19. J.D.Ferry, Viscoelastic Properties of Polymers (J.Wiley, New York 1980)
20. K.Binder, in Molecular Level Calculations of the Structure and Properties of Non–Crystalline Polymers (J.Bicerano, ed.) Marcel Dekker, New York, in press
21. K.Binder, in Computer Simulation of Polymers, ed.by E.A.Colbourn, Longman, Harlow, U.K., in press
22. K.Binder, Makromol.Chem, in press
23. F.T.Wall, S.Windwer, and P.J.Gans, in Methods in Computational Physics, Vol.1 (eds.B.Alder, S.Fernbach and M.Rotenberg, Academic Press, New York 1963)
24. A.Baumgärtner, Ann.Rev.Phys.Chem.35, 419 (1984), and in Ref.18, Chap.5
25. K.Kremer and K.Binder, Computer Phys.Repts.7, 259 (1988)
26. I.Carmesin and K.Kremer, Macromolecules 21, 2819 (1988); J.Phys.(Paris) 51, 915 (1990)
27. H.P.Wittmann and K.Kremer, Comp.Phys.Comm.61, 309 (1990)
28. W.Paul, K.Binder, D.W.Heermann, and K.Kremer, J.Phys.(Paris) II1, 37 (1991)
29. W.Paul, K.Binder, D.W.Heermann, and K.Kremer, J.Chem.Phys., in press
30. H.–P.Deutsch and K.Binder, J.Chem.Phys.94, 2294 (1991)
31. H.–P.Deutsch and K.Binder, Europhys.Lett.(submitted)
32. H.–P.Deutsch, preprint
33. H.–P.Deutsch and K.Binder, preprint
34. H.–P.Wittmann, K.Kremer and K.Binder, J.Chem.Phys.(submitted)
35. W.Paul, K.Binder, K.Kremer, and D.W.Heermann, Macromolecules
36. J.Baschnagel, K.Binder, W.Paul, M.Laso, U.W.Suter, I.Batoulis, W.Jilge and T.Bürger, J.Chem.Phys., in press
37. H.–P.Deutsch and R.Dickman, J.Chem.Phys.93, 8983 (1990)
38. A.Lopez–Rodriguez, H.P.Wittmann and K.Binder, Macromolecules 23, 4327 (1990)
39. J.C.Guillou and J.Zinn–Justin, Phys.Rev.B21, 3976 (1980)
40. K.Kremer and G.S.Grest, J.Chem.Phys.92, 5057 (1990)
41. D.S.Pearson, G.Verstrate, E.von Meerwall, and F.C.Schilling, Macromolecules 20, 1133 (1987)
42. M.J.Huggins, J.Chem.Phys.9, 440 (1941); P.J.Flory, J.Chem.Phys.10, 51 (1942)
43. A.Sariban and K.Binder, J.Chem.Phys.86, 5853 (1987)
44. A.Sariban and K.Binder, Macromolecules 21, 711 (1988)
45. K.S.Schweizer and J.G.Curro, J.Chem.Phys.88, 7242 (1988); 94, 3986 (1991)
46. K.S.Schweizer and J.G.Curro, Chem.Phys.149, 105 (1990)
47. K.Binder, Ferroelectrics 73, 43 (1987)
48. V.Privman (ed.) Finite Size Scaling and Numerical Simulation of Statistical Systems (World Scientific, Singapore 1990)
49. A.M.Ferrenberg and R.H.Swendsen, Phys.Rev.Lett.61, 2635 (1988); 63, 1195 (1989)
50. A.Sariban and K.Binder, Macromolecules 24, 578 (1991)
51. F.S.Bates, M.Muthukumar, G.D.Wignall, and L.J.Fetters, J.Chem.Phys.89, 535 (1988)
52. K.Binder, Colloid & Polymer Sci.266, 871 (1988)
53. H.Vogel, Phys.Z.22, 642 (1921); G.S.Fulcher, J.Ann.Ceram.Soc.8, 339 (1925)
54. R.Kohlrausch, Ann.Phys.(Leipzig) 12, 393 (1874); G.Williams and D.C.Watts, Trans.Faraday Soc.66, 80 (1980)
55. J.Jäckle, Rep.Progr.Phys.49, 171 (1986)

Simulating Spatiotemporal Chaos with Coupled Map Lattices

K. Kaneko

Department of Pure and Applied Sciences, University of Toyko,
Komaba, Meguro-ku, Tokyo 153, Japan

Abstract. Dynamical behavior in spatially extended systems is modelled by coupled map lattices (CML). The predominancy of CML over traditional methods is emphasized. Some examples of qualitative universality classes are briefly given. Applications of CMLs are given which cover pattern formation, crystal growth, boiling, convection, excitable media, neural dynamics, and information processing.

1 Introduction

For this quarter century, studies in low-dimensional chaos have exploded in almost all branches of sciences. Chaos has been an important step towards science of dynamical complexity. Low-dimensional chaos, however, is not the ultimate origin for dynamical complexity.

When we look back at the history of low-dimensional chaos, we have to admit that the low-dimensional chaos is still rather special in the nature, even though it is much more common in nonlinear systems than integrable systems with regular dynamical behaviors. In experiments for low-dimensional chaos, system parameters are chosen so that only few modes are excited (e.g., laser, spin-wave chaos), or the system is set up so that spatial degrees of freedom are suppressed (examples are chemical chaos in which a system is fully stirred; Rayleigh-Benard convection in a small aspect ratio, where only few rolls are allowed). Thus low-dimensional chaos appears only in restricted situation in nonlinear nonequilibrium systems. Generally speaking, nonlinear nonequilibrium systems tend to show spatiotemporal chaos, i.e., chaotic dynamics with spatial degrees of freedom of many active modes. Modelling and characterization of spatiotemporal chaos are important in the study of turbulence in a general sense, not only in fluid dynamics but also in solid-state physics, optics, chemical reaction with diffusion, pattern formation problems, and in biology. Examples of spatiotemporal chaos are seen in Benard convection, convection in liquid crystals, Taylor flow, open flow in fluid systems, chemical reaction with diffusion, some solid-state systems such as Josephson junction array, charge density wave, and spin wave turbulence, and some biological networks.

It has been 10 years since the author introduced a simple model with essential features of spatiotemporal chaos; **Coupled Map Lattice (CML)**.

Springer Proceedings in Physics, Vol. 70 **Computational Approaches in Condensed-Matter Physics**
Editors: S. Miyashita · M. Imada · H. Takayama

Studies in CML have been expanding not only in the field of spatiotemporal chaos and pattern formation but also in the fields of biology, mathematics and engineering.

A CML is a dynamical system with a discrete time, discrete space, and continuous state [1-6]. Our strategy of modelling dynamical phenomena in spatially extended systems by CML is based on the following steps:

(A) Choose a (set of) field variable(s) on a lattice. This set of variable(s) is not in a microscopic, but in a semi-macroscopic level (e.g., temperature, fluid velocity field, concentration of some chemical substances).

(B) Decompose the phenomena into independent units (e.g., convection, reaction, diffusion, local chaos and so on).

(C) Replace each unit by a simple parallel dynamics on a lattice: the dynamics consists of nonlinear transformation of the field variable on each lattice point and/or a coupling term among suitably chosen neighbors.

(D) Carry out each unit dynamics ("procedure") successively.

We often encounter with a type of spatiotemporal chaos created by local nonlinear dynamics and spatial diffusion, as is seen in a chemical reaction diffusion system, Benard convection with a large aspect ratio, and so on. For studies of this type of spatiotemporal chaos, we choose local chaos and spatial diffusion as separate procedures. By adopting nonlinear mapping $x(i) \to x'(i) = f(x(i))$ for local chaos and the discretized diffusion process $x'(i) \to (1-\epsilon)x'(i) + (\epsilon/2)[x'(i+1) + x'(i-1)]$, we arrive at the following CML;

$$x_{n+1}(i) = (1-\epsilon)f(x_n(i)) + \frac{\epsilon}{2}[f(x_n(i+1)) + f(x_n(i-1))] \qquad (1)$$

where n is a discrete time step and i is a lattice point ($i = 1, 2, \cdots, N$=system size) with a periodic boundary condition.

As the local map $f(x)$, we can choose any 1-dimensional map (with chaotic behavior). The above model has extensively been studied as a prototype model for spatiotemporal chaos. We often take the logistic map

$$f(x) = 1 - ax^2, \qquad (2)$$

since the map has already been investigated in detail. Features in the logistic map lattice, however, can be seen in a wide class of mappings and in other types of couplings, and are thought to give a universal class in spatiotemporal chaos.

In the model (1), the independent procedures in (B) are local transformation (eq.(2)) and the diffusion process, which are separated parallel procedures. The model consists of the sequential repetition of these two procedures. By introducing the variable right after the local transformation ($y_n(i) = f(x_n(i))$), the following equivalent form with (1) is obtained:

$$y_{n+1}(i) = f((1-\epsilon)y_n(i) + \epsilon/2[y_n(i+1) + y_n(i-1)]), \qquad (3)$$

which form may be more familiar with some researchers (in the field of neural networks and some spin systems).

Let us compare our CML with other models for spatially extended systems, in particular, partial differential equation (PDE) and cellular automata (CA) (see Table below).

Table

Model	State	Space	Time	Level
CA	Discrete	Discrete	Discrete	Micro
CML	Continuous	Discrete	Discrete	Semi-Macro
PDE	Continuous	Continuous	Continuous	Macro

In PDE, analytic methods have been developed to reduce equations from some equations such as the Navier-Stokes with the aids of perturbative techniques (e.g., reductive perturbation). One problem here is that such reduction is possible only in restricted parameter regimes, and that one tends to use this reduction beyond its validity. It is often believed that some PDEs are derived from the first principle through some coarse-graining. This is not necessary true; even the Navier-Stokes equation has no clear derivation from a molecular level. Still, PDEs have some merits; parameters there can often correspond to measurable quantities easily, such as the viscosity, conductivity, and so on. Problems in PDE are that (1) numerical integration requires much time, thus unsuitable for heuristic studies; and that (2) neither dynamical system theory nor lattice statistical mechanics is straightforwardly applied.

CA are originated in the construction of digital computers, thus architecturally suitable for such computers. Simulations of CA are quite fast, with the use of special-purpose machine for CA, which can easily be designed. Some notions in lattice statistical mechanics are easily applied (for example, kinetic Ising model essentially belongs to this level of modelling). In CA, however, applications of dynamical systems theory are difficult, since they have continuity neither in a state nor in a parameter. Another drawback in CA is that it cannot create information in the sense of Shannon, as long as the lattice size is finite. Thus an aperiodic state is not possible in a finite lattice. In CA, each cell corresponds to a microscopic variable. Thus we need a very large lattice size to have a realistic simulation.

On the other hand, we can get a suitably statistical behavior within a small lattice system in CML, since a field variable in CML is a coarse-grained quantity of a semi-macroscopic level. Following the philosophy of statistical physics, a suitably coarse-grained model provides a more economic and higher level description of the nature. This semi-macroscopic description in CML also leads to numerical efficiency in simulations, which enables us heuristic studies. Discreteness in space and time is suitable for the present architecture of computers. In CML, straightforward applications of dynamical systems theory are possible, with some connection with the statistical mechanics on a lattice system.

2 Qualitative Universality Classes in CMLs

In modeling in physics, we often tend to believe that a model must have one-to-one correspondence with a phenomenon in concern. For example, if one tries to model turbulence, one often adopts molecular dynamics or a PDE which has a fair basis on the phenomena at the level of microscopic process. If one succeeds in reproducing the phenomena from this type of model system, then what can one learn? One dangerous trap in computational physics is that one may be still at the same level as the direct observation of the complex phenomenon itself, if the simulation is too successful. We cannot **understand** the complex phenomena through this type of model.

In the studies of CML, we take a constructive approach. By combining procedures, we try to find the phenomenology of complex spatiotemporal behavior. If this approach gives a novel class of notion, which reproduces some natural phenomena, we can **understand** the complex behavior, even if (or *because*) the model equation itself is not completely derived from a microscopic level. This approach is based on the belief in the existence of universality class in physics. A model cannot be exacly same as the nature herself, anyway, and we have to assume that there is some universality of phenomenology, independent of details of the modelling. In CML studies, we search for a novel qualitative universality class, without bothering the details of phenomenology. Through this approach, we understand how such phenomenology appears, in what class it is commonly seen, and what the essence of the phenomena is. Only through this approach we can understand why some type of complex behavior is common in nature, irrespective of details, and then we can predict what class of systems leads to such behavior. We believe that this approach is the only route to the understanding of complex systems.

In physics, quantitative universality is justified with the use of renormalization group, as is seen in critical phenomena, onset of chaos, and field theory. Similar justification may be possible for CML, and the search for it is strongly recommended.

Even without this quantitative basis, however, our CML approach can have a basis in a wider context; qualitative universality. It is important to predict and classify qualitative phenomenology, independent of the details of a model. Although mathematical foundation of qualitative universality is left for the future, this qualitative picture is important, to develop further our scope of physics to complex systems.

In the present section, we briefly discuss some qualitative universality classes discovered in CMLs. Most results here are based on the simulation of diffusively coupled logistic lattice (eq.(1),(2)). These universality classes, however, hold for a wide range of CMLs, and some of them are also found in experiments and in partial differential equation systems.

(I) Frozen random pattern [4]

At the parameter for period-doubling of logistic map, the corresponding coupled logistic lattice (3) forms kinks with different phases of oscillation [2]. By the doublings, domains of various sizes are formed. With the further increase of nonlinearity $a(> 1.401...)$, the system starts to show a chaotic behavior. Domains of various sizes are formed. The frozen random pattern leads to spatial bifurcation. Even if the model is homogeneous in space, attractors can have strong spatial dependence. The final spatial structure depends on initial conditions. In a large domain, the motion is chaotic, while it is almost period-8 at smaller domains, period-4, for much smaller domains, and period-2 for the smallest ones.

(II) Pattern selection with suppression of chaos [4]

In the frozen random pattern, domains of arbitrarily large sizes can exist, depending on initial conditions. As the nonlinearity is further increased ($a > 1.55$), larger domains start to be unstable and split into smaller domains. Final domain sizes are independent of initial conditions. Through the transient process, domains of few special sizes are selected. After the selection, the position of a domain is frozen and does not move in space. Selected are such domain sizes that the dynamics in them is less chaotic (motion with shorter periods). The diffusion tries to homogenize a system, while the chaotic motion makes the system inhomogeneous due to the sensitive dependence on initial conditions. These two tendencies conflict with each other. In an (initially given) large domain, chaos is so strong that the domain splits into smaller ones. Once a domain structure is formed with suppression of chaos, the conflict is resolved and the domain structure is stabilized.

(III) Spatiotemporal Intermittency [2]

In the pattern selection, spatially ordered structure is stable. Transition from such ordered pattern to fully developed spatiotemporal chaos occurs via spatiotemporal intermittency (STI). In STI there are laminar motion and turbulent bursts in spacetime. Each spacetime pixel can be classified into laminar and bursts. Since the introduction of STI in 1984 [2], studies on STI have been growing both experimentally and theoretically. So far, there are two types of STI.

In the first type of STI[2,8], there is no spontaneous creation of bursts. If a site and its neighbors are laminar, the site remains laminar in the next step. Before the onset of STI, there is a stable state with spatial homogeneity and temporal periodicity. A typical example is given by the coupled logistic lattice at the period-3 window for the local map. Possible relationships of type-I STI with directed percolation have been intensively investigated [8].

In the second type of STI, there exists [7,4] spontaneous creation of turbulent bursts, as long as some coarse-grained reduction of states is used. There is some probability of creation of bursts even if all the states of a site and its neighbors are laminar. STI here cannot be represented by a cellular automaton or a related percolation model. This STI is observed with the transition with a spatial pattern. A typical example of this STI is observed at the transition region from pattern selection to fully developed spatiotemporal chaos in the coupled logistic lattice (3). In the type-II STI, the temporal change corresponding to the selected pattern has a very long memory. This long-term correlation leads to selective-flicker noise: The dynamical form factor $P(k,\omega)$ (power of Fourier transform of the spacetime pattern $x_n(i)$) exhibits the $\omega^{-\beta}$ noise ($\beta \approx 1.9$), only for the wavenumber $k \approx k_p$, the wavenumber of selected pattern [4].

Type-II spatiotemporal intermittency is a common route to full developed spatiotemporal chaos, as has recently been observed in various experiments, including Benard convection with a large aspect ratio[17,18], Faraday instability of wave [20], a 2-dimensional electric convection of liquid crystal [19], and viscous rotating fluid [21]. The power law distribution of laminar time is also confirmed in these experiments, as well as the above selective flicker-like noise for $P(k,\omega)$.

(IV) Travelling Wave [9]

When the coupling ϵ is large (larger than .42), domain structures in (I) and (II) are no longer fixed in space, but can move with some velocity. At the parameter corresponding to the frozen state (I), the motion of a domain is rather irregular, while a regular travelling wave is seen in the pattern selection regime. We also note that attractors with different wave velocities coexist, depending on the spatial asymmetry of a selected wave pattern.

(V) Quasistationary supertransients [12]

In spatiotemporal chaos, we often encounter with very long transients. Indeed, in a class of STC, transient time before falling on an attractor diverges exponentially or faster with the system size. In the transient regime, STC is quasistationary. Any quantifier characterizing the dynamics fluctuates around its mean value, without showing any decay. This quasistationary state is almost indistinguishable from attractors. The transient time increase is so fast that we cannot observe the attractor, which may be periodic. This supertransient is the source of type-I spatiotemporal intermittency and the stability of fully developed spatiotemporal chaos [12]. Supertranisent is a generic behavior in a coupled system of local maps with topological chaos but without a chaotic attractor.

Extension of our diffusively coupled map lattice to a two-dimensional space is quite straightforward. We have again seen a similar transition sequence of pattern dynamics. Extension to spatially asymmetric coupling models is relevant to the study of open fluid flow (such as pipe flow), where the dynamics is strongly influenced by the upperflow. An extreme case is given by the CML with one-way coupling; $x_{n+1}(i) = (1-\epsilon)f(x_n(i)) + \epsilon f(x_n(i-1))$. The model gives similar pattern dynamics changes as well as spatial period-doubling, and selective amplification of noise[10,11].

These universality classes are quantitatively characterized with the use of statistical and dynamical quantifiers. So far, we have applied (1) power spectrum in space and time [4]; (2) pattern distribution and pattern entropy [4];

(3) Lyapunov spectrum and corresponding eigen vectors [2,4]; (4) co-moving Lyapunov spectra [11,4,13]; (5) sub-space-time Lyapunov spectra [15]; (6) propagation speed of disturbance [4]; and (7) co-moving mutual information flow [4]. Theoretical formulation for these quantifiers has been started, with the use of rigorous statistical mechanics [14] and self-consistent approximation [15,16], but most problems are left for the future.

3 Applications of Coupled Map Lattices

Most dynamical phenomena in spatially extended systems are described by the combination of some elementary local dynamics. These phenomena are constructed by successive procedures with local dynamics and local spatial coupling. Here we give few examples of applications of CMLs.

(I) Pattern formation (spinodal decomposition)

When a system is quenched from a disordered state (at high-temperatures) to an ordered state (at low-temperatures), spatial pattern formation proceeds, as is typically known as spinodal decomposition. Traditionally this problem has been studied by the time-dependent Ginzburg-Landau equation or by the kinetic Ising model with Monte Carlo method.

Oono and Puri have proposed a CML for this problem. Their model is based on the coarse graining in a cell with a large number of sites from the kinetic Ising model. The local dynamics of a cell leads to a map with two stable fixed points, e.g.,

$$f(x) = \tanh(\beta x). \tag{4}$$

The same diffusive coupling form as in the previous section is adopted to include the ordering process (to align the order parameter of two neighboring regions; "ferro-coupling"). This approach is easily extended to the pattern formation with some conservation law or with a vector order parameter.

Applications are also possible to the simulation of crystal growth. We take the order parameter x corresponding to the density ($x = 1$ is solid, $x = -1$ is gas), and the temperature field t on a lattice. As procedures, we take the phase separation dynamics, diffusion of temperature, and the latent heat. The diffusion type coupling for x and t is straightforward from eq.(1), while the solidification process is represented by the map $x_n(i,j) \rightarrow f(t_n(i,j), x_n(i,j))$, with $f(t,x) = tanh(\beta(x-t))$. The latent heat, on the other hand, is described by the procedure $t_n(i,j) \rightarrow t'_n(i,j) = t_n(i,j) + c(f(t_n(i,j), x_n(i,j)) - x_n(i,j))$. With a suitable choice of parameters, growth of a "solid" region is observed ([6]; see also [23]).

(II) Excitable media

Another possible application of CML lies in the pattern dynamics with an excitable state and relaxation from it. We have seen such examples in reaction diffusion within excitable media, and in some biological problems such as heart rhythm and neural response.

A simple 1-dimensional map with an excitable state is introduced by Nagumo and Sato[24]. They have used the map $f(x) = b \times (x - H(x)) + c$,where $H(x)$ is the Heaviside's step function ($H(x) = 1$ for $x > 0$ and $H(x) = 0$ for $x < 0$). Here, $x > 0$ corresponds to the " fired state". The constant term "c" comes from an external stimulus applied on a single neuron. By replacing this constant "c" by stimuli from other lattice points, we can construct a CML model for excitable media. For example, let us assume that each element emits a pulse if $x > 0$ ("fired"). Then we obtain the following CML for a 2-dimensional lattice;

$$x_{n+1}(i,j) = b \times (x_n(i,j) - H(x_n(i,j)))+$$
$$d \times (H(x_n(i+1,j)) + H(x_n(i-1,j)) + H(x_n(i,j+1)) + H(x_n(i,j-1))).$$

The model leads to spiral formation and travelling wave as are observed in experiments with excitable media[6]. An important application of excitable media to biology is the heart rhythm, where the turbulent spiral formation is seen as the firbrilliation [25].

(III) Dripping handrail and boiling chaos

Models with a threshold dynamics and diffusion also appear in a variety of fields, including dripping handrail, boiling, and so on. As an example of a 1-d lattice, let us take a dripping handrail [5,12]. Water is constantly poured by rain onto a handrail, leading to a constant increase of water at the rail up to a threshold. If the amount exceeds the threshold, the water drops from the handrail. This threshold dynamics is given by the local map $f(x) = x + c(mod\ 1)$.There is diffusion along the handrail, which is again modelled by the diffusive coupling (eq. (1)). The model exhibits the supertransients in §2 [12]. A snapshot pattern may be seen in icicles at the handrail.

An extension to a 2-dimensional system is straightforward. The 2-dimensional model may be regarded as a simple model for boiling, since the boiling process consists of the increase of temperature up to a threshold and diffusion.

Yanagita has recently constructed a CML model for boiling, adopting the procedures of phase transition to bubbles, their floating by buoyancy, and heat diffusion. The model clearly reproduces the nucleus-film transition, observed in experiments of boiling phenomena [26].

(IV) Thermal convection of fluid

Convection of fluid is one of the most important experimental systems in spatiotemporal chaos. The dynamics in convective fluid consists of following processes;transport of fluid and heat by Lagrangean scheme, diffusion of velocity by viscosity, buoyancy, and heat diffusion. Recently a CML model is proposed by Yanagita and the author[27], with the use of field variable $\vec{v}(i,j)$ (velocity) and $E(i,j)$ (internal energy). Our procedures consist of Lagrangian and Eulerian parts. Eulerian part consists of the procedures for buoyancy force, heat diffusion and viscosity, which are easily carried out in the same manner as the previous examples.

Lagrangean scheme expresses the advection of velocity and temperature. We set quasi-particle at each lattice site (i, j). The particle has a velocity $\vec{v}(i, j)$ and changes its position to $(i+\delta i, j+\delta j)$ where $\delta i = v_x(i, j), \delta j = v_y(i, j)$. Field variables (velocity and internal energy here) are carried by this particle. If there is no lattice point at the new position, the transported velocity and energy are allocated to four neighboring sites with the weights according to the distance between the particle and the lattice sites.

Simulations of the model reproduce almost all phenomena in Bénard convection experiments. With the increase of temperature difference between top and bottom plates, we have seen the formation of convection rolls, onset of chaos, and period-doubling route to chaos. In a large aspect ratio, spatiotemporal intermittency (STI) [17,18] is well reproduced. With the increase of temperature gradient, the model shows a transition between soft and hard turbulence, as is characterized by the temperature distribution change from Gaussian to exponential [27]. The roll formation process in a three-dimensional convection also agrees with experimental observation. Indeed, our simulation is the first computer demonstration with agreement to experiments for STI, soft-hard turbulence transition, and three-dimensional roll formation.

(V) Josephson junction or charge density wave as coupled pendula

Problems of coupled pendula are frequently seen in the solid-state physics. Typical examples include a Josephson junction array and charge density wave (CDW). Since a forced pendulum can be modelled by the circle map $f(x) = x + (K/2\pi)sin(2\pi x) + \Omega$ (see e.g., [3]), it is natural to use a coupled circle map lattice for the above problems. In the coupled circle lattice, soliton turbulence [5] is observed as a variant of spatiotemporal intermittency, where a laminar region includes stable propagation of kinks. See also [28,29] for applications of these models.

(VI) Neural dynamics

In studies of recent neural networks, dynamics with few stable fixed points or with discrete states (at the CA level) is often used (e.g., McCllough-Pitts neuron). In real physiology, the neural dynamics is more complex, including a chaotic behavior for a small ensemble of neurons or for a single neuron. In this sense, recent neural networks are oversimplified. Of course, they may capture some essential features of the neural dynamics, but may miss some important points, too. In the standpoint of statistical mechanics, it must be much better to start from a higher-level dynamics than the neural networks with very simple elements, in order to understand a higher-level function in the brain. Thus it is natural and important to ask the following question: What happens if we use more complex elements (with a chaotic response) with simple global couplings instead, as an abstraction from the neural dynamics?

On this purpose, we have introduced a globally coupled map (GCM) as a mean-field-type extension (Fushimi-Temperley-type model) of CML, with a coupling to all other elements. The GCM corresponding to our diffusively coupled map lattice is written as

$$x_{n+1}(i) = (1-\epsilon)f(x_n(i)) + (\epsilon/N)\sum_{j=1}^{N} f(x_n(j)). \tag{5}$$

The above model has a remarkably rich behavior (partly similar but much richer than Sherrington-Kirkpatrick model as a mean-field model for spin-glass) including the formation of synchronized clusters, transition of clustering, hierarchical clusters, chaotic itinerancy over attractor ruins, the breakdown of the law of large numbers due to the hidden coherence [32]. Relevance of these phenomenology to neural dynamics is also discussed [32,33], while experimental studies are started in nonlinear optics and lipid membrane[34,35].

(VII) Coupled Map Engineering

Relevance of low-dimensional chaos to biological information processing has recently been emphasized [39,37,3]. High-dimensional chaos has much larger potential ability for the information processing. Since most biological networks include many (chaotic) dynamical elements, we expect that such ability may play an important role in the existing biological systems. So far the following possible applications of CMLs to engineering problems are proposed:

(1) Spatiotemporal chaos as a source of "structured noise": Stochastic process is often used for optimization. Spatiotemporal chaos keeps some structure in spacetime. Thus faster search keeping some spatiotemporal structure is possible if we use spatiotemporal chaos instead of random noise (see also [37,38] for the use of chaos).

(2) Separated procedures as different functions: By relating each procedure in our CML with some processing task, it may be possible to compromise multiple different requests by a suitable choice of CML. Price et al. [31] have applied CML to the image data processing, where the following two opposite requirements must be compromised; preservation of relevant structure (e.g., patterns with clear edges) to keep the information in the image, and smoothing to remove small noise. In CML, the former is attained by a (suitably chosen) local mapping, while the latter is easily achieved by the diffusion procedure.

Other application include[1]: (3) Memory storage at many attractors with different domain structures: (4) Switching among attractors by inputs to elements: (5) Spontaneous transition among local structures in spatiotemporal intermittency: (6) Information creation and transmission by chaotic travelling wave: (7) Selective amplification of input signal by CML with one-way coupling: (8) Dynamic categorization with hierarchical structure of domains or clusters.

So far we have adapted to live with digital computers, since the invention of Turing's tape, Ulam-Neumann's cellular automaton, and the machine of von Neumann's type. Coupled map lattices may provid a different type of machine; analogue parallel processors (see also [36]). It will be important to reveal the ability of CML machine which the conventional computer does not have.

I would like to thank Tatsuo Yanagita for discussions and information on CML modelling and simulations for convection, boiling, and fluid flow. This work is partially supported by Grant-in-Aids for Scientific Research from the Ministry of Education, Science, and Culture of Japan, and by a cooperative research program in Institute for Statistical Mathematics.

References

[1] *Theory and Applications of Coupled Map Lattices*, ed. K. Kaneko, Manchester Univ. Press, 1992 and reference cited therein.
[2] K. Kaneko, Prog. Theor. Phys. 72 (1984) 480, 74 (1985) 1033:
[3] K. Kaneko, *Collapse of Tori and Genesis of Chaos in Dissipative Systems*, (World Sci. Pub., 1986; based on Ph.D. thesis of 1983)
[4] K. Kaneko, Physica 23D (1986) 436; 34D (1989) 1; 36D (1989) 60
[5] J. P. Crutchfield and K. Kaneko, "Phenomenology of Spatiotemporal Chaos", in *Directions in Chaos* (World Scientific, 1987)
[6] K. Kaneko "Simulating Physics with Coupled Map Lattices" in *Formation, Dynamics, and Statistics of Patterns* (ed. K. Kawasaki, A. Onuki, and M. Suzuki, World. Sci. 1990)
[7] J. D. Keeler and J. D. Farmer, Physica 23D (1986) 413
[8] H. Chate and P. Manneville, Physica 32D (1988) 409; see also P. Grassberger and T. Schreiber, Physica 50 D (1991) 177
[9] K. Kaneko, Travelling Wave in Coupled Map Lattices, preprint (1991)
[10] K. Kaneko, Phys. Lett. 111A (1985) 321
[11] R.J. Deissler and K. Kaneko, Phys. Lett. 119A (1987) 397
[12] J.P. Crutchfield and K. Kaneko, Phys. Rev. Lett. 60 (1988) 2715; K. Kaneko, Phys. Lett. 149A (1990) 105
[13] G. Mayer-Kress and K. Kaneko, J. Stat. Phys. 54 (1989) 1489
[14] L.A. Bunimovich and Ya.G. Sinai, Nonlinearity 1 (1989) 491
[15] K. Kaneko, Phys. Lett. 139 A (1989) 47; K. Kaneko, Prog. Theor. Phys. Suppl. 99 (1989) 263
[16] see also J.M. Houlrik, I. Webman, and M. H. Jensen, Phys. Rev. 41A (1990) 4210
[17] S. Ciliberto and P. Bigazzi, Phys. Rev. Lett. 60 (1988) 286
[18] F. Daviaud, M. Dubois, P. Berge, Europhys. Lett. 9 (1989) 441
[19] S. Nasuno, M. Sano, Y. Sawada, in Cooperative Dynamics in Complex Systems, (ed. H. Takayama, Springer, 1989)
[20] J. Gollub and R. Ramshankar, in *New Perspectives in Turbulence* (eds. S. Orszag and L. Sirovich, Springer)
[21] Y. Couder, reported at the summer school "Pattern, Order, and Chaos" (Como, 1990)
[22] Y.Oono and S. Puri, Phys. Rev. Lett. 58 (1986)836; Phys. Rev. 38 A (1988) 1542
[23] D.A. Kessler, H. Levine, and W.N. Reynolds, Phys. Rev. 42A (1990) 6125
[24] J. Nagumo and S. Sato, Kybernetik 10 (1972) 155
[25] D. Chialvo, private communication (1989); H.Ito and L. Glass, Phys. Rev. Lett. 66 (1991)671
[26] T. Yanagita, preprint (1991), submitted to Phys. Rev. Lett.
[27] T. Yanagita and K. Kaneko, preprint (1991), submitted to Phys. Rev. Lett.

[28] P. Alstrom and R. K. Ritala, Phys. Rev. 35A (1987) 300: S. Coppersmith, Phys. Rev. 38 A (1988) 375
[29] P.Hadley and K. Wiesenfeld, Phys. Rev. Lett. 62 (1989) 1335
[30] J.F. Magnan, et al. talk at the SIAM conf. on dynamical systems (Florida, 1990)
[31] C.B. Price, P. Wambacq, and A. Oosterlinck, SPIE Application of Digital Image Processing XI, 1990, San Diego
[32] K. Kaneko, Phys. Rev. Lett. 63 (1989)219; Physica 41 D (1990)137;Phys. Rev. Lett. 65 (1990) 1391; J. Phys. A, 24 (1991) 2107; Physica D (1991), in press
[33] I. Tsuda and K. Kaneko, in preparation; H. Nozawa et al., in preparation, NICON-system (1991)
[34] C. Bracikowski and R. Roy, Chaos 1 (1991) 49: see also J-L Chern and K. Otsuka (preprint NTT Res.Lab. 1991) for simulation of globally coupled optical array.
[35] M. Saito et al., preprint (1991) submitted J. Phys. Soc. Jpn.
[36] A.V. Holden, J.V. Tucker, and B.C. Thompson, Physica 49 D (1991) 240
[37] W. Freeman and C. A. Skarda, Brain Res. Rev. 10 (1985) 147; W. Freeman, Brain Res. Rev. 11 (1986) 259
[38] P. Davis, ATR preprint (1989)
[39] I. Tsuda, Prog. Theor. Phys. Suppl. 79 (1984) 241; I. Tsuda, *Chaos-teki-Noukan*, (1990, Science-Pub., in Japanese)

Dynamical Behavior of Thermostated Harmonic Oscillators

S. Nosé

Department of Physics, Faculty of Science and Technoloy,
Keio University, 3-14-1 Hiyoshi, Kohoku-ku, Yokohama 223, Japan

Abstract. Behavior of a system consisting of identical classical harmonic oscillators coupled to Nosē-Hoover thermostat is studied. Regular recurrences of initial conditions are observed in this system.

The canonical distribution of a classical system described with a Hamiltonian H(p,q) is attained by coupling the system with Nosē-Hoover thermostat[1], whose equations of motion are

$$\frac{dq_i}{dt} = \frac{\partial H}{\partial p_i}, \quad \frac{dp_i}{dt} = -\frac{\partial H}{\partial q_i} - \zeta p_i, \quad \frac{d\zeta}{dt} = (\Sigma_i\, p_i \frac{\partial H}{\partial p_i} - gkT)/Q \ . \quad (1)$$

An additional term $-\zeta p_i$ expresses coupling with a thermostat. The change of the variable ζ is governed by the last equation. g is the number of degrees of freedom of the system and Q is a parameter controlling the speed of energy transfer between the system and the thermostat. Nosē-Hoover thermostat produces the canonical distribution if an extended system consisting of a physical system and a thermostat is ergodic[1]. This condition is generally satisfied in a large system. However, a simple system of one harmonic oscillator coupled to Nosē-Hoover thermostat shows chaotic behavior only in some portion of phase space[2]. We study a system of two or more harmonic oscillators. Recurrence phenomenon is observed in various conditions if the frequencies of the oscillators are all equal (the identical model).

The equations of motion for the identical model in energy (H_i) - angle (ϕ_i) variables in a scaled unit (ω_i=1.0, m_i=1.0, kT=1.0) are

$$\frac{dH_i}{dt} = -2\zeta H_i \cos^2\phi_i \ , \quad \frac{d\phi_i}{dt} = 1 + \zeta \sin\phi_i \cos\phi_i \ , \quad (2)$$

$$Q\frac{d\zeta}{dt} = \Sigma_i\, 2H_i \cos^2\phi_i - gkT \ . \quad (3)$$

Several properties of the system can be derived analytically.

(1) The equations of motion are invariant under orthogonal transformations of the cartesian coordinates and the momenta. Therefore , all the trajectories starting from different initial conditions relating by an orthogonal transformation are identical.

Springer Proceedings in Physics, Vol. 70 Computational Approaches in Condensed-Matter Physics
Editors: S. Miyashita · M. Imada · H. Takayama

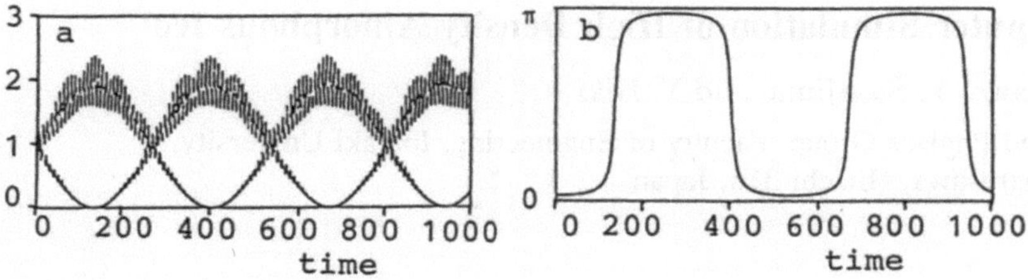

Figure 1. The time evolution of a) the energy H_i and b) the difference of angles $\Delta\phi$. Two oscillators, Q=4.0, and $\Delta\phi_0=\pi/36$.

(2) The differences of angles are

$$\frac{d(\phi_i-\phi_j)}{dt} = \zeta \cos(\phi_i+\phi_j)\sin(\phi_i-\phi_j) . \tag{4}$$

If $\Delta\phi=\phi_i-\phi_j = 0 \pmod{\pi}$, $\Delta\phi$ is always kept at zero. Overtaking of angles is forbidden and the order of the angles are conserved, $\phi_1 < \phi_2 < \phi_3 < \ldots\ldots < \phi_g < \phi_1+\pi \mod(\pi)$ in the identical model.

(3) In two limiting cases, (i) $\phi_i=\phi_j$, (ii) $H_i=1.0$, and $\phi_i=i\pi/g$, all oscillators move synchronously.

The calculations with two oscillators are carried out by 4th order Runge-Kutta algorithm. The initial conditions of this calculation are $H_i=1.0$, $\zeta=0$. Control parameters are the mass parameter Q and the difference of angles $\Delta\phi_0$. The time evolution of the energy of each oscillator and $\Delta\phi$ are depicted in fig. 1 (Q=4.0, $\Delta\phi_0=\pi/36$, $\Delta t=0.05$). The change of the energy consists of very rapid oscillation due to the coupling with the thermostat and slow beat relating to the energy transfer between two oscillators. At about t=540, the system almost returns to the initial configuration. $\Delta\phi$ changes very quickly from a small initial value to a value close to π near the point when the energy of one of oscillators is almost zero.

Several conclusions obtained from this calculation are as follows. The frequency Ω of the beat changes as $A\sin(\Delta\phi_0)$. $\Delta\phi$ and the beat of H_i are limited between $\Delta\phi_0$ and $\pi-\Delta\phi_0$, and between $1-\cos(\Delta\phi_0)$ and $1+\cos(\Delta\phi_0)$, respectively. Ω decreases with Q.

References

[1] S.Nosé, Prog.Theor.Phys.Suppl. 103 , 1 (1991). See also references therein.
[2] H.A.Posch, W.G.Hoover, and F.J.Vesely, Phys.Rev. A33, 4253 (1986)

Computer Simulation of High Density Amorphous Ice

S. Ozawa, Y. Sasajima, and Y. Hiki

Applied Physics Group, Faculty of Engineering, Ibaraki University, Nakanarusawa, Hitachi 316, Japan

Abstract Formation process of high density amorphous ice has been studied by the Monte Carlo computer experiment. Ice crystal Ih was uniformly compressed (10 - 30 % reduction in linear dimensions) and relaxed at 77 K. It has been shown that the overall shape of pair distribution functions for the relaxed state was similar to that for the liquid state.

1. Introduction

During the last decade, amorphous states of ice have been realized experimentally by many workers [1]. The typical methods to produce amorphous ice are 1) deposition of water vapor onto cooled substrate (low density amorphous), and 2) compression of ice crystal under 10 kbar pressure at 77 K (high density amorphous). By the latter method bulk specimens can be obtained, which is useful to investigate various properties of amorphous ice. We here carried out Monte Carlo (MC) computer experiments to produce high density amorphous ice, expecting to understand the atomistic process of amorphous state formation.

2. Method

The initial molecular configuration for the MC experiment is the proton-ordered ice Ih. The system is uniformly compressed and then relaxed at 77 K by the MC annealing procedure which permits translational and rotational movements of water molecules at a rate proportional to the Boltzmann factor $\exp(-\Delta E/k_{\mathrm{B}}T)$. The energy change ΔE is calculated by using the CC intermolecular potential [2] which contains 9 long-range Coulomb and 13 short-range exponential terms. The stability of normal pressure ice under this interacting potential has been checked [3]. The adopted supplemental techniques are the replica method for the periodic boundary condition (each replica contains 96 water molecules), the Ewald's method for the summation of slowly converging Coulomb terms, and the bookkeeping method to save the CPU time.

Springer Proceedings in Physics, Vol. 70 **Computational Approaches in Condensed-Matter Physics**
Editors: S. Miyashita · M. Imada · H. Takayama

3. Results

An example of the data obtained in the high pressure MC experiment (15% reduction) is shown in Fig. 1. The uniformly compressed ice Ih structure gradually changes into a random structure (see (a) - (c)). In order to study the process more quantitatively, the pair distribution functions (PDF) for these states have been calculated (see (d) - (f)). As the time elapses, the PDF peaks become diffused, and the position of the first peak shifts towards the large distance side. It is worthwhile to note that the shape of PDF in (e) is similar to that of experimentally observed liquid state PDF. The time of the relaxation sensitively depends on the compression rate. In the 10 % compression MC experiment, we could not observe the liquid-like PDF within a reasonable CPU time.

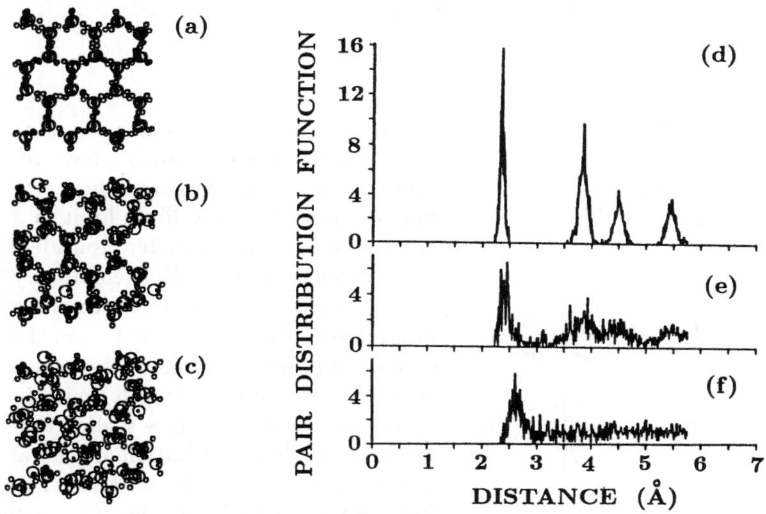

Fig. 1 Molecular configurations at different stages of relaxation process, (a) 3125 (b) 9375 and (c) 18750 Monte Carlo Steps; and the pair distribution function for each of these states, (d),(e) and (f).

References

1) Proc. Symp. Phys. and Chem. of Ice, in J. de Phys., **48** (1986).
2) V. Carravetta and E. Clementi, J. Chem. Phys., **81**, 2646 (1984).
3) S. Ozawa, Y. Sasajima and Y. Hiki, Proc. PCI-91, Sapporo, in press.

Statistical Mechanics of Polymer Networks

K. Ohno

Institute for Materials Research, Tohoku University,
Katahira, Sendai 980, Japan

Abstract. By using the scaling theory for the total number of configurations of general polymer networks, the critical exponents associated with the short distance behavior of the end(A)-to-end(B) distance (r^{AB}) distribution function $g_{\mathcal{G}}(r^{AB}) \sim (r^{AB})^{\theta_{\mathcal{G}}}$ of a polymer network $\mathcal{G}$ is obtained as $\theta_{\mathcal{G}} = [\gamma(n_A)+\gamma(n_B)-\gamma(n_A+n_B)-1]/\nu$, where n_A (or n_B) denotes the number of branching at the end-point A (or B), and $\gamma(n)$ denotes the configuration-number-exponent for n-arm star polymers. Similar expressions are also obtained for a surface-grafted polymer networks.

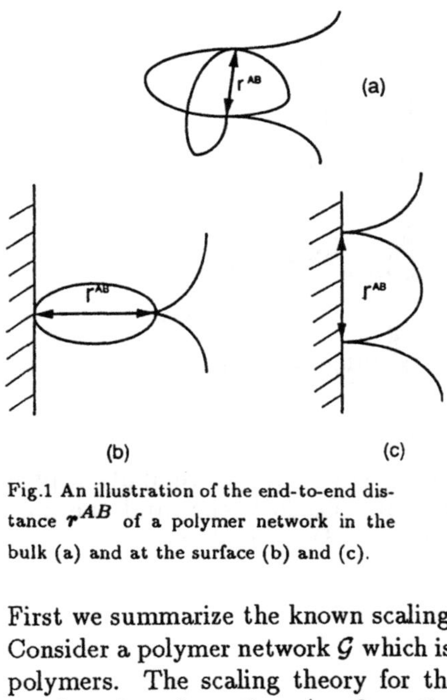

Fig.1 An illustration of the end-to-end distance r^{AB} of a polymer network in the bulk (a) and at the surface (b) and (c).

A variety of polymer networks, e.g. star polymers and comb polymers, have recently been synthesized and their statistical properties in a dilute limit in a good solvent have attracted considerable current interest both theoretically and experimentally. Here we are mainly interested in the universal feature of the statistical properties of networks, which is realized in the large length limit. In this paper, we derive a new scaling theory for the radial distribution of general polymer networks constructed from long linear polymers with equal length l, both for the case of a free network in the bulk and for a surface-grafted network with some ends adsorbed.

First we summarize the known scaling relations of general polymer networks[1]. Consider a polymer network $\mathcal{G}$ which is composed of f mutually connected linear polymers. The scaling theory for the exponent $\gamma_{\mathcal{G}}$ associated with the total number of configurations (configuration-number-exponent) of a general polymer network $\mathcal{G}$ in the bulk is expressed as follows; for the polymer network with n_h h-functional units (h-fold branching point) in the bulk and totally composed of f linear polymers with the same length, $\gamma_{\mathcal{G}}$ is given by $\gamma_{\mathcal{G}} = \alpha-1-f+\sum_{h=1}^{\infty} n_h \Delta_h$, where the exponent Δ_f is related to the configuration-number-exponents $\gamma(f)$

Springer Proceedings in Physics, Vol. 70 Computational Approaches in Condensed-Matter Physics
Editors: S. Miyashita · M. Imada · H. Takayama

of a f-arm star polymer via $\gamma(f) = \alpha - 1 + (\gamma - \alpha)f/2 + \Delta_f$. Similarly for a surface-grafted polymer network $\mathcal{G}$ (at a "repulsive" or "marginal" wall) with n_h h-functional units in the bulk and n'_h h-functional units at the surface (and totally composed of f linear polymers with the same length), the configuration-number-exponent $\gamma_\mathcal{G}$ is given by $\gamma_\mathcal{G} = \alpha - 1 - f + \nu + \sum_{h=1}^{\infty}[n_h\Delta_h + n'_h\Delta'_h]$. Here the exponent Δ'_f is related to the configuration-number-exponent $\gamma_s(f)$ of the center-adsorbed f-arm star polymers as $\gamma_s(f) = \alpha - 1 + \nu + (\gamma - \alpha)f/2 + \Delta'_f$, with the exponent α given by $\alpha = 2 - \nu d$. The reader should note here that the exponent $\gamma_s(f)$ has different values for repulsive and marginal walls[2].

We recently discussed the short distance behavior of the end-to-end distance distribution function of a free or surface-grafted star polymers[2]. The aim of this paper is to generalize this discussion to the case of general polymer networks.

As was stated in (3.21) of Ref.[1], the end(A)-to-end(B) (an end may be a branching point) distribution function for general free networks has the scaling form $g(r^{AB}) = \phi(r^{AB}/L^\nu)/L^{\nu d}$, where L is the total chain length $l_1 + l_2 + \cdots + l_f$. We are interested in the short distance ($r^{AB} \sim 0$) behavior of this function. Let us assume that the scaling function $\phi(x)$ behaves as $\phi(x) \sim x^{\theta_\mathcal{G}}$ for $x \to 0$. The exponent $\gamma'_\mathcal{G}$ when r^{AB} is fixed to be quite small is given by $\gamma'_\mathcal{G} = \gamma_\mathcal{G} + \Delta_{n_A+n_B} - \Delta_{n_A} - \Delta_{n_B}$. This relation holds because the n_A-functional unit combines with the other n_B-functional unit to produce the $(n_A + n_B)$-functional unit (see Fig.1a). Therefore we have $\mathcal{N}(r^{AB} \to 0) \sim L^{\gamma_\mathcal{G}+\Delta_{n_A+n_B}-\Delta_{n_A}-\Delta_{n_B}-1}\mu^L$, as the total number of configuration of this topology. This behavior should be compared with the same quantity with general value of r^{AB}, $\mathcal{N}(r^{AB}) = \mathcal{N}g(r^{AB}) = L^{\gamma_\mathcal{G}-1-\nu d}\mu^L\phi(r^{AB}/L^\nu) \sim L^{\gamma_\mathcal{G}-1-\nu[d+\theta_\mathcal{G}]}\mu^L(r^{AB})^{\theta_\mathcal{G}}$, for $r^{AB} \to 0$, where $\mathcal{N} \sim L^{\gamma_\mathcal{G}-1}\mu^L$ denotes the usual total number of configuration of the network $\mathcal{G}$. The comparison of the power of L yields $\theta_\mathcal{G} = [\gamma(n_A) + \gamma(n_B) - \gamma(n_A + n_B) - 1]/\nu$.

Next we consider the case of the surface-grafted polymer networks. There are two ways of approaching one end $\mathbf{r}^A = (\mathbf{r}_\parallel^A, z^A)$ to the other $\mathbf{r}^B = (\mathbf{r}_\parallel^B, 0)$ which is fixed onto the surface. First we set $r_\parallel^{AB} = |\mathbf{r}_\parallel^A - \mathbf{r}_\parallel^B| \to 0$ with finite z^A (see Fig.1b). Then, repeating a similar argument as above, we have $g(r_\parallel^{AB} = 0, z^A) \sim (z^A)^{\theta_{\perp\mathcal{G}}}$ with $\theta_{\perp\mathcal{G}} = [\gamma(n_A) + \gamma_s(n_B) - \gamma_s(n_A + n_B) - 1]/\nu$. The second is the case where z^A is very small compared with $r_\parallel^{AB}$ (see Fig.1c). In this case, we have $g(r_\parallel^{AB}, z^A = 0) \sim (r_\parallel^{AB})^{\theta_{\parallel\mathcal{G}}}$ with $\theta_{\parallel\mathcal{G}} = [\gamma_s(n_A) + \gamma_s(n_B) - \gamma_s(n_A + n_B) - 1]/\nu$. One may regard that these formulae are the simple generalization of the corresponding formulae for star polymers.

[1] K. Ohno and K. Binder, J. de Physique (Paris) **49**, 1329 (1988). [2] K. Ohno and K. Binder, J. Chem. Phys. **95** (1991); J. Stat. Phys. **64**, 781 (1991).

Studies on Metastable States by Ising Machine m-TIS2

H. Tomita and S. Miyashita

Department of Physics, CLAS, Kyoto University, Kyoto 606, Japan

Abstract. We investigated the relaxation of the metastable state which appears in the ordered state of Ising system. We propose a new definition of the spinodal point concerning the dynamical nature of relaxation of the magnetization.

Although the absence of the spinodal point in the thermodynamic limit has been confirmed, we still find metastable behaviors in the kinetic Ising model [1,2]. Thus we investigate the system-size effect and the statistical property of the relaxation carefully using the Ising machine *m-TIS2* [3]. We study the two-dimensional Ising model, $\mathcal{H}/k_{\mathrm{B}}T = -K\sum_{<ij>} s_i s_j - H\sum_i s_i$. The cases with lattice size $L = 64 \sim 720$ ($N = L \times L$), temperature $T = 0.8T_{\mathrm{c}}$ ($K = 0.55086$, the spontaneous magnetization $m_{\mathrm{s}} = 0.95441$) are examined. The initial configuration is the all up-spin state and its relaxation under a negative field is simulated $100 \sim 1000$ times for each set of parameters.

We have a well-known notion of the mean field spinodal point (MFSP). This can be seen in the very early stage of the relaxation of the magnetization $m(t)$ starting from completely random configurations. Rapid relaxations to a local minimum state with $m \simeq +m_{\mathrm{s}}$ are found. After this initial stage there appears a flat part of $m(t)$ for small field $|H| \leq H_{\mathrm{C}} \sim 4K$. Let us define its lifetime t_1 by $m(t_1) = 0$. This increases with decreasing $|H|$ and apparently seems to diverge at some critical value $H^* \sim K/L$ [1,2]. This may be called thermodynamic spinodal point (THSP), because it is caused by the bistable nature of the thermodynamic constrained free-energy of a finite system, which has a central barrier $\Delta F/k_{\mathrm{B}}T \sim KL^{d-1}$ for $|H| \ll H^*$ due to the interface energy. The new notion proposed here is a dynamical spinodal point (DSP) which differs from both of MFSP and THSP.

When $|H|$ decreases, a steep change of the relative variance $r = \sqrt{\langle (t_1 - \langle t_1 \rangle)^2 \rangle}/\langle t_1 \rangle$ from $r \simeq 0$ to $r \simeq 1$ is found before THSP. This fact suggests a switch-over in the relaxation mechanism: Throughout all regions including $r \simeq 1$ where t_1 largely

Springer Proceedings in Physics, Vol. 70 Computational Approaches in Condensed-Matter Physics
Editors: S. Miyashita · M. Imada · H. Takayama

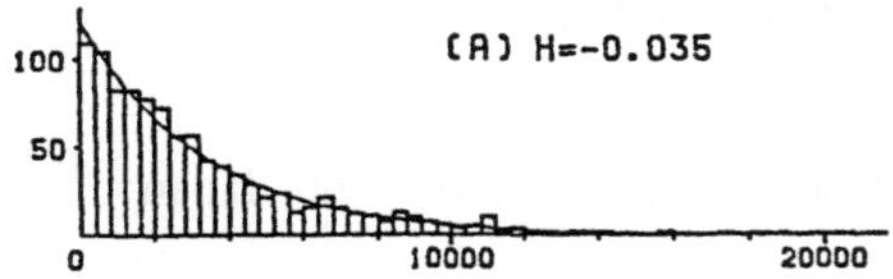

Fig.1 Histogram of 1000 samples for the lifetime t_1 in $L = 128$ system. The values of the relative variance r are (A)0.996, (B)0.817 and (C)0.409, respectively.

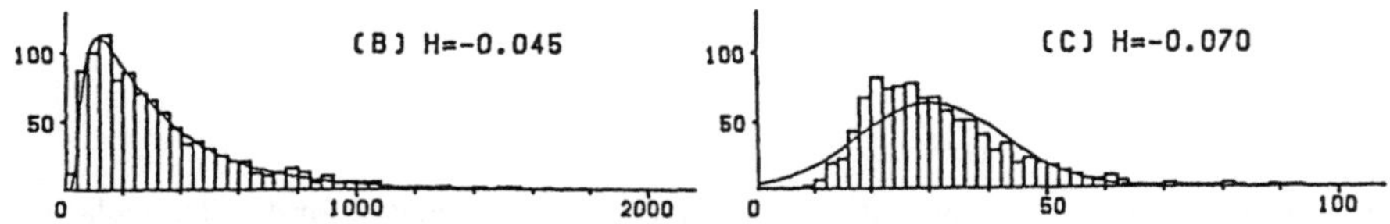

fluctuates, shapes of $m(t)$ are found to be remarkably similar, i.e. all sample curves are superposed into a unique curve for given H and L if time t of each is shifted by $t_1 - \langle t_1 \rangle$. This fact implies Suzuki's scaling [4]. Thus the main process of relaxation seems to be well described by dynamic potential as $dm/dt = -d\phi/dm$. The difference in the relaxation mechanism becomes clearer if the lifetime distributions are analyzed in detail with use of this potential picture. Typical cases are shown in Fig.1. (A)*Metastable*: The fitting curve drawn is exponential one, which is characteristic of the Poissonian decay through a potential barrier. Note that $r \simeq 1.0$ in this region. (B)*Critical*, i.e. DSP: The dynamical barrier just disappears. The fitting curve is obtained by solving a Fokker-Planck equation for unstable potential with only one fitting parameter. (C)*Non-critical*: Little deformation is expected during running down the monotonic potential slope. A Gaussian fitting curve is drawn. It should be noted that the critical value $H^*_{\rm DSP} \simeq 0.045$ is well distinguishable from $H^*_{\rm THSP} \sim K/L = 0.0043$ for $K = 0.55086$ and $L = 128$.

References

[1] E. Stoll and T. Schneider, Phys. Rev. **A6** (1972), 429.

[2] K. Binder and H.Müller-Krumbhaar, Phys. Rev. **B9** (1974), 2328.

[3] N. Ito, M. Taiji and M. Suzuki, J. Phys. Soc. Jpn. **56** (1987), 4218 and also see their article in this issue.

[4] M. Suzuki, Prog. Theor. Phys. **56** (1976), 77 and 477; See also a review article in Adv. Chem. Phys. **46** (1981), 195.

Raman Correlation Function of Percolating Networks

T. Nakayama and K. Yakubo

Department of Applied Physics, Hokkaido University, Sapporo 060, Japan

Abstract. The Raman correlation function $C(q,\omega)$ is calculated for $d=2$ percolating networks by direct simulations. An array-processing supercomputer is efficiently utilized by applying a novel numerical algorithm. We have found that $C(q,\omega)$ has the universal behavior scaled by the characteristic frequency.

1. Raman Correlation Function

The inelastic light scattering experiments yield a great deal of information on the dynamics of fractal structures[1,2]. The interpretation of experimental data, however, have been controversial. The motivation for this paper arises from such an obscurity. We have numerically calculated the Raman correlation function $C(q,\omega)$ of $d=2$ percolating networks.

The Raman intensity $I(\mathbf{q},\omega)$ with a frequency shift $(\omega/2\pi)$ and a wavenumber shift $(\mathbf{q})$ is proportional to the Raman tensor $I_{\alpha\beta}(\mathbf{q},\omega)$, which is the space and time Fourier transform of the correlation function of the dielectric susceptibility $\epsilon_{\alpha\beta}(\mathbf{r},t)$. The Greek subscripts α and β indicate Cartesian components. Since the dielectric susceptibility $\epsilon_{\alpha\beta}$ is related with the elastic strain tensor $e_{\alpha\beta}$, the Raman tensor is expressed by $I_{\alpha\beta}(\mathbf{q},\omega) = ((n(\omega)+1)/\omega)\tilde{C}_{\alpha\beta}(\mathbf{q},\omega)$. Here $n(\omega)$ is the Bose population factor and $\tilde{C}_{\alpha\beta}(\mathbf{q},\omega)$ is defined as

$$\tilde{C}_{\alpha\beta}(\mathbf{q},\omega) = \sum_{\lambda} \delta(\omega-\omega_\lambda) \left| \sum_i e^{\lambda}_{\alpha\beta}(\mathbf{r}_i) \exp(i\mathbf{q}\cdot\mathbf{r}_i) \right|^2 ,$$

where $e^{\lambda}_{\alpha\beta}$ is the strain of the mode λ.

The frequency dependence of $\tilde{C}_{\alpha\beta}(\mathbf{q},\omega)$ for fractons is analyzed by several authors[1-3]. All analyses are based on two key assumptions: (i) $\tilde{C}_{\alpha\beta}(\mathbf{q},\omega)$ could be described by "ensemble averaged" fracton wavefunction $\bar{\psi}_{fr}$. $\bar{\psi}_{fr}$ has spherical symmetry around its localization center. (ii) The profile of $\bar{\psi}_{fr}$ could be described by an envelope function neglecting spatial oscillation of fracton modes. We find that the second assumption is very misleading as well as the first one as pointed out previously by Nakayama et al.[4] and Alexander[3].

2. Numerical Results and Conclusion

The systems are site-percolating networks consisting of N atoms with unit mass and linear springs ($K=1$) connecting two nearest neighbor atoms. The vibrational displacement of each atom has only one component, so the Raman tensor becomes scalar. The numerical technique employed here is based on the mechanical resonance to extract pure eigenmodes. Using this algorithm, it becomes possible to investigate lattice vibrations in large systems containing as many as $\sim 10^6$ atoms. In addition, the algorithm has an advantage to be easily vectorized for implementation on an array-processing supercomputer.

Springer Proceedings in Physics, Vol. 70 Computational Approaches in Condensed-Matter Physics
Editors: S. Miyashita · M. Imada · H. Takayama

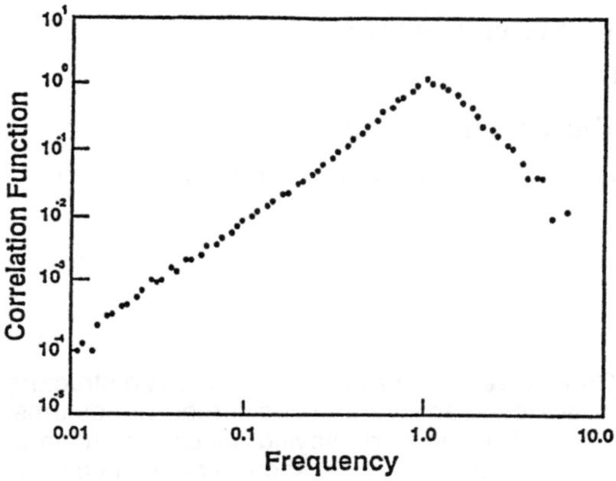

Fig.1 The ω-dependence of the Raman correlation function $C(q,\omega)$ of two-dimensional percolating networks at p_c. The correlation function is calculated by averaging over five percolating networks formed on 500×500 square lattices. The maximum site number of our systems is $N = 110793$ and the minimum $N = 93382$. The abscissa is scaled by the dimensionless frequency ω/ω_0. The value of $C(q,\omega)$ is scaled by its maximum value.

The calculated result of the dynamical correlation function $C(q,\omega)(\equiv \tilde{C}(q,\omega)/\mathcal{D}(\omega))$ is shown in Fig. 1. Figure 1 indicates the ω-dependence of $C(q,\omega)$ of $d = 2$ percolating networks at $p_c(=0.593)$ for 250 wavenumbers $(2\pi/250 \leq q \leq 2\pi)$. The abscissa represents the frequency ω/ω_0, where ω_0 is the frequency at which $C(q,\omega)$ has the largest value for each *fixed wavenumber*. Filled circles indicate the averaged value over data within narrow range close to ω/ω_0. It should be emphasized that $C(q,\omega)$ can be scaled by the characteristic frequency ω_0 for any wavenumbers. The result also shows that the frequency dependence of $C(q,\omega)$ is expressed as $C(q,\omega) \propto \omega^{2\pm0.1}$ in the frequency regime of $\omega \ll \omega_0$ and $C(q,\omega) \propto \omega^{-3\pm0.1}$ for $\omega \gg \omega_0$. The previous analyses for the dynamical correlation function have predicted that $C(q,\omega)$ vanishes for $\omega \ll \omega_0$ and $C(q,\omega) \propto \omega^{\alpha}$ for $\omega \gg \omega_0$, where α is positive. $C(q,\omega)$ obtained here, however, has obviously the peak at $\omega = \omega_0$. The discrepancy between our simulated results and previous analyses is due to the fact that spatial oscillations of fracton modes are neglected in previous analyses.

In conclusion, we have calculated the dynamical correlation function $C(q,\omega)$ for large percolating networks. We have found that $C(q,\omega)$ exhibits the universal behavior scaled by the characteristic frequency. Our results conflict with previously proposed theories. The origin of this discrepancy is due to the misleading assumption on fracton modes.

This work was supported in part by a Grand-in-Aid of Scientific Research on Priority Areas, Computational Physics as a New Frontiers in Condensed Matter, Research of the Ministry of Education, Science and Culture, Japan. We thank to R.Orbach for his fruitful discussion on this subject.

References

1. A. Boukenter, B. Champagnon, E. Duval, J. Dumas, J. F. Quinson, and J. Serugetti, Phys. Rev. Lett. **57**, 2391 (1986).
2. See, for example, E. Courtens and R. Vacher, Proc. R. Soc. Lond. A**423**, 55 (1989), and references therein.
3. S. Alexander, Phys. Rev. B**40**, 7953 (1989).
4. T. Nakayama, K. Yakubo, and R. Orbach, J. Phys. Soc. Jpn. **58**, 1891 (1989).
5. K. Yakubo and T. Nakayama, Phys. Rev. **40**, 517 (1989).
6. K. Yakubo, T. Nakayama, and H. J. Maris, J. Phys. Soc. Jpn. **60** 3249 (1991).

Dynamics of Charge Density Wave Systems with Multidomain Structure

H. Matsukawa, N. Koshita, Y. Takada, and C. Sakiyama

Department of Physics, Nara Women's University, Nara 630, Japan

Abstract. A model for dynamics of Charge Density Wave with multidomain structure is studied. It consists from sinusoidally coupled overdamped forced pendulums. We investigate it numerically under dc and dc+ac fields paying special attention to switching and chaotic response. It is shown that the model shows chaos in contrast to a linearly coupled model.

Incommensurate charge density waves (CDW's) in quasi-one-dimensional conductors are typical examples of cooperative nonlinear dynamical systems with randomness[1]. Above a dc electric threshold field E_T the sliding motion of CDW occurs, which associates with a dc current with a spontaneous current oscillation, the so called narrow band noise (NBN). Its fundamental frequency ω_{NBN} is proportional to the dc CDW current. Below E_T CDW is pinned by impurities in a sample. Under dc+ac field NBN interferes with the ac field and yields nonlinear dynamical phenomena such as mode locking (ML), in which ω_{NBN} is drawn into a rational value of the external ac frequency ω_{ex} and then the dc CDW current keeps a constant value in certain finite range of the dc field. Most of these phenomena are well described by Fukuyama-Lee-Rice model[2], which treats CDW as a classical deformable object under impurity potentials and describes its behavior by its phase variable[3]. The qualitative aspects of them are, however, reproduced by linearly coupled overdamped forced pendulums with only two degrees of freedom[4]. In the latter the phase of each pendulum and coupling correspond to the phase at each impurity site and elastic coupling of CDW, respectively. The gravitational force, which corresponds to each impurity pinning force, changes its direction randomly for each domain due to random distribution of impurities.

In most samples the increment of dc current at E_T is sharp but continuous. But in some samples switching phenomena occur, in which the dc current jumps at E_T with hysteresis. In these samples chaotic response also occurs under dc+ac field[5]. Both phenomena never happen in the FLR and linearly coupled pendulums models[3,4]. Hall et al. showed experimentally that in these samples CDW is divided into domains and in each domain CDW moves with different dc velocity[5]. Such motion is impossible in these models, since the force of the spring which connects pendulums grows with their phase difference.

In order to investigate the effects of multidomain structure we here study sinusoidally coupled pendulums model numerically. Its equation of motion is expressed as follows,

$$\phi_1 = a_1\sin(\phi_1) + b\sin(\phi_2-\phi_1) + \varepsilon_{dc} + \varepsilon_{ac}\sin(\omega_{ex}t) \quad ,$$
$$\phi_2 = a_2\sin(\phi_2+\delta) + b\sin(\phi_1-\phi_2) + \varepsilon_{dc} + \varepsilon_{ac}\sin(\omega_{ex}t) \quad .$$

Here $\phi_{1,2}$ and $a_{1,2}$ are phases and magnitudes of pinning potentials of domain 1 and 2, respectively, δ is a misfit of favorable phases of pinning force, b is coupling

Springer Proceedings in Physics, Vol. 70 **Computational Approaches in Condensed-Matter Physics**
Editors: S. Miyashita · M. Imada · H. Takayama

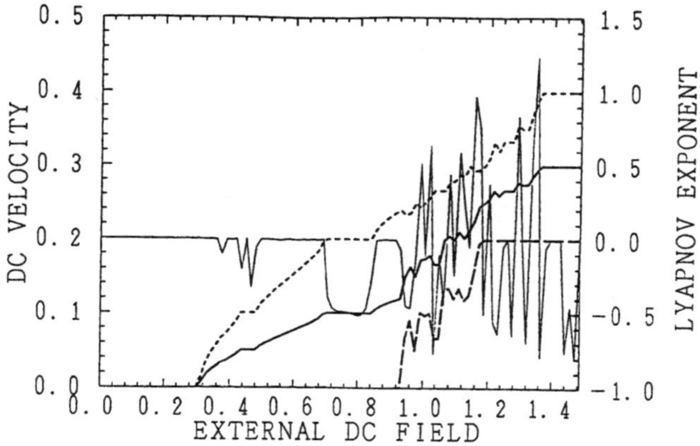

Fig.1 dc velocities of domain 1 and 2(broken and dotted lines), their average(bold line) and maximum Lyapunov exponent(thin line) in a sample with a_1=1.5, a_2=0.2, b=0.5, $\delta=\pi$, ε_{ac}=0.5 and $\omega_{ex}=0.2\pi$. In the present unit dc current equals to ω_{NBN}/π.

strength and ε_{dc} and ε_{ac} are magnitudes of external dc and ac fields, respectively. In this model domain 1 and 2 can move with different dc velocities. The mean field version of this model under dc field was already studied and it is known that it shows switching phenomena[6].

In fig.1 we show dc velocities of two domains and their average and maximum Lyapunov exponent under dc+ac field as a function of dc field. An interesting point is that under large field domain 1 shows 1/1 ML and 2 does 2/1 ML and then the whole system apparently shows 2/3 ML. Below this range the maximum Lyapunov exponent takes positive values, which indicate chaotic response. Within our study we can find such response only in the sample where the magnitudes of the pinning potentials of two domains are much different and then the threshold fields of them are also different as in the present sample. At the threshold field of domain 1 the increment of total CDW current is so sharp and looks like switching. Such behavior is also observed under dc field alone. Any hysteresis is, however, not observed in our study. We have also investigated a model consisting from 30 pendulums sinusoidally coupled in one dimension. In this case chaotic response is also observed even when the magnitude of all impurity potentials is set to be equal. But switching is not. It is interesting to study a similar model with more degrees of freedom and in higher dimensions in relation to the mean field model.

References

[1] G.Gruner, Rev.Mod.Phys.60,1129(1988) .

[2] H.Fukuyama, J.Phys.Soc.Jpn.41,513(1976); H.Fukuyama and P.A.Lee, Phys.Rev. B17,535(1978); P.A.Lee and T.M.Rice, ibid. B19,3970(1979).

[3] H.Matsukawa and H.Takayama, J.Phys.Soc.Jpn.56,1507(1987); H.Matsukawa, ibid.56,1522(1987), 57,3463(1988) and Synthetic Metals 29,F343(1989).

[4] H.Matsukawa and H.Takayama, Synthetic Metals19,7(1987); M.Inui and S.Doniach, Phys.Rev.B35,6244(1987).

[5] R.P.Hall and A.Zettl, Solid State Commun.50,813(1984); R.P.Hall, M.Sherwin and A.Zettl, Phys.Rev.B29,7076(1984); R.P.Hall, M.F.Hundley and A.Zettl,ibid. B56, 2399(1986).

[6] S.H.Strogatz, C.M.Marcus, R.M.Westervelt and R.E.Mirollo, Phys. Rev.Lett. 61, 2380(1988).

Part VII

Computer Facilities

Computational Physics on Connection Machine Supercomputers

E. Loh, Jr.

Thinking Machines Corporation, 245 First Street, Cambridge, MA 02142, USA

Abstract. Developments in computer architectures are changing the way we do computational physics. Data-parallel programming languages eliminate unnecessary serialization, which obscures the parallelism in nature, and allow scientists to program computers at higher levels. Massively parallel hardware designs allow both higher performance and better performance-per-price. In this talk, I will discuss data-parallel programming, Connection Machine (CM™) design, and selected applications on Connection Machine supercomputers from computational condensed-matter physics.

1. Introduction

Unfortunately, mathematical models of our physical world are rarely solvable analytically. Analytic techniques often assume extreme parameter values, which are generally uninteresting for real-life problems. Alternatively, problems necessarily entail large amounts of data, perhaps to describe the conformation of a molecule or the shape of a crystal defect. Thus, researchers must rely increasingly on numerical tools.

Computer technology by itself cannot unlock the universe's secrets for us. Our understanding of the physical world will always depend most on the quality of our insights. As our reliance upon numerical methods continues to increase, development of algorithms – fast fourier transforms, conjugate gradient solutions, implicit methods, preconditioners, Lanczos techniques, and so on – will also play a vital role in theoretical physics. Nevertheless, computer technologies, including both hardware and software advances, also play an essential role and have proven order-of-magnitude improvements every few years.

In this paper, we discuss how massively parallel computer architectures are changing the way we do computational physics. Since large-scale computational problems in condensed-matter physics generally involve large numbers of degrees-of-freedom, all obeying the same physics, the serialization that occurs in conventional computer programs is spurious and tedious. Data-parallel programming languages use the parallelism in nature and simplify programming tasks. The elements of data-parallel languages will be presented, using illustrations from the recently adopted Fortran standard, Fortran 90.

Connection Machine CM-2™and CM-200™models are discussed as examples of massively parallel computer architectures. [1] These supercomputers are based on thousands of processors working together, providing unprecedented performance, price performance, and scalability.

Springer Proceedings in Physics, Vol. 70 Computational Approaches in Condensed-Matter Physics
Editors: S. Miyashita · M. Imada · H. Takayama

Finally, a few examples, chosen from condensed-matter physics performed on Connection Machine supercomputers, are used to illustrate scientific work being performed on massively parallel machines. [2]

2. Data-Parallel Programming

Data parallelism is a style of programming in which instructions are applied to many pieces of data at once. In the computer, one might imagine processors working independently on similar tasks, with all processors synchronizing from time to time to exchange data. In a numerical simulation, one might imagine similar, local processes taking place throughout simulational space with occasional breaks for global communications.

Certain communications primitives are provided, both to facilitate programming and to provide access to communications patterns that have been optimized on the parallel computer. Such patterns include local communications and communications on rectangular grids.

2.1 Example from Fortran 90

Most massively parallel computers require users to write programs for the various processors and explicitly pass data among the processors as necessary. Connection Machine supercomputers, however, support data-parallel extensions of the common, high-level languages Fortran, C, and Lisp, called CM Fortran™, C*®, and *Lisp™, respectively. Here, we will illustrate the features of data-parallel languages using examples from Fortran. While CM Fortran is essentially the same as Fortran 90, the new ANSI Fortran standard, some differences exist since CM Fortran was developed before the standard was adopted in 1991.

Let us begin by examing a short code fragment for solving a Poisson equation using successive relaxation. The problem is to solve

$$\nabla^2\phi = f$$

with $\phi = \phi_0$ on some boundary. In each iteration, one calculates the residue $r = f - \nabla^2\phi$, updates $\phi \rightarrow \phi + \alpha\ r$ in the interior of the domain for some suitably chosen α, and checks for convergence. The Laplacian is commonly differenced as

$$\nabla^2\phi = \frac{\phi(+\hat{\mathbf{x}}) - 2\phi + \phi(-\hat{\mathbf{x}})}{dx^2} + \frac{\phi(+\hat{\mathbf{y}}) - 2\phi + \phi(-\hat{\mathbf{y}})}{dy^2}\ .$$

A short Fortran code fragment that effects successive relaxation is

```
SUBROUTINE RELAX(PHI, F, PHI0, INTERIOR, NX, NY)

LOGICAL, ARRAY(NX,NY) :: INTERIOR
REAL, ARRAY(NX,NY) :: PHI, F, PHI0, RESIDUE
COMMON /PARAMS/ DX, DY, ALPHA, EPSILON
```

```
      PHI=PHI0

      DO
         RESIDUE=F
&         -(CSHIFT(PHI,1,1)-2.*PHI+CSHIFT(PHI,1,-1))/DX**2
&         -(CSHIFT(PHI,2,1)-2.*PHI+CSHIFT(PHI,2,-1))/DY**2
         WHERE (INTERIOR) PHI=PHI+ALPHA*RESIDUE
         ERROR=MAXVAL(ABS(RESIDUE),MASK=INTERIOR)
         IF (ERROR.LT.EPSILON) EXIT
      END DO

      RETURN
      END
```

2.2 Elemental Operations

The code fragment illustrates a number of basic features of data-parallel programming languages.

The basic data structure is the array, a rectangular grid of data values in some, possibly high-dimensional, space. This grid is unrelated to the number of processors in the computer or their configuration, though compiler directives may be used to control the layout of these grid values on the processors to optimize performance. In general, machine-specific software lays out the data to distribute the workload among all processors and this process is transparent to the programmer.

Array syntax is used to invoke operations on entire arrays or subsections of arrays at once, rather than embedding statements deep within nested **DO** loops. For example, **PHI=PHI0** assigns the elements of **PHI0** to the respective elements of **PHI**. This requires the two arrays to be conformable – that is, they must be of the same size and shape. Conformable arrays need not all have their data laid out identically on the machine's processors. The emphasis, again, is on making the computer architecture as transparent as possible to the programming scientist.

Nevertheless, not every element need participate in every statement. The process of specifying which elements are active in any operation is called contextualization. This may be effected with **WHERE** statements or with masks that are passed as arguments to intrinsic functions.

Control flow in data parallel programs remains serial and is managed with statements such as **DO** and **IF**.

2.3 Communications

The key challenge for massively parallel computers, however, is communications. By allowing any processor to communicate with any other processor at any time, full versatility is allowed. Unfortunately, communicating solely in this general fashion makes programming difficult and debugging a nightmare. Thus, data parallelism adds important communication primitives. Computer design should not only optimize communications performance in general, but performance for data-parallel primitives in particular.

Reduction operations combine data values. For example, **MAXVAL(A)** returns the maximum value of all elements of the array **A**. Fortran intrinsic functions that reduce arrays to lower dimensional structures include **MAXVAL**, **MINVAL**, **SUM**, **COUNT**, **ANY**, and **ALL**.

On massively parallel computers, such reductions are first formed on-processor and then collected across processors. For example, to find the maximum value in an array, each processor computes the maximum of values residing on the node. These maxima are then combined in a binary fashion, with each two processors determining their maximum, then each four determining theirs, and so on. This means that once each processor has determined its maximum, one must pay a $log_2(P)$ premium in running time, during which not all of the P processors in the system can remain busy. This premium is small compared with the total cost of the operation when there is a lot of data per processor. Further, the overall speedup due to the parallel hardware generally more than compensates for this cost.

Meanwhile, such a binary combination generally improves arithmetic precision. In a serial, floating-point accumulation, for example, latter values get added into a large accumulator with low precision. In a hierarchical accumulation, on the other hand, partial sums of similar order are added to each other.

The opposite of reduction is broadcast, which distributes data from one or more processors out to the rest of the processors. Broadcasts, being the counterparts of reductions, also pay a $log_2(P)$ premium for parallelism.

Permutations are one-to-one communication patterns and can be used for general communications. Each processor can send (receive) a value to (from) any other processor. Special cases such as near neighbor communications on rectangular grids may be important for problems with localized physics. These patterns are easily invoked with Fortran intrinsics. For example, **CSHIFT(PHI,DIM=2,SHIFT=-1)** fetches the elements of **PHI** from locations displaced by -1 along dimension 2, producing **PHI(I,J-1)** in element **(I,J)**.

Parallel-prefix ("scan") operations are another powerful data-parallel communications primitive. These operations compute partial reductions along selected axes in rectangular grids. For example, one may choose to replace every element a_i by $\sum_{j=1}^{i} a_j$. On a serial computer, this sort of recursive operation is best implemented by "scanning" the elements, starting with $i = 1$, and accumulating partial sums along the way. Very general parallel-prefix operations can be specified by selecting elements along the scan where the partial results are to be reset. Such recursion appears to be an inherently serial operation, but the same logarithmic combination techniques used for global reductions can be used for such partial reductions.

2.4 Data Parallelism in Computational Condensed-Matter Physics

Data parallelism arises in computational physics for a variety of reasons. Analytic calculations may become intractable if there are too many parameters that describe the problem: the structure of a unit cell, an empirically determined interaction potential, and so on.

In condensed-matter physics, however, the chief origin of data parallelism is even more natural and gives rise to many more degrees-of-freedom – we are interested in many-body problems and are unable to solve these problems analytically, outside of extreme parameter regimes. The resulting computational solutions, as the talks at this symposium amply demonstrate, invariably deal with many degrees-of-freedom. Sometimes these degrees-of-freedom are "real", such as the spin degrees-of-freedom in Ising and classical-spin-glass simulations, or artificially introduced as a part of the computational algorithm, as with fermion Monte Carlo.

Not only is computational condensed-matter physics data-intensive, but it is also *control* thin. In automobile-engine design, in contrast, the computational processes that take place in the various parts of the engine are all different. Computer programs must include code that describe these various processes, which take place independently of each other. Such problems pose special challenges for the data-parallel approach. In contrast, we can generally represent problems in condensed-matter physics quite compactly. The physics is "homogeneous", with the same model being applicable throughout the simulation space, rather than different processes being active at different sites. For example, in most Monte Carlo algorithms, the rules for updating a degree-of-freedom are the same throughout the lattice. A Brillouin Zone integrand is the same at each point in k-space. The prime difficulty in such problems is simply that these models or solution algorithms must be applied to increasingly more degrees-of-freedom to study large lattices, low temperatures, or detailed crystal structures. Thus, the data-parallel approach, in which a simple stream of instructions is executed in parallel on many pieces of data, is most natural.

2.5 Parallel-Computer Designs

Choice of programming languages should be dictated by the physics, rather than the computer architecture. Thus, data-parallel programming languages should generally be used in computational condensed-matter physics not only for massively parallel computers but for conventional, serial and vector, computers as well. Three points that specifically concern massively parallel computers, however, should be made.

First, data-parallel languages are well-suited for programming massively parallel computers. The contructs of data parallelism, which we will describe below, map well onto typical hardware capabilities.

Next, these languages are needed to control the complexity massively parallel designs present. One could easily construct a "computer" with arbitrarily powerful capacity simply by linking many processors together. (Let us ignore, for the moment, issues of financial costs and comunication bottlenecks!) Programming such a beast would be incredibly difficult. Taking full advantage of the capacity and versatility of this system would be a daunting task. Whether one is interested in programming a multiple-instruction, multiple-data (MIMD) system like the one we just described or a single-instruction, multiple-data (SIMD) system like a CM-2 or CM-200 supercomputer, a restricted machine model is needed to simplify the programming task for the physicist. Low-level programming allows computationally sophisticated scientists to extract greater performance on individual subroutines on specific computers, but a basic machine model is required to simplify the general programming task.

Finally, the data-parallel programming paradigm will survive several stages of massively parallel computer designs. Old-style programs have to be vectorized, multitasked, and so on as computer-design strategies evolve. In contrast, a data-parallel program does not require significant recoding to accommodate increases in available computer power provided either by increased numbers of processors or increased power-per-processor.

2.6 Consequences for Computer Programming

Data-parallel languages ease other challenges for the programmer as well. The process of abstraction from mathetical formulae to computer code is simplified since formulae have to been written only once, rather than many times within **DO** loops for each degree-of-freedom in the model. Computer codes should not serialize what happens in parallel in nature.

Since data-parallel programs do not introduce spurious serialization, they are more compact and easier to debug and maintain.

Finally, in programming massively parallel computers, there is an additional difficulty in debugging codes in that different processors may be executing different programs and may be at different instructions at the time that a bug occurs. The simplified control stream in data-parallel programs – all processors executing similar tasks, but each one on its own data – reduces this debugging problem.

3. CM-2/CM-200 Architecture

Massively parallel computer architecture affords computational physicists a number of attractive advantages. Most of all, fundamental limits in nature – characterized by the speed of light, Planck's constant, and so on – place performance constraints on a single computer processor. Clever technological choices, such as using gallium arsenide instead of silicon or immersing electronic components directly in coolant, can push back the performance limits that designers realize, but nature, all the same, limits the ultimate performance these designers will achieve. The next strategy, then, is to combine a number of processors together. Such a strategy is likely to be our only path to orders-of-magnitude increases in performance over today's single processors.

Further, the cost-effectiveness of midlevel-performance components is better than that for high-end ones. Thus, massive parallelism offers a cost advantage over few-processor systems for computer speed, memory capacity, and memory access.

Technology is advancing faster for these midlevel components as well, meaning that massively parallel systems can ride the wave of evolving technology. Indeed, the CM-2 and CM-200 models use commercial chips for both computation and memory and the CM-2 has moved through more than an order-of-magnitude increase in memory capacity in its four-year lifetime as memory technology has evolved.

Massively parallel systems are scalable, making systems available both to those with low budgets and to those with high requirements.

Finally, homogeneity of components simplifies trouble-shooting and hardware maintenance.

In this section, we will discuss the architecture of the CM-2 and CM-200 models, manufactured by Thinking Machines Corporation, of Cambridge, Massachusetts. One can think of the basic hardware designs of the two systems as the same, but with a clock speed of 7 MHz for the CM-2 and 10 MHz for the CM-200.

Connection Machine hardware design first focused on bit processing. As computational condensed-matter theorists, however, we are more interested in numerical computing. Therefore, this section will describe CM systems in terms of their floating-point nodes. For example, we will describe a fully configured system as being composed of 2,048 nodes, rather than 65,536 bit-serial processors, which is the convention.

3.1 Sequencer

In data-parallel programs, there is a narrow control stream. That is, the sequence of instructions to be executed is very simple, but the instructions must be broadcast to many processors. In CM-2 and CM-200 systems, each partition of processing nodes is under the control of a single, microcoded sequencer, which decodes instructions from the computer program and broadcasts them to the individual processors.

3.2 Processing Node

Each processing node in CM supercomputers has a floating-point chip for either 32-bit or 64-bit floating-point arithmetic. On-chip registers are utilized by compilers to minimize memory accesses. Currently, Connection Machine systems utilize commercially available chips that run at 14 Mflps on the CM-2 and at 20 Mflps on the CM-200 on chained multiply/add operations.

Each processing node also has its own memory. Currently, CM systems utilize commerical memory chips and can have 4 MByte per processing node. Each memory module has its own error-correction code (ECC) for fault tolerance.

There is also a custom-designed floating-point and memory interface chip. Perhaps the most important function of this chip for numerical computing is its indirect-addressing hardware. Since CM supercomputers are distributed-memory machines, table lookups are, conceivably, time-consuming operations, requiring data fetches from across the entire system. For small-to-moderate table sizes, then, codes generally replicate a copy of the table for each processing node. The indirect-addressing hardware allows independent table lookups on the various computing nodes.

Another custom-designed chip provides 32 bit-serial processors per node and the interface to the communications network.

3.3 Communications Network

The communications network in the CM-2 and CM-200 models has the same topology as a binary hypercube in d dimensions, with a processing nodes at each of the 2^d vertices and a communications wire along each of the $d2^{d-1}$

edges. Data messages passed along these wires are accompanied by parity bits to detect faults in communications.

A router directs messages along the communications wires for generalized communications patterns. Due to the high connectivity of the network, a message never needs to travel more than d wires to reach any node, assuming no traffic jams in the router. Typical bandwidth for general communications patterns is approximately 0.2 (0.3) MByte/s per node on the CM-2 (CM-200).

System software lays data out in the hypercube in such a way as to optimize local communications on rectangular grids. Nearest-neighbor communications are either on-node or one wire hop. Due, once again, to the high connectivity of the network, communication between grid points separated by a power-of-two along any axis, important for hierarchical reductions, broadcasts, and scans, is never more than two wire hops.

Multiwire communications software utilizes all d wires emanating from a node to achieve the full bandwidth of the hypercube – approximately 0.5 (0.7) MByte/s per node on the CM-2 (CM-200).

Other communication patterns are also efficient on the hypercube. A fast-fourier-transform butterfly, for example, yields 2.5 (3.6) MByte/s per node on the CM-2 (CM-200).

3.4 Other System Characteristics

Systems can have from $128 = 2^7$ to $2,048 = 2^{11}$ nodes. On a fully configured CM-200 system, then, one can realize a peak aggregate performance of 40 Gflps. Such a system would have 8 GByte of memory.

Though these peak figures are, of course, theoretical maxima rather than typical, realized performances, system performance is still quite high. An oil-reservoir simulation, a recent entry to the Gordon Bell competition, ran at 14 Gflps on a CM-2 and at 20 Gflps on a CM-200. The LINPACK benchmark runs at 9 Gflps in double precision and at 17 Gflps in single precision on a CM-200. Performance on fast fourier transforms (FFT) can range from 3-10 Gflps, depending on the operation, and matrix multiplies are on the order of 10 Gflps. Users typically expect and achieve performances of a few Gflps.

Processors may all be used together for one job or they may be partitioned for different control streams. A partition may also be timeshared among several jobs.

The programming environment uses a UNIX® operating system, providing familiar editing, file systems, debugging, security, and so on.

Graphics hardware and software are integrated into the CM system.

4. Selected Applications from Condensed-Matter Physics

Approximately sixty CM systems have been installed worldwide, including systems in the United States, Europe, Japan, and Australia. They are in government labs, private industry, universities, and supercomputing centers. These systems are used in a variety of applications. In this section, we mention a few condensed-matter applications performed on Connection Machine supercomputers.

4.1 Langevin / Molecular Dynamics

Coherent localized and extended defects, such as vortices, dislocations, domain walls, and grain boundaries, play important roles in modifying both equilibrium (e.g., strength) and non-equilibrium (e.g., response) properties of materials. Spiral growth patterns, which are an important example and appear in many contexts in nonlinear science, have been studied in large-system simulations on a CM-2. [3] In the study, growth of a crystal surface was modeled by a dynamic solid-on-solid model, analogous to a discrete two-dimensional sine-Gordon equation. The competing influences of both homogeneous (thermal) nucleation and Frank-Read sources of spiral growth patterns were incorporated. The dynamics were simulated using a classical Langevin molecular dynamics scheme. The model is a fine example of the data-parallel paradigm. The physics is completely homogeneous and the interactions are purely local. Meanwhile, great computing power is needed in this study to gather good statistics on the temperature-dependent phenomena.

4.2 Quantum Spin Monte Carlo

With the initial discoveries of the high-temperature superconducting compounds, interest in the two-dimensional Heisenberg antiferromagnet was dramatically rekindled. An interesting feature in the Raman spectrum of undoped La_2CuO_4 is the broad peak around 3000 cm^{-1}. The ratio of the peak width to the averaged frequency is parameter free within the 2d Heisenberg model and checks the applicability of the model to the undoped compound. Unfortunately, calculations within conventional spin-wave theory or exact diagonalizations on small clusters provide poor agreement with the experimentally measured ratio. Variational Monte Carlo studies were undertaken to determine if this was due to shortcomings in the solutions or inapplicability of the model. [4] Monte Carlo configurations were generated on a combination of workstations and conventional supercomputers. Measurements of the Raman moments, the intensive part of the calculation, was performed on a CM-2. Consistency between numerical results and the 3000 cm^{-1} peak in the experimental spectrum argues for applicability of the Heisenberg model to the base compound.

4.3 Car-Parrinello

The Si (111) surface, with the (7x7) reconstruction, has been one of the most controversial and widely studied surfaces of any solid. Its enormous complexity has defied any complete and realistic theoretical treatment of its properties. Using the large memory and powerful performance of a CM-2 computer, *ab initio* calculations of the geometric and electronic structure of the surface can finally be performed using the Car-Parrinello method. [5] At a kinetic-energy cutoff of 6 Ry, the calculation uses a basis of 23,000 planewaves. At 8 Ry, there are 36,000 planewaves, requiring 122 MWords of memory. Such an application can run in a one-quarter partition of a CM-2 computer, on which one iteration would take less than two hours.

4.4 Lanczos

There has been a great deal of debate over the relative contributions of electron-electron and electron-phonon interactions to the rich phenomenology of conjugated polymers. Tight-binding Hamiltonians, such as the Peierls-Hubbard model, incorporate these terms. Due to the low dimensionality of the compounds, numerical calculations on even short (up to 16 sites) chains quickly extrapolate to produce the infinite-chain limit and so Lanczos diagonalizations have been used to study such systems with great success. The computational challenge in large-system calculations is to manipulate wavefunctions in the extremely large Hilbert spaces generated by the many-body degrees-of-freedom. Often, symmetries of the Hamiltonian are exploited to reduce this Hilbert space, but that approach has extremely limited utility for conjugated polymers, for which lattice distortions are important. Nevertheless, the large memory of the CM-2 has been used routinely to study chains up to 14 sites, for which the Hilbert space has 12 million states. [6] Indeed, in-core diagonalizations of 16-site systems, with Hilbert spaces of 166 million states, are possible on the CM-2.

5. Conclusion

While physical insight will continue to be the key to a better theoretical understanding of nature, we continue to rely increasingly on computer technology to help us solve mathematical models.

Massive parallelism offers us a means to break past fundamental limits to single-processor performance. Data-parallel programming languages enable us to harness the complexity of such new computer architectures and to describe data-intensive, but physically homogeneous, systems more eloquently to the computer.

The advent of Connection Machine supercomputers has proven the suitability of data-parallel language to describe physical systems and the capability of massively parallel systems to surpass the performance of conventional supercomputer architectures. The challenge of the upcoming years is to unlearn the serial mindset we have been taught to use in conventional programming. Otherwise, we need only wait for massively parallel technology to run its inevitable course toward Teraflop computing.

References

1. See also Connection Machine Model CM-2 Technical Summary, Version 6.0, November 1990, Thinking Machines Corp., Cambridge, Mass.

2. See also B.M. Boghosian, Computers in Physics **4**, 14 (1990).

3. F. Falo, A.R. Bishop, P.S. Lomdahl, B. Horowitz, Phys. Rev. B **43**, 8081 (1991).

4. Z. Liu, E. Manousakis, Phys. Rev. B **43**, 13246 (1991).

5. J. D. Joannopoulos, K. Brommer, B. E. Larson, M. Needels, to be published.

6. J.T. Gammel, D.K. Campbell, E. Loh, S. Mazumdar, S.N. Dixit, to appear in J. Materials Research.

Ising Machine m-TIS2

N. Ito[1], *M. Taiji*[2], *and M. Suzuki*[2]

[1]Computing and Information Systems Center, Japan Atomic Energy Research Institute, Tokai-mura, Naka-gun, Ibaraki 319-11, Japan
[2]Department of Physics, University of Tokyo, Hongo, Bunkyo-ku, Tokyo 113, Japan

Abstract. The characteristics and specifications of m-TIS2 are given. The m-TIS2 is a dedicated machine for the Monte Carlo simulation of the Ising models.

From the beginning of 1980s, several dedicated computer systems for Ising Monte Carlo simulation have been designed and constructed[1]. And they produced many important results. We have constructed a new special purpose computer system for the Ising Monte Carlo simulation. It is the advanced machine of our previous m-TIS[2] and it is named m-TIS2. The m-TIS was constructed in 1986 and was used in several studies[3]. The host computer of m-TIS2 is the personal computer[4]. There are twelve m-TIS2s presently which are installed in several research groups in Japan.

The m-TIS and m-TIS2 have a unique architecture as dedicated computer. The Ising machines which were made before the m-TIS and m-TIS2 are installed the spin memories which hold the configurations of the system, that is, spin, bond and other configurations. The rolls of the host computer are the initialization of the machine and the analysis of the physical behavior. The Ising machine updates the given spin configuration following the initialized Hamiltonian independently of the host computer. After several Monte Carlo steps, the spin configuration and physical quantities are transferred from the machine to the host and they are analyzed to estimate physical behavior. Therefore these Ising machines are "loosely coupled"with the host computer. On the other hand our machines are "strongly coupled"with the host computer. They do not hold all the configurations in them. They hold only the spin states and other information which are necessary to update several spins. The whole configurations are in the memory storage devices of the host computer. Personal computers with Intel's CPU, 8086, 80286 and 80386, are their hosts. Therefore they are designed so that the 16 spins which are coded in one computer word are updated at every operation. Several words are transferred from the host and one word which contains updated 16 spins is transferred from the machine after one operation.

This strongly-coupled architecture has some advantages. Firstly, the hardware becomes simple because the address generation circuit and configuration memory devices are not necessary. Instead of them, several 16bits shift registers with special functions are installed. These registers are called the local-information registers. Secondly, the memory management, interrupt treatment and other functions necessary for system management are treated by the host computer and its operating system. This makes no problem because the machine returns the control privilege within 16 system clocks at the latest.

There is, however, disadvantage in this architecture. It is the appearance of data transfer bottle-neck between the host and the machine. For example, about four words should be transferred from and to the host for square-lattice nearest-neighbor-coupling Ising model. Because the clock speed of m-TIS2 is 8MHz, the transfer rate necessary to maintain the peak performance is 2M words per second. For the more complicated Hamiltonian, the required transfer rate becomes larger.

Table 1: The models which can be treated by the present m-TIS2 are listed. The number in the column "uniform"denotes the interaction range, for example, 3 means the interactions with up to the third neighbor spins. The $\pm J$ and $\pm h$ mean random bond and random field models, respectively. Uniform field can be charged in any model.

lattice	uniform	$\pm J$	$\pm h$	$\pm J$ and $\pm h$
square	3	○	○	○
hexagonal	3	○	○	○
triangular	3	○	○	○
cubic	1	○	○	○
hypercubic				
$d = 4$	1	×	○	×
$d = 5$	1	×	○	×
$d = 6$	1	×	×	×

The following features are extended in the m-TIS2 compared with the m-TIS. (1) The number of local-information registers is increased. This extends the models which can be simulated. Examples of the models which can be simulated by m-TIS2 are given in Table 1. (2) The 32bit-wide random numbers are used. It was 16bit-wide in m-TIS. (3) The address of the transition probability memory is 16bit-wide. Therefore 16bit information can be involved for updating one spin. It was 13bit in m-TIS. (4) There are twelve physical-quantity counters. There was only one in m-TIS. The m-TIS2 can measure physical quantities up to twelve bits, for example, the energy and the magnetization.

Several technical improvements are also made. All the circuits except the interface board are in one circuit-printed board. The local-information registers, random-number generator, physical-quantity counters, system-clock generator and control circuits are implemented in three LSIs. The XILINX's LCA 3090 are used. The LCA configurations are written from the host computer at the beginning of the simulation. We can select several configurations of local-information registers and random-number generation algorithms by changing the configuration files. The details are to be given in [4].

This work is supported by the Research Fund of the Ministry of Education, Science and Culture.

References

[1] H. J. Hilhorst, A. F. Bakker, C. Bruin, A. Compagner and A. Hoogland, J. Stat. Phys. **34** (1984) p. 987.

[2] M. Taiji, N. Ito and M. Suzuki, Rev. Sci. Instrum., **59** (1988) p.2483.

[3] N. Ito, M. Taiji and M. Suzuki, J. Phys. Soc. Jpn. **56** (1987) p.4218. N.Ito, M.Taiji and M.Suzuki, J. de Physique Colloque **C8** Supplement (1988)p.1397. N. Ito and M. Suzuki, J. Phys. Soc. Jpn. **60** (1991) p. 1978.

[4] M. Taiji, N. Ito and M. Suzuki, in preparation.

Index of Contributors

Springer Proceedings in Physics

Managing Editor: H. K. V. Lotsch

MIX
Papier aus verantwortungsvollen Quellen
Paper from responsible sources
FSC® C105338

If you have any concerns about our products,
you can contact us on
ProductSafety@springernature.com

In case Publisher is established outside the EU,
the EU authorized representative is:
Springer Nature Customer Service Center GmbH
Europaplatz 3, 69115 Heidelberg, Germany

Printed by Libri Plureos GmbH
in Hamburg, Germany